D1481753

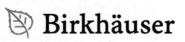

Bernhard Krötz • Omer Offen • Eitan Sayag
Editors

Representation Theory, Complex Analysis, and Integral Geometry

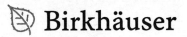 Birkhäuser

Editors

Bernhard Krötz
Institut für Analysis
Leibniz Universität Hannover
Welfengarten 1
30167 Hannover
Germany
kroetz@math.uni-hannover.de

Omer Offen
Department of Mathematics
Technion – Israel Institute of Technology
Haifa 32000
Israel
offen@tx.technion.ac.il

Eitan Sayag
Department of Mathematics
Ben-Gurion University of the Negev
Be'er Sheva 84105
Israel
sayage@math.bgu.ac.il

ISBN 978-0-8176-4816-9   e-ISBN 978-0-8176-4817-6
DOI 10.1007/978-0-8176-4817-6
Springer New York Dordrecht Heidelberg London

Library of Congress Control Number: 2011942315

Mathematics Subject Classification (2010): 11F30, 11F37, 11M36, 11S90, 17B08, 20H10, 22D10, 22Exx, 32N10, 33C52, 42B35, 43A75, 43A85, 43A90, 46E35, 53C35, 53D20

www.birkhauser-science.com

Preface

This volume is an outgrowth of a special summer term on "Harmonic analysis, representation theory, and integral geometry", hosted by the Max Plank Institute for Mathematics (MPIM) and the then newly founded Hausdorff Research Institute for Mathematics (HIM) in Bonn in 2007. It was organized and led by S. Gindikin and B. Krötz with the help of O. Offen and E. Sayag. The purpose of this book is to make an essential part of the activity from the summer term available to a wider audience.

The book contains research contributions on the following themes: connecting periods of Eisenstein series on orthogonal groups and double Dirichlet series (Gautam Chinta and Omer Offen); vanishing at infinity of smooth functions on symmetric spaces (Bernhard Krötz and Henrik Schlichtkrull); a formula involving all the Rankin–Selberg convolutions of holomorphic and non-holomorphic cusp forms (Jay Jorgenson and Jürg Kramer); a scheme of a new proof for the so-called Helgason conjecture on a Riemannian symmetric space $X = G/K$ of the non-compact type (Simon Gindikin); an algorithm for the computation of special unipotent representations attached to certain regular K-orbits on a flag variety of the dual group (Dan Ciubotaru, Kyo Nishiyama, and Peter E. Trapa); applications of symplectic geometry, particularly moment maps, to the study of arithmetic issues in invariant theory (Marcus J. Slupinski and Robert J. Stanton); and restrictions of representations of $SL_2(\mathbb{C})$ to $SL_2(\mathbb{R})$ treated in a geometric way, thus providing a useful introduction to this research area (Birgit Speh and T. N. Venkataramana).

In addition, the volume contains three papers of an expository nature that should be considered a bonus. The first, by Joseph Bernstein, is a course for beginners on the representation theory of Lie algebras; experts can also benefit from this. Although Feigin and Zelevinski published an expanded version of these notes, the original from 1976, which is much more suitable for beginners, had never been published. The second contribution, by Jacques Faraut, introduces the work of Okounkov and Olshanski on the asymptotics of spherical functions on symmetric spaces of a large rank. The third, by Yuri A. Neretin, is an introduction to the Stein–Sahi complementary series.

Acknowledgments

We thank the invited lecturers and participants for creating a stimulating atmosphere of cooperation and communication without which this volume would not have been possible. We thank the referees for their efficient and helpful reports. We also express our gratitude to MPIM and HIM for providing us with a wonderful working environment.

Hannover, Germany   Bernhard Krötz
Haifa, Israel   Omer Offen
Be'er-Sheva, Israel   Eitan Sayag

Contents

Contributors

Joseph Bernstein Department of Mathematics, Tel-Aviv University, Tel-Aviv, Israel

Gautam Chinta Department of Mathematics, The City College of CUNY, New York, NY, USA

Dan Ciubotaru Department of Mathematics, University of Utah, Salt Lake City, UT, USA

Jacques Faraut Équipe "Analyse Algébrique", Institut de Mathématiques, Université Pierre et Marie Curie, Paris Cedex, France

Simon Gindikin Department of Mathematics, Hill Center, Rutgers University, Piscataway, NJ, USA

Jay Jorgenson Department of Mathematics, City College of New York, New York, NY, USA

Jürg Kramer Institut für Mathematik, Humboldt-Universität zu Berlin, Berlin, Germany

Bernhard Krötz Institut für Analysis, Leibniz Universität Hannover, Hannover, Germany

Yuri A. Neretin Mathematics Department, University of Vienna, Vienna, Austria

Mathematical Physics Group, Institute for Theoretical and Experimental Physics, Moscow, Russia

TFFA, Mechanics and Mathematics Department, Moscow State University, Moscow, Russia

Kyo Nishiyama Department of Mathematics, Graduate School of Science, Kyoto University, Sakyo, Kyoto, Japan

Omer Offen Mathematics Department, Technion – Israel Institute of Technology, Haifa, Israel

Henrik Schlichtkrull Department of Mathematics, University of Copenhagen, København, Denmark

Marcus J. Slupinski IRMA, Université Louis Pasteur (Strasbourg), Strasbourg Cedex, France

B. Speh Department of Mathematics, Cornell University, Ithaca, NY, USA

Robert J. Stanton Department of Mathematics, Ohio State University, Columbus, OH, USA

Peter E. Trapa Department of Mathematics, University of Utah, Salt Lake City, UT, USA

T.N. Venkataramana Tata Institute for Fundamental Research, Mumbai, India

On Function Spaces on Symmetric Spaces

Bernhard Krötz and Henrik Schlichtkrull

Abstract Let $Y = G/H$ be a semisimple symmetric space. It is shown that the smooth vectors for the regular representation of G on $L^p(Y)$ vanish at infinity.

Keywords Smooth vectors • Decay of matrix coefficients • RiemannLebesgue lemma • Symmetric spaces

Mathematics Subject Classification (2010): 43A85, 43A90, 46E35

1 Vanishing at Infinity

Let G be a connected unimodular Lie group, equipped with a Haar measure dg, and let $1 \leq p < \infty$. We consider the left regular representation L of G on the function space $E_p = L^p(G)$.

Recall that $f \in E_p$ is called a *smooth vector for* L if and only if the map

$$G \to E_p, \quad g \mapsto L(g)f$$

is a smooth E_p-valued map.

Write $\mathfrak{g}$ for the Lie algebra of G and $\mathcal{U}(\mathfrak{g})$ for its enveloping algebra. The following result is well known, see [3].

B. Krötz (✉)
Max-Planck-Institut für Mathematik, Vivatsgasse 7, D-53111 Bonn, Deutschland

Institut für Analysis, Leibniz Universität Hannover, Welfengarten 1, 30167 Hannover, Germany
e-mail: kroetz@mpim-bonn.mpg.de; kroetz@math.uni-hannover.de

H. Schlichtkrull
Department of Mathematics, University of Copenhagen, Universitetsparken 5,
DK-2100 København, Denmark
e-mail: schlicht@math.ku.dk

B. Krötz et al. (eds.), *Representation Theory, Complex Analysis, and Integral Geometry*,
DOI 10.1007/978-0-8176-4817-6_1, © Springer Science+Business Media, LLC 2012

Theorem 1. *The space of smooth vectors for L is*

$$E_p^\infty = \{f \in C^\infty(G) \mid L_u f \in L^p(G) \text{ for all } u \in \mathcal{U}(\mathfrak{g})\}.$$

Furthermore, $E_p^\infty \subset C_0^\infty(G)$, the space of smooth functions on G which vanish at infinity.

Our concern is with the corresponding result for a homogeneous space Y of G. By that we mean a connected manifold Y with a transitive action of G. In other words,

$$Y = G/H$$

with $H \subset G$ a closed subgroup. We shall require that Y carries a G-invariant positive measure dy. Such a measure is unique up to scale and commonly referred to as Haar measure. With respect to dy, we form the Banach spaces $E_p := L^p(Y)$. The group G acts continuously by isometries on E_p via the left regular representation:

$$[L(g)f](y) = f(g^{-1}y) \qquad (g \in G, y \in Y, f \in E_p).$$

We are concerned with the space E_p^∞ of smooth vectors for this representation. The first part of Theorem 1 is generalized as follows, see [3], Theorem 5.1.

Theorem 2. *The space of smooth vectors for L is*

$$E_p^\infty = \{f \in C^\infty(Y) \mid L_u f \in L^p(Y) \text{ for all } u \in \mathcal{U}(\mathfrak{g})\}.$$

We write $C_0^\infty(Y)$ for the space of smooth functions vanishing at infinity. Our goal is to investigate an assumption under which the second part of Theorem 1 generalizes, that is,

$$E_p^\infty \subset C_0^\infty(Y). \tag{1}$$

Notice that if H is compact, then we can regard $L^p(G/H)$ as a closed G-invariant subspace of $L^p(G)$, and (1) follows immediately from Theorem 1.

Likewise, if $Y = G$ regarded as a homogeneous space for $G \times G$ with the left×right action, then again (1) follows from Theorem 1, since a left×right smooth vector is obviously also left smooth.

However, (1) is false in general as the following class of examples shows. Assume that Y has finite volume but is not compact, e.g. $Y = \mathrm{Sl}(2, \mathbb{R})/\mathrm{Sl}(2, \mathbb{Z})$. Then the constant function $\mathbf{1}_Y$ is a smooth vector for E^p, but it does not vanish at infinity.

2 Proof by Convolution

We give a short proof of (1) for the case $Y = G$, based on the theorem of Dixmier and Malliavin (see [2]). According to this theorem, every smooth vector in a Fréchet representation (π, E) belongs to the Gårding space, that is, it is spanned by vectors of the form $\pi(f)v$, where $f \in C_c^\infty(G)$ and $v \in E$. Let such a vector $L(f)g$, where $g \in E_p = L^p(G)$ be given. Then by unimodularity

$$[L(f)g](y) = \int_G f(x)g(x^{-1}y)\,dx = \int_G f(yx^{-1})g(x)\,dx. \qquad (2)$$

For simplicity, we assume $p = 1$. The general case is similar. Let $\Omega \subset G$ be compact such that $|g|$ integrates to $< \epsilon$ over the complement. Then for y outside of the compact set $\operatorname{supp} f \cdot \Omega$, we have

$$yx^{-1} \in \operatorname{supp} f \Rightarrow x \notin \Omega,$$

and hence

$$|L(f)g(y)| \leq \sup |f| \int_{x \notin \Omega} |g(x)|\,dx \leq \sup |f|\,\epsilon.$$

It follows that $L(f)g \in C_0(G)$.

Notice that the assumption $Y = G$ is crucial in this proof, since the convolution identity (2) makes no sense in the general case.

3 Semisimple Symmetric Spaces

Let $Y = G/H$ be a semisimple symmetric space. By this, we mean:

- G is a connected semisimple Lie group with finite center.
- There exists an involutive automorphism τ of G such that H is an open subgroup of the group $G^\tau = \{g \in G \mid \tau(g) = g\}$ of τ-fixed points.

We will verify (1) for this case. In fact, our proof is valid also under the more general assumption that G/H is a reductive symmetric space of Harish–Chandra's class, see [1].

Theorem 3. *Let $Y = G/H$ be a semisimple symmetric space, and let $E_p = L^p(Y)$ where $1 \leq p < \infty$. Then*

$$E_p^\infty \subset C_0^\infty(Y).$$

Proof. A little bit of standard terminology is useful. As customary we use the same symbol for an automorphism of G and its derived automorphism of the Lie algebra $\mathfrak{g}$. Let us write $\mathfrak{g} = \mathfrak{h} + \mathfrak{q}$ for the decomposition in τ-eigenspaces according to eigenvalues $+1$ and -1.

Denote by K a maximal compact subgroup of G. We may and shall assume that K is stable under τ. Write θ for the Cartan-involution on G with fixed point group K, and write $\mathfrak{g} = \mathfrak{k} + \mathfrak{p}$ for the eigenspace decomposition for the corresponding derived involution. We fix a maximal abelian subspace $\mathfrak{a} \subset \mathfrak{p} \cap \mathfrak{q}$.

The simultaneous eigenspace decomposition of $\mathfrak{g}$ under ad $\mathfrak{a}$ leads to a (possibly reduced) root system $\Sigma \subset \mathfrak{a}^*\backslash\{0\}$. Write $\mathfrak{a}_{\mathrm{reg}}$ for $\mathfrak{a}$ with the root hyperplanes removed, i.e.:

$$\mathfrak{a}_{\mathrm{reg}} = \{X \in \mathfrak{a} \mid (\forall \alpha \in \Sigma)\, \alpha(X) \neq 0\}.$$

Let $M = Z_{H \cap K}(\mathfrak{a})$ and $W_H = N_{H \cap K}(\mathfrak{a})/M$.

Recall the polar decomposition of Y. With $y_0 = H \in Y$ the base point of Y it asserts that the mapping

$$\rho : K/M \times \mathfrak{a} \to Y, \quad (kM, X) \mapsto k\exp(X) \cdot y_0$$

is differentiable, onto and proper. Furthermore, the element X in the decomposition is unique up to conjugation by W_H, and the induced map

$$K/M \times_{W_H} \mathfrak{a}_{\mathrm{reg}} \to Y$$

is a diffeomorphism onto an open and dense subset of Y.

Let us return now to our subject proper, the vanishing at infinity of functions in E_p^∞. Let us denote functions on Y by lowercase roman letters, and by the corresponding uppercase letters their pull backs to $K/M \times \mathfrak{a}$, for example $F = f \circ \rho$. Then f vanishes at infinity on Y translates into

$$\lim_{\substack{X \to \infty \\ X \in \mathfrak{a}}} \sup_{k \in K} |F(kM, X)| = 0. \tag{3}$$

We recall the formula for the pull back by ρ of the invariant measure dy on Y. For each $\alpha \in \Sigma$ we denote by $\mathfrak{g}^\alpha \subset \mathfrak{g}$ the corresponding root space. We note that $\mathfrak{g}^\alpha$ is stable under the involution $\theta\tau$. Define p_α, resp. q_α, as the dimension of the $\theta\tau$-eigenspace in $\mathfrak{g}^\alpha$ according to eigenvalues $+1, -1$. Define a function J on $\mathfrak{a}$ by

$$J(X) = \left| \prod_{\alpha \in \Sigma^+} [\cosh \alpha(X)]^{q_\alpha} \cdot [\sinh \alpha(X)]^{p_\alpha} \right|.$$

With $d(kM)$ the Haar-measure on K/M and dX the Lebesgue-measure on $\mathfrak{a}$ one then gets, up to normalization:

$$\rho^*(dy) = J(X)\, d(k, X) := J(X)\, d(kM)\, dX.$$

We shall use this formula to relate certain Sobolev norms on Y and on $K/M \times \mathfrak{a}$. Fix a basis $X_1, \ldots, X_n$ for $\mathfrak{g}$. For an n-tupel $\mathbf{m} = (m_1, \ldots, m_n) \in \mathbb{N}_0^n$, we define elements $X^{\mathbf{m}} \in \mathcal{U}(\mathfrak{g})$ by

$$X^{\mathbf{m}} := X_1^{m_1} \cdot \ldots \cdot X_n^{m_n}.$$

These elements form a basis for $\mathcal{U}(\mathfrak{g})$. We introduce the L^p-Sobolev norms on Y,

$$S_{m,\Omega}(f) := \sum_{|\mathbf{m}| \le m} \left[\int_{\Omega} |L(X^{\mathbf{m}})f(y)|^p\, dy \right]^{1/p}$$

where $\Omega \subset Y$, and where $|\mathbf{m}| := m_1 + \ldots + m_n$. Then a function $f \in C^\infty(Y)$ belongs to E_p^∞ if and only if $S_{m,Y}(f) < \infty$ for all m.

Likewise, for $V \subset \mathfrak{a}$ we denote

$$S_{m,V}^*(F) := \sum_{|\mathbf{m}| \le m} \left[\int_{K \times V} |L(Z^{\mathbf{m}})F(kM, X)|^p\, J(X)\, d(k, X) \right]^{1/p}.$$

Here Z refers to members of some fixed bases for $\mathfrak{k}$ and $\mathfrak{a}$, acting from the left on the two variables, and again $\mathbf{m}$ is a multiindex.

Observe that for $Z \in \mathfrak{a}$ we have for the action on $\mathfrak{a}$,

$$[L(Z)F](kM, X) = [L(Z^k)f](k \exp(X) \cdot y_0),$$

where $Z^k := \mathrm{Ad}(k)(Z)$ can be written as a linear combination of the basis elements in $\mathfrak{g}$, with coefficients which are continuous on K. It follows that for every m there exists a constant $C_m > 0$ such that for all $F = f \circ \rho$,

$$S_{m,V}^*(F) \le C_m S_{m,\Omega}(f), \tag{4}$$

where $\Omega = \rho(K/M, V) = K \exp(V) \cdot y_0$.

Let $\epsilon > 0$ and set

$$\mathfrak{a}_\epsilon := \{ X \in \mathfrak{a} \mid (\forall \alpha \in \Sigma)\, |\alpha(X)| \ge \epsilon \}.$$

Observe that there exists a constant $C_\epsilon > 0$ such that

$$(\forall X \in \mathfrak{a}_\epsilon) \quad J(X) \ge C_\epsilon. \tag{5}$$

We come to the main part of the proof. Let $f \in E_p^\infty$. We shall first establish that

$$\lim_{\substack{X \to \infty \\ X \in \mathfrak{a}_\epsilon}} F(eM, X) = 0. \tag{6}$$

It follows from the Sobolev lemma, applied in local coordinates, that the following holds for a sufficiently large integer m (depending only on p and the dimensions of K/M and $\mathfrak{a}$). For each compact symmetric neighborhood V of 0 in $\mathfrak{a}$, there exists a constant $C > 0$ such that

$$|F(eM, 0)|$$

$$\leq C \sum_{|\mathbf{m}| \leq m} \left[\int_{K/M \times V} |[L(Z^{\mathbf{m}})F](kM, X)|^p \, d(k, X) \right]^{1/p} \tag{7}$$

for all $F \in C^\infty(K/M \times \mathfrak{a})$. We choose V such that $\mathfrak{a}_\epsilon + V \subset \mathfrak{a}_{\epsilon/2}$.

Let $\delta > 0$. Since $f \in E^p$, it follows from (4) and the properness of ρ that there exists a compact set $B \subset \mathfrak{a}$ with complement $B^c \subset \mathfrak{a}$, such that

$$S_{m,B^c}^*(F) \leq C_m S_{m,\Omega}(f) < \delta, \tag{8}$$

where $\Omega = K \exp(B^c) \cdot y_0$.

Let $X_1 \in \mathfrak{a}_\epsilon \cap (B + V)^c$. Then $X_1 + X \in \mathfrak{a}_{\epsilon/2} \cap B^c$ for $X \in V$. Applying (7) to the function

$$F_1(kM, X) = F(kM, X_1 + X),$$

and employing (5) for the set $\mathfrak{a}_{\epsilon/2}$, we derive

$$|F(eM, X_1)|$$

$$\leq C \sum_{|\mathbf{m}| \leq m} \left[\int_{K/M \times V} |[L(Z^{\mathbf{m}})F_1](kM, X)|^p \, d(k, X) \right]^{1/p}$$

$$\leq C' \sum_{|\mathbf{m}| \leq m} \left[\int_{K/M \times B^c} |[L(Z^{\mathbf{m}})F](kM, X)|^p \, J(X) \, d(k, X) \right]^{1/p}$$

$$= C' S_{m,B^c}^*(F) \leq C'\delta,$$

from which (6) follows.

In order to conclude the theorem, we need a version of (6) which is uniform for all functions $L(q)f$, for q in a fixed compact subset Q of G.

Let $\delta > 0$ be given, and as before let $B \subset \mathfrak{a}$ be such that (8) holds. By the properness of ρ, there exists a compact set $B' \subset \mathfrak{a}$ such that

$$QK \exp(B) \cdot y_0 \subset K \exp(B') \cdot y_0.$$

We may assume that B' is W_H-invariant. Then for each $k \in K$, $X \notin B'$ and $q \in Q$ we have that

$$q^{-1}k \exp(X) \cdot y_0 \notin K \exp(B) \cdot y_0, \tag{9}$$

since otherwise we would have

$$k \exp(X) \cdot y_0 \in qK \exp(B) \cdot y_0 \subset K \exp(B') \cdot y_0$$

and hence $X \in B'$.

We proceed as before, with B replaced by B', and with f, F replaced by $f_q = L_q f$, $F_q = f_q \circ \rho$. We thus obtain for $X_1 \in \mathfrak{a}_\epsilon \cap (B' + V)^c$,

$$|F_q(eM, X_1)| \leq CS^*_{m,(B')^c}(F_q) \leq C\, C_m S_{m,\Omega'}(f_q)$$

where $\Omega' = K \exp((B')^c) \cdot y_0$.

Observe that for each X in $\mathfrak{g}$ the derivative $L(X)f_q$ can be written as a linear combination of derivatives of f by basis elements from $\mathfrak{g}$, with coefficients which are uniformly bounded on Q. We conclude that $S_{m,\Omega'}(f_q)$ is bounded by a constant times $S_{m,Q^{-1}\Omega'}(f)$, with a uniform constant for $q \in Q$. By (9) and (8), we conclude that the latter Sobolev norm is bounded from the above by δ.

We derive the desired uniformity of the limit (6) for $q \in Q$,

$$\lim_{\substack{X \to \infty \\ X \in \mathfrak{a}_\epsilon}} \sup_{q \in Q} |F_q(eM, X)| = 0. \tag{10}$$

Finally, we choose an appropriate compact set Q. Let $C_1, \ldots, C_N \subset \mathfrak{a}$ be the closed chambers relative to Σ. For each chamber C_j, we choose $X_j \in C_j$ such that $X_j + C_j \subset \mathfrak{a}_\epsilon$. It follows that

$$\mathfrak{a} = \bigcup_{j=1}^{N} (-X_j + \mathfrak{a}_\epsilon). \tag{11}$$

Set $a_j = \exp(X_j) \in A$ and define

$$Q := \bigcup_{j=1}^{N} a_j K.$$

Note that for $q = a_j k$ we have

$$F_q(eM, X) = F(k^{-1}M, X - X_j).$$

Let $\delta > 0$ be given. It follows from (10) that there exists $R > 0$ such that $|F_q(eM, Y)| < \delta$ for all $q \in Q$ and all $Y \in \mathfrak{a}_\epsilon$ with $|Y| \geq R$. For every $X \in \mathfrak{a}$ with $|X| \geq R + \max_j |X_j|$, we have $X \in -X_j + \mathfrak{a}_\epsilon$ for some j and $|X + X_j| \geq R$. Hence for all $k \in K$,

$$|F(kM, X)| = |F_q(eM, X + X_j)| < \delta,$$

where $q = a_j k^{-1}$. Thus,

$$\lim_{X \to \infty} F(kM, X) = 0,$$

uniformly over $k \in K$, as was to be shown.   □

Remark. Let $f \in L^2(Y)$ be a K-finite function which is also finite for the center of $\mathcal{U}(\mathfrak{g})$. Then it follows from [4] that f vanishes at infinity. The present result is more general, since such a function necessarily belongs to E_2^∞.

References

[1] E. van den Ban, *The principal series for a reductive symmetric space, II*, Jour. Funct. Anal. **109** (1992), 331–441.

[2] J. Dixmier and P. Malliavin, *Factorisations de fonctions et de vecteurs indéfiniment différentiables*, Bull. Sci. Math. **102** (1978), 307–330.

[3] N. S. Poulsen, *On C^∞-vectors and intertwining bilinear forms for representations of Lie groups*, J. Funct. Anal. **9** (1972), 87–120.

[4] Z. Rudnick and H. Schlichtkrull, *Decay of eigenfunctions on semisimple symmetric spaces*, Duke Math. J. **64** (1991), no. **3**, 445–450.

A Relation Involving Rankin–Selberg
L-Functions of Cusp Forms and Maass Forms

Jay Jorgenson and Jürg Kramer

Abstract In previous articles, an identity relating the canonical metric to the hyperbolic metric associated with any compact Riemann surface of genus at least two has been derived and studied. In this article, this identity is extended to any hyperbolic Riemann surface of finite volume. The method of proof is to study the identity given in the compact case through degeneration and to understand the limiting behavior of all quantities involved. In the second part of the paper, the Rankin–Selberg transform of the noncompact identity is studied, meaning that both sides of the relation after multiplication by a nonholomorphic, parabolic Eisenstein series are being integrated over the Riemann surface in question. The resulting formula yields an asymptotic relation involving the Rankin–Selberg L-functions of weight two holomorphic cusp forms, of weight zero Maass forms, and of nonholomorphic weight zero parabolic Eisenstein series.

Keywords Automorphic forms • Eisenstein series • L-functions • Rankin–Selberg transform • Heat kernel

Mathematics Subject Classification (2010): 11F30, 32N10, 20H10, 11M36

J. Jorgenson (✉)
Department of Mathematics, City College of New York, Convent Avenue at 138th Street,
New York, NY 10031, USA
e-mail: jjorgenson@mindspring.com

J. Kramer
Institut für Mathematik, Humboldt-Universität zu Berlin, Unter den Linden 6,
D-10099 Berlin, Germany
e-mail: kramer@math.hu-berlin.de

B. Krötz et al. (eds.), *Representation Theory, Complex Analysis, and Integral Geometry*,
DOI 10.1007/978-0-8176-4817-6_2, © Springer Science+Business Media, LLC 2012

1   Introduction

1.1   Background

Beginning with the article [13], we derived and studied a basic identity, stated in
(1) below, coming from the spectral theory of the Laplacian associated with any
compact hyperbolic Riemann surface. In the subsequent papers, this identity was
employed to address a number of problems, including the following: Establishing
precise relations between analytic invariants arising in the Arakelov theory of
algebraic curves and hyperbolic geometry (see [13]), proving the noncompleteness
of a newly defined metric on the moduli space of algebraic curves of a fixed
genus (see [14]), deriving bounds for canonical and hyperbolic Green's functions
(see [15]), and obtaining bounds for Faltings's delta function with applications
associated with Arakelov theory (see [16]). In this article, we expand our application
of the results from [13] to analytic number theory. In brief, we first generalize
the identity (1) to general noncompact, finite volume hyperbolic Riemann surfaces
without elliptic fixed points; this relation is stated in equation (2) below. We then
compute the Rankin–Selberg convolution with respect to (2), and show that the
result yields a new relation involving Rankin–Selberg L-functions of cusp forms of
weight two and Maass forms, as well as the scattering matrix of the nonholomorphic
Eisenstein series of weight zero.

1.2   The Basic Identity

Let X denote a compact hyperbolic Riemann surface, necessarily of genus $g \geq 2$.
Let $\{f_j\}$ be a basis of the g-dimensional space of cusp forms of weight two, which
we assume to be orthonormal with respect to the Petersson inner product. Then
we set

$$\mu_{\mathrm{can}}(z) = \frac{1}{g} \cdot \frac{i}{2} \sum_{j=1}^{g} |f_j(z)|^2 \mathrm{d}z \wedge \mathrm{d}\bar{z}$$

for any point $z \in X$. Let Δ_{hyp} denote the hyperbolic Laplacian acting on the space of
smooth functions on X, and $K(t; z, w)$ the corresponding heat kernel; set $K(t; z) = K(t; z, z)$. We use μ_{shyp} to denote the $(1, 1)$-form of the constant negative curvature
metric on X such that X has volume one, and μ_{hyp} to denote the $(1, 1)$-form of the
metric on X with constant negative curvature equal to -1. With this notation, the
key identity of [13] states

$$\mu_{\mathrm{can}}(z) = \mu_{\mathrm{shyp}}(z) + \frac{1}{2g} \int_0^\infty \Delta_{\mathrm{hyp}} K(t; z) \, \mathrm{d}t \mu_{\mathrm{hyp}}(z) \qquad (z \in X). \qquad (1)$$

The first result in this paper is to generalize (1) to general noncompact, finite volume hyperbolic Riemann surfaces without elliptic fixed points. Specifically, if X is such a noncompact, finite volume hyperbolic Riemann surface of genus g with p cusps and no elliptic fixed points, then

$$\mu_{\mathrm{can}}(z) = \left(1 + \frac{p}{2g}\right)\mu_{\mathrm{shyp}}(z) + \frac{1}{2g}\int_0^\infty \Delta_{\mathrm{hyp}} K(t;z)\, dt\, \mu_{\mathrm{hyp}}(z) \qquad (z \in X).$$
(2)

The proof of (2) we present here is to study (1) for a degenerating family of hyperbolic Riemann surfaces and to use known results for the asymptotic behavior of the canonical metric form μ_{can} (see [12]), the hyperbolic heat kernel (see [18]), and small eigenvalues and eigenfunctions of the Laplacian (see [21]).

In [2], the author extends the identity (2) to general finite volume quotients of the hyperbolic upper half-plane, allowing for the presence of elliptic elements. The proof does not employ degeneration techniques, as in this paper, but rather follows the original method of proof given in [13] and [15]. The article [2] is part of the Ph.D. dissertation completed under the direction of the second named author of the present article.

1.3   The Rankin–Selberg Convolution

For the remainder of this article, we assume $p > 0$. Let P denote a cusp of X and $E_{P,s}(z)$ the associated nonholomorphic Eisenstein series of weight zero. In essence, the purpose of this article is to evaluate the Rankin–Selberg convolution with respect to (2), by which we mean to multiply both sides of (2) by $E_{P,s}(z)$ and to integrate over all $z \in X$.

By means of the uniformization theorem, there is a Fuchsian group of the first kind $\Gamma \subseteq \mathrm{PSL}_2(\mathbb{R})$ such that X is isometric to $\Gamma \backslash \mathbb{H}$. Furthermore, we can choose Γ so that the point $i\infty$ in the boundary of $\mathbb{H}$ projects to the cusp P, which we assume to have width b. Writing $z = x + iy$, well-known elementary considerations then show that the expression

$$\int_X E_{P,s}(z)\mu_{\mathrm{can}}(z)$$

$$= \int_X E_{P,s}(z)\left(\left(1 + \frac{p}{2g}\right)\mu_{\mathrm{shyp}}(z) + \frac{1}{2g}\int_0^\infty \Delta_{\mathrm{hyp}} K(t;z)\, dt\, \mu_{\mathrm{hyp}}(z)\right)$$

is equivalent to

$$\int_{y=0}^\infty \int_{x=0}^b y^s \mu_{\mathrm{can}}(z)$$

$$= \int_{y=0}^\infty \int_{x=0}^b y^s \left(\left(1 + \frac{p}{2g}\right)\mu_{\mathrm{shyp}}(z) + \frac{1}{2g}\int_0^\infty \Delta_{\mathrm{hyp}} K(t;z)\, dt\, \mu_{\mathrm{hyp}}(z)\right). \quad (3)$$

The majority of the computations carried out in this article are related to the evaluation of (3). To be precise, for technical reasons we consider the integrals in (3) multiplied by the factor $2gb^{-1}\pi^{-s}\Gamma(s)\zeta(2s)$, where $\Gamma(s)$ is the Γ-function and $\zeta(s)$ is the Riemann ζ-function.

1.4 The Main Result

Having posed the problem under consideration, we can now state the main result of this article after establishing some additional notation.

The cusp forms f_j, being invariant under the map $z \mapsto z + b$, allow a Fourier expansion of the form

$$f_j(z) = \sum_{n=1}^{\infty} a_{j,n} e^{2\pi i n z/b}.$$

Following notations and conventions in [4], we let

$$\widetilde{L}(s, f_j \otimes \overline{f}_j) = G_\infty(s) \cdot L(s, f_j \otimes \overline{f}_j),\tag{4}$$

where

$$G_\infty(s) = (2\pi)^{-2s-1}\Gamma(s)\Gamma(s+1)\zeta(2s),$$

$$L(s, f_j \otimes \overline{f}_j) = \sum_{n=1}^{\infty} \frac{|a_{j,n}|^2}{(n/b)^{s+1}}.$$

As shown in [4], the Rankin–Selberg L-function $\widetilde{L}(s, f_j \otimes \overline{f}_j)$ is holomorphic for $s \in \mathbb{C}$ with $\mathrm{Re}(s) > 1$, admits a meromorphic continuation to all $s \in \mathbb{C}$, and is symmetric under $s \mapsto 1 - s$.

Let φ_j be a nonholomorphic weight zero form which is an eigenfunction of Δ_{hyp} with eigenvalue $\lambda_j = s_j(1 - s_j)$, hence $s_j = 1/2 + ir_j$. From [11], we recall the expansion

$$\varphi_j(z) = \alpha_{j,0}(y) + \sum_{n\neq0} \alpha_{j,n} W_{s_j}(nz/b),$$

where

$$\alpha_{j,0}(y) = \alpha_{j,0} y^{1-s_j},$$

$$W_{s_j}(w) = 2\sqrt{\cosh(\pi r_j)}\sqrt{|\mathrm{Im}(w)|}K_{ir_j}(2\pi|\mathrm{Im}(w)|)e^{2\pi i \mathrm{Re}(w)} \quad (w \in \mathbb{C}),$$

and $K.(\cdot)$ denotes the classical K-Bessel function. Again, following notations and conventions in [4], we let

$$\widetilde{L}(s, \varphi_j \otimes \overline{\varphi}_j) = G_{r_j}(s) \cdot L(s, \varphi_j \otimes \overline{\varphi}_j),$$

where

$$G_{r_j}(s) = s(1-s)\pi^{-2s}\Gamma^2\left(\frac{s}{2}\right)\Gamma\left(\frac{s}{2}+ir_j\right)\Gamma\left(\frac{s}{2}-ir_j\right)\zeta(2s),$$

$$L(s, \varphi_j \otimes \overline{\varphi}_j) = \sum_{n \neq 0} \frac{|\alpha_{j,n}|^2}{(n/b)^{s-1}}.$$

As shown in [4], the Rankin–Selberg L-function $\widetilde{L}(s, \varphi_j \otimes \overline{\varphi}_j)$ is holomorphic for $s \in \mathbb{C}$ with $\mathrm{Re}(s) > 1$, admits a meromorphic continuation to all $s \in \mathbb{C}$, and is symmetric under $s \mapsto 1-s$. Observe that our completed L-function $\widetilde{L}(s, \varphi_j \otimes \overline{\varphi}_j)$ differs from the L-function defined in [4] because of the appearance of the multiplicative factor $s(1-s)$ in the definition of $G_{r_j}(s)$.

Similarly, one can define completed Rankin–Selberg L-functions associated with the nonholomorphic Eisenstein series $E_{P,s}(z)$ for any cusp P on X having a Fourier expansion of the form

$$E_{P,s}(z) = \delta_{P,\infty}y^s + \phi_{P,\infty}(s)y^{1-s} + \sum_{n \neq 0} \alpha_{P,s,n}W_s(nz/b)$$

with $\phi_{P,\infty}(s)$ denoting the (P, ∞)-th entry of the scattering matrix.

With all this, the main result of this article is the following theorem. For any $\varepsilon > 0$ and $s \in \mathbb{C}$ with $\mathrm{Re}(s) > 1$, define the Θ-function

$$\Theta_\varepsilon(s) = \sum_{\lambda_j > 0} \frac{\cosh(\pi r_j)e^{-\lambda_j \varepsilon}}{2\lambda_j}\widetilde{L}(s, \varphi_j \otimes \overline{\varphi}_j)$$

$$+ \frac{1}{8\pi}\sum_{P \text{ cusp}}\int_{-\infty}^{\infty}\frac{\cosh(\pi r)e^{-(r^2+1/4)\varepsilon}}{r^2+1/4}\widetilde{L}(s, E_{P,1/2+ir} \otimes \overline{E}_{P,1/2+ir})\,dr$$

and the universal function

$$F_\varepsilon(s) = \frac{\zeta(s)b^{s-1}}{2\pi^2}\int_0^{\infty}\frac{r\sinh(\pi r)e^{-(r^2+1/4)\varepsilon}}{r^2+1/4}G_r(s)\,dr.$$

Then the L-function relation involving Rankin–Selberg L-functions of cusp forms and Maass forms

$$\lim_{\varepsilon \to 0} \left(\Theta_\varepsilon(s) - F_\varepsilon(s) \right)$$

$$= \sum_{j=1}^{g} \widetilde{L}(s, f_j \otimes \overline{f}_j) - 4\pi \zeta(s) b^{s-1} G_\infty(s) - \pi^{-s} \frac{2s}{s+1} \Gamma(s) \zeta(2s) \phi_{\infty,\infty} \left(\frac{s+1}{2} \right)$$

$$(5)$$

holds true. By taking $\varepsilon > 0$ in (5), one has an error term which is $o(1)$ as ε approaches zero. This error term is explicit and given in terms of integrals involving the hyperbolic heat kernel.

A natural question to ask is to what extent the relation of L-functions (5) implies relations between the Fourier coefficients of the holomorphic weight two forms and the Fourier coefficients of the Maass forms under consideration. In general, extracting such information from a limiting relationship such as (5) could be very difficult. However, as stated, our analysis yields an explicit expression for the error term by rewriting (5) for a fixed $\varepsilon > 0$, which allows for additional considerations. The problem of using (5) to study possible relations among the Fourier coefficients is currently under investigation.

1.5   General Comments

If X is the Riemann surface associated with a congruence subgroup, then the series $\phi_{\infty,\infty}(s)$ can be expressed in terms of Dirichlet L-functions associated with even characters with conductors dividing the level (see [8] or [10]). With these computations, one can rewrite (5) further so that one obtains an expression involving Rankin–Selberg L-functions associated with cusp forms of weight two, Maass forms, nonholomorphic Eisenstein series, and classical zeta functions. However, the relation stated in (5) holds for any finite volume hyperbolic Riemann surface without elliptic fixed points. In order to eliminate the restriction that X has no elliptic fixed points, one needs to revisit the proof of (2), and possibly (1), in order to allow for elliptic fixed points. As stated above, this project currently is under investigation in [2]; however, we choose to focus in this paper on deriving (5) with the simplifying assumption that X has no elliptic fixed points in order to draw attention to the presence of an L-function relation coming from the basic identity (2). We will leave for future work the generalization of (2) to arbitrary finite volume hyperbolic Riemann surfaces, which may have elliptic fixed points, and derive the relation analogous to (5).

From Riemannian geometry, theta functions naturally appear as the trace of a heat kernel, and the small time expansion of the heat kernel has a first-order term which is somewhat universal and a second-order term which involves integrals of a curvature of the Riemannian metric. In this regard, (5) suggests that the sum of Rankin–Selberg L-functions

$$\sum_{j=1}^{g} \widetilde{L}(s, f_j \otimes \overline{f}_j)$$

represents some type of curvature integral relative to the theta function $\Theta_\varepsilon(s)$. Further investigation of this heuristic observation is warranted.

1.6   Outline of the Paper

In Sect. 2, we recall necessary background material and establish additional notation. In Sect. 3, we prove (2) and further develop the identity (2) using the spectral expansion of the heat kernel $K(t; z, w)$. In Sect. 4, we evaluate the integrals in (3) using the revised analytic expressions of (2), and in Sect. 5, we gather the computations from Sect. 4 and prove (5).

2   Notations and Preliminaries

2.1   Hyperbolic and Canonical Metrics

Let Γ be a Fuchsian subgroup of the first kind of $\mathrm{PSL}_2(\mathbb{R})$ acting by fractional linear transformations on the upper half-plane $\mathbb{H} = \{z \in \mathbb{C} \mid z = x + iy, \, y > 0\}$. We let X be the quotient space $\Gamma \backslash \mathbb{H}$ and denote by g the genus of X. We assume that Γ has no elliptic elements and that X has $p \geq 1$ cusps. We identify X locally with its universal cover $\mathbb{H}$.

In the sequel μ denotes a (smooth) metric on X, i.e., μ is a positive $(1, 1)$-form on X. In particular, we let $\mu = \mu_{\text{hyp}}$ denote the hyperbolic metric on X, which is compatible with the complex structure of X, and has constant negative curvature equal to -1. Locally, we have

$$\mu_{\text{hyp}}(z) = \frac{i}{2} \cdot \frac{dz \wedge d\overline{z}}{y^2}.$$

We write $\text{vol}_{\text{hyp}}(X)$ for the hyperbolic volume of X; recall that $\text{vol}_{\text{hyp}}(X)$ is given by $2\pi(2g - 2 + p)$. The scaled hyperbolic metric $\mu = \mu_{\text{shyp}}$ is simply the rescaled hyperbolic metric $\mu_{\text{hyp}}/\text{vol}_{\text{hyp}}(X)$, which measures the volume of X to be one.

Let $S_k(\Gamma)$ denote the $\mathbb{C}$-vector space of cusp forms of weight k with respect to Γ equipped with the Petersson inner product

$$\langle f, g \rangle = \frac{i}{2} \int_X f(z) \overline{g(z)} \, y^k \, \frac{dz \wedge d\bar{z}}{y^2} \qquad (f, g \in S_k(\Gamma)).$$

By choosing an orthonormal basis $\{f_1, \ldots, f_g\}$ of $S_2(\Gamma)$ with respect to the Petersson inner product, the canonical metric $\mu = \mu_{\mathrm{can}}$ of X is given by

$$\mu_{\mathrm{can}}(z) = \frac{1}{g} \cdot \frac{i}{2} \sum_{j=1}^{g} |f_j(z)|^2 \, dz \wedge d\bar{z}.$$

We denote the hyperbolic Laplacian on X by Δ_{hyp}; locally, we have

$$\Delta_{\mathrm{hyp}} = -y^2 \left(\frac{\partial^2}{\partial x^2} + \frac{\partial^2}{\partial y^2} \right). \qquad (6)$$

The discrete spectrum of Δ_{hyp} is given by the increasing sequence of eigenvalues

$$0 = \lambda_0 < \lambda_1 \leq \lambda_2 \leq \ldots$$

2.2 Modular Forms, Maass Forms, and Eisenstein Series

Throughout we assume, as before, that the cusp width of the cusp $i\infty$ equals b. In Sect. 1.4, we established the notation for holomorphic cusp forms of weight two and Maass forms with respect to Γ, as well as the corresponding Rankin–Selberg L-functions, so we do not repeat the discussion here.

The eigenfunctions for the continuous spectrum of Δ_{hyp} are provided by the Eisenstein series $E_{P,s'}$ (associated with each cusp P of X) with eigenvalue $\lambda = s'(1 - s')$, hence $s' = 1/2 + ir$ ($r \in \mathbb{R}$). They have Fourier expansions of the form

$$E_{P,s'}(z) = \alpha_{P,s',0}(y) + \sum_{n \neq 0} \alpha_{P,s',n} W_{s'}(nz/b),$$

where

$$\alpha_{P,s',0}(y) = \delta_{P,\infty} y^{s'} + \phi_{P,\infty}(s') y^{1-s'},$$

$$W_{s'}(w) = 2\sqrt{\cosh(\pi r)} \sqrt{|\mathrm{Im}(w)|} K_{ir}(2\pi |\mathrm{Im}(w)|) e^{2\pi i \mathrm{Re}(w)} \qquad (w \in \mathbb{C});$$

here $\delta_{P,\infty}$ is the Kronecker delta and $\phi_{P,\infty}(s')$ is the (P, ∞)-th entry of the scattering matrix (see [11]). For example, the function $\phi_{\infty,\infty}(s')$ is given by a Dirichlet series of the form

$$\phi_{\infty,\infty}(s') = \sqrt{\pi}\,\frac{\Gamma(s'-1/2)}{\Gamma(s')}\sum_{n=1}^{\infty}\frac{a_n}{\mu_n^{2s'}}, \tag{7}$$

where the quantities a_n and μ_n are explicitly given in [11], p. 60.

For $s \in \mathbb{C}$, $\mathrm{Re}(s) > 1$, we define the completed Rankin–Selberg L-function attached to $E_{P,s'}$ by

$$\widetilde{L}(s, E_{P,s'} \otimes \overline{E}_{P,s'}) = G_r(s) \cdot L(s, E_{P,s'} \otimes \overline{E}_{P,s'}), \tag{8}$$

where

$$G_r(s) = s(1-s)\pi^{-2s}\Gamma^2\left(\frac{s}{2}\right)\Gamma\left(\frac{s}{2}+ir\right)\Gamma\left(\frac{s}{2}-ir\right)\zeta(2s),$$

$$L(s, E_{P,s'} \otimes \overline{E}_{P,s'}) = \sum_{n \neq 0}\frac{|\alpha_{P,s',n}|^2}{(n/b)^{s-1}}.$$

2.3 Hyperbolic Heat Kernel and Variants

The hyperbolic heat kernel $K_{\mathbb{H}}(t; z, w)$ $(t \in \mathbb{R}_{>0};\ z, w \in \mathbb{H})$ on $\mathbb{H}$ is given by the formula

$$K_{\mathbb{H}}(t; z, w) = K_{\mathbb{H}}(t; \rho) = \frac{\sqrt{2}e^{-t/4}}{(4\pi t)^{3/2}}\int_{\rho}^{\infty}\frac{re^{-r^2/(4t)}}{\sqrt{\cosh(r)-\cosh(\rho)}}\,dr,$$

where $\rho = d_{\mathrm{hyp}}(z, w)$ denotes the hyperbolic distance from z to w. The hyperbolic heat kernel $K(t; z, w)$ $(t \in \mathbb{R}_{>0};\ z, w \in X)$ on X is obtained by averaging over the elements of Γ, namely

$$K(t; z, w) = \sum_{\gamma \in \Gamma} K_{\mathbb{H}}\big(t; z, \gamma(w)\big).$$

The heat kernel on X satisfies the equations

$$\left(\frac{\partial}{\partial t} + \Delta_{\mathrm{hyp},z}\right)K(t; z, w) = 0 \qquad (w \in X),$$

$$\lim_{t \to 0}\int_X K(t; z, w)\,f(w)\,\mu_{\mathrm{hyp}}(w) = f(z) \quad (z \in X)$$

for all C^{∞}-functions f on X. As a shorthand, we write $K(t; z) = K(t; z, z)$.

With the notations from Sect. 2.2, we introduce the modified heat kernel function

$$K^{\mathrm{cusp}}(t;z) = K(t;z) - \sum_{0 \le \lambda_j < 1/4} |\alpha_{j,0}|^2 y^{2-2s_j} e^{-\lambda_j t}$$

$$- \frac{1}{4\pi} \sum_{P \text{ cusp}} \int_{-\infty}^{\infty} |\delta_{P,\infty} y^{1/2+ir} + \phi_{P,\infty}(s) y^{1/2-ir}|^2 e^{-(r^2+1/4)t} dr. \quad (9)$$

Denoting by Γ_∞ the stabilizer of the cusp ∞, we can define the following partial heat kernel functions

$$K_0(t;z) = \sum_{\gamma \in \Gamma \backslash \Gamma_\infty} K_{\mathbb{H}}(t;z,\gamma(z)), \quad (10)$$

$$K_\infty(t;z) = \sum_{\gamma \in \Gamma_\infty} K_{\mathbb{H}}(t;z,\gamma(z)) \quad (11)$$

giving rise to the decomposition

$$K(t;z) = K_0(t;z) + K_\infty(t;z).$$

3   The Fundamental Identity

In this section, we derive the identity (2) by studying the relation (1) for a degenerating family of compact hyperbolic Riemann surfaces. The corresponding statement is proven in Lemma 3.1. In the remainder of the section, we manipulate the terms in (2) assuming $p > 0$ in order to obtain an equivalent formulation of the relation which then will be suited for our computations in the subsequent sections. Specifically, we first express the heat kernel on the underlying Riemann surface in terms of its spectral expansion, which involves Maass forms and nonholomorphic Eisenstein series, and we remove the terms associated with the constant terms in the Fourier expansions of the Maass forms and the nonholomorphic Eisenstein series (see Proposition 3.3). We then express the heat kernel as a periodization over the uniformizing group and remove the contribution from the parabolic subgroup associated with a single cusp (see Lemma 3.8 as well as the preliminary computations and remarks). The main result of this section is Theorem 3.9.

Lemma 3.1. *With the above notations, we have*

$$\mu_{\mathrm{can}}(z) = \left(1 + \frac{p}{2g}\right) \mu_{\mathrm{shyp}}(z) + \frac{1}{2g} \int_0^{\infty} \Delta_{\mathrm{hyp}} K(t;z) \, dt \, \mu_{\mathrm{hyp}}(z). \quad (12)$$

Proof. The proof of identity (12) in case X is compact, i.e. $p = 0$, for any $g \geq 2$ is given in [13] as well as the appendix to [16]. We will now prove (12) by induction on p by considering degenerating sequences of finite volume hyperbolic Riemann surfaces. More specifically, we assume that (12) holds for any hyperbolic Riemann surface of genus g with p cusps, and then prove the relation for hyperbolic Riemann surfaces of any genus with $p+1$ cusps. Whereas the method of proof can be viewed as standard perturbation theory, we choose to include all details in order to determine all constants, specifically the multiplicative factor of μ_{hyp} in (2).

If X has genus g and $p + 1$ cusps, then, following the methodology of [12] and [18], one can construct a degenerating family $\{X_\ell\}$ with the following properties:

- For $\ell > 0$, each surface X_ℓ has genus $g + 1$ and p cusps,
- the degenerating family has precisely one pinching geodesic of length ℓ approaching zero,
- the limiting surface X_0, which necessarily has two components, is such that X is isometric to one of the two components.

Let X and X' be the two components of X_0 with hyperbolic volumes $v = \text{vol}_{\text{hyp}}(X)$ and $v' = \text{vol}_{\text{hyp}}(X')$, respectively; by construction, X' has genus one and one cusp. The hyperbolic volume of X_ℓ equals $v + v'$, and the induction hypothesis for X_ℓ reads (using an obvious change in notation)

$$2(g + 1)\mu_{\text{can}, X_\ell}(z) = \big(2(g + 1) + p\big)\mu_{\text{shyp}, X_\ell}(z)$$

$$+ \int_0^\infty \Delta_{\text{hyp}, X_\ell} K_{X_\ell}(t; z)\, dt\; \mu_{\text{hyp}, X_\ell}(z). \qquad (13)$$

We now determine the limiting value of (13) through degeneration. Throughout, we will let $z \in X_\ell$ be any point which limits to a point $z \in X$.

From [12], we have that

$$\lim_{\ell \to 0} \big(2(g + 1)\mu_{\text{can}, X_\ell}(z)\big) = 2g\mu_{\text{can}, X}(z). \qquad (14)$$

From [1], we recall that

$$\lim_{\ell \to 0} \big(\mu_{\text{hyp}, X_\ell}(z)\big) = \mu_{\text{hyp}, X}(z),$$

which leads to

$$\lim_{\ell \to 0} \big((2(g + 1) + p)\mu_{\text{shyp}, X_\ell}(z)\big) = \frac{2(g + 1) + p}{v + v'}\mu_{\text{hyp}, X}(z). \qquad (15)$$

Let now λ_{1,X_ℓ} denote the smallest nonzero eigenvalue of the hyperbolic Laplacian $\Delta_{\mathrm{hyp},X_\ell}$ on X_ℓ, with corresponding eigenfunction φ_{1,X_ℓ}. From [18], we have that

$$\lim_{\ell \to 0} \left(K_{X_\ell}(t;z) - \frac{1}{v+v'} - \varphi_{1,X_\ell}^2(z)e^{-\lambda_{1,X_\ell}t} \right) = K_X(t;z) - \frac{1}{v}$$

with uniformity of the convergence for all $t > 0$ (see [18], Lemma 3.2). The proof given in [18] extends (see Remark 3.2) to show that

$$\lim_{\ell \to 0} \Delta_{\mathrm{hyp},X_\ell} \left(K_{X_\ell}(t;z) - \frac{1}{v+v'} - \varphi_{1,X_\ell}^2(z)e^{-\lambda_{1,X_\ell}t} \right) = \Delta_{\mathrm{hyp},X} \left(K_X(t;z) - \frac{1}{v} \right),$$
(16)

with a corresponding uniformity result, which allows us to arrive at the conclusion that

$$\lim_{\ell \to 0} \left(\int_0^\infty \Delta_{\mathrm{hyp},X_\ell} K_{X_\ell}(t;z)\, dt - \frac{\Delta_{\mathrm{hyp},X_\ell}\varphi_{1,X_\ell}^2(z)}{\lambda_{1,X_\ell}} \right) = \int_0^\infty \Delta_{\mathrm{hyp},X} K_X(t;z)\, dt.$$
(17)

By substituting the limit computations (14), (15), and (17) into (13), we are led to

$$2g\mu_{\mathrm{can},X}(z) = \int_0^\infty \Delta_{\mathrm{hyp},X} K_X(t;z)\, dt\, \mu_{\mathrm{hyp},X}(z)$$

$$+ \left(\frac{2(g+1)+p}{v+v'} + \lim_{\ell \to 0} \left(\frac{\Delta_{\mathrm{hyp},X_\ell}\varphi_{1,X_\ell}^2(z)}{\lambda_{1,X_\ell}} \right) \right) \mu_{\mathrm{hyp},X}(z),$$

so we are left to prove that

$$\frac{2(g+1)+p}{v+v'} + \lim_{\ell \to 0} \left(\frac{\Delta_{\mathrm{hyp},X_\ell}\varphi_{1,X_\ell}^2(z)}{\lambda_{1,X_\ell}} \right) = \frac{2g+(p+1)}{v}.$$
(18)

The construction of the degenerating family $\{X_\ell\}$ from [18] begins by constructing a degenerating family of compact Riemann surfaces with distinguished points, after which one obtains a degenerating family of finite volume hyperbolic Riemann surfaces by employing the uniformization theorem. As a result, there is an underlying real parameter u, which describes the degenerating family $\{X_\ell\}$. An asymptotic relation between u and ℓ is established in [21]; for our purposes, it suffices to use that $\ell \to 0$ as $u \to 0$, and conversely. With all this, it is proven in [21] that one has the asymptotic expansion

$$\lambda_{1,X_\ell} = \alpha_1 u + O(u^2) \quad \text{as} \quad u \to 0$$
(19)

for some constant α_1. In addition, one has from [21] the asymptotic expansions

$$\varphi_{1,X_\ell}(z) = c_{0,X}(z) + c_{1,X}(z)u + O(u^2) \quad \text{as} \quad u \to 0 \qquad (z \in X), \qquad (20)$$

and

$$\varphi_{1,X_\ell}(z) = c_{0,X'}(z) + c_{1,X'}(z)u + O(u^2) \quad \text{as} \quad u \to 0 \qquad (z \in X'). \qquad (21)$$

In [18], it is proven that small eigenvalues and small eigenfunctions converge through degeneration; hence, the functions $c_{0,X}$ and $c_{0,X'}$ are constants. More precisely, since φ_{1,X_ℓ} is orthogonal to the constant functions on X_ℓ and has L^2-norm one, we have the relations

$$c_{0,X}v + c_{0,X'}v' = 0 \quad \text{and} \quad c_{0,X}^2 v + c_{0,X'}^2 v' = 1,$$

from which we immediately derive

$$c_{0,X} = \pm \left(\frac{v'}{v(v+v')} \right)^{1/2} \quad \text{and} \quad c_{0,X'} = \mp \left(\frac{v}{v'(v+v')} \right)^{1/2}. \qquad (22)$$

The uniformity of the convergence of heat kernels through degeneration from [18] and the convergence of hyperbolic metrics through degeneration from [1], allow one to conclude that, since φ_{1,X_ℓ} is an eigenfunction of $\Delta_{\mathrm{hyp},X_\ell}$ with eigenvalue λ_{1,X_ℓ}, the asymptotic expansions (19) and (20) yield the relation (keeping in mind that the function $c_{0,X}$ is constant)

$$\Delta_{\mathrm{hyp},X} c_{1,X}(z) = \alpha_1 c_{0,X}. \qquad (23)$$

In the same way, we derive from (20) the asymptotic expansion

$$\Delta_{\mathrm{hyp},X_\ell} \varphi_{1,X_\ell}^2(z) = \Delta_{\mathrm{hyp},X} c_{0,X}^2(z) + \Delta_{\mathrm{hyp},X}\left(2c_{0,X}(z)c_{1,X}(z)\right)u + O(u^2)$$

$$= 2c_{0,X}\Delta_{\mathrm{hyp},X} c_{1,X}(z)u + O(u^2) \quad \text{as} \quad u \to 0. \qquad (24)$$

Using (19), (22), (23), and (24), we arrive at

$$\lim_{\ell \to 0} \left(\frac{\Delta_{\mathrm{hyp},X_\ell}\varphi_{1,X_\ell}^2(z)}{\lambda_{1,X_\ell}} \right) = \lim_{u \to 0} \left(\frac{2c_{0,X}\Delta_{\mathrm{hyp},X} c_{1,X}(z)u + O(u^2)}{\alpha_1 u + O(u^2)} \right)$$

$$= 2c_{X,0}^2 = \frac{2v'}{v(v+v')}.$$

Recalling the formulae

$$v = 2\pi\left(2g - 2 + (p+1)\right) \quad \text{and} \quad v' = 2\pi,$$

we finally compute

$$\frac{2(g+1)+p}{v+v'} + \frac{2v'}{v(v+v')} = \frac{v(v/(2\pi)+3)}{v(v+v')} + \frac{2v'}{v(v+v')}$$

$$= \frac{1}{2\pi}\frac{v^2+3vv'+2v'^2}{v(v+v')} = \frac{1}{2\pi}\frac{v+2v'}{v} = \frac{2g+(p+1)}{v},$$

which completes the proof of claim (18) and hence the proof of the lemma.   □

Remark 3.2. We describe here how one can extend the arguments from [18] and references therein to prove formula (16); we continue to use the notation from the proof of Lemma 3.1. The pointwise convergence

$$\lim_{\ell \to 0} \Delta_{\mathrm{hyp},X_\ell} K_{X_\ell}(t;z) = \Delta_{\mathrm{hyp},X} K_X(t;z) \tag{25}$$

follows immediately from [17], Theorem 1.3 (iii). Using the inverse Laplace transform, one concludes from (25) the convergence of small eigenvalues and small eigenfunctions (see, for example, [9] for complete details) to conclude that (16) holds pointwise for all $t > 0$. Theorem 1.3 in [17] states further conditions under which the convergence in (25) is uniform, which immediately implies that the convergence in (16) holds for fixed z and t lying in any bounded, compact subset of $t > 0$, so it remains to prove uniform convergence for t near zero and near infinity. The uniformity of the convergence near zero is established as part of the proof of Theorem 1.3 in [17] since the identity term does not contribute to the realization of the heat kernel through group periodization. What remains is to prove uniformity of the convergence in (16) as t approaches infinity. For this, the method of proof of Lemma 3.2 in [18] applies. More specifically, one writes the function

$$\Delta_{\mathrm{hyp},X_\ell}\left(K_{X_\ell}(t;z) - \frac{1}{v+v'} - \varphi_{1,X_\ell}^2(z)e^{-\lambda_{1,X_\ell}t}\right)$$

as the Laplace transform of a measure as in [18], p. 649. In this case, the measure is not bounded, but standard bounds for the sup-norm of L^2-eigenfunctions of the Laplacian imply that the measure is bounded by a positive measure, which suffices to apply the method of proof of Lemma 3.2 in [18]. With all this, one concludes the pointwise convergence asserted in (16) and integrable, uniform bounds for all $t > 0$, from which (17) follows.

Proposition 3.3. *With the above notations, in particular using the form (7) for the function* $\phi_{\infty,\infty}(s')$, *we have*

$$\mu_{\mathrm{can}}(z) = \frac{1}{4\pi g}\mu_{\mathrm{hyp}}(z) + \frac{1}{2g}\int_0^\infty \Delta_{\mathrm{hyp}} K^{\mathrm{cusp}}(t;z)\,dt\,\mu_{\mathrm{hyp}}(z)$$

$$+ \frac{1}{g}\sum_{\mu_n < 1/y}\frac{2a_n\mu_n y^3}{\sqrt{1-(\mu_n y)^2}}\mu_{\mathrm{hyp}}(z). \tag{26}$$

We point out that the sum in (26) vanishes if $y \gg 0$.

Proof. The proof is based on formula (12) from Lemma 3.1 and consists in substituting the integrand $K(t;z)$ by $K^{\mathrm{cusp}}(t;z)$. We compute

$$\Delta_{\mathrm{hyp}} K(t;z) = \Delta_{\mathrm{hyp}} K^{\mathrm{cusp}}(t;z) - \sum_{0\le\lambda_j<1/4}|\alpha_{j,0}|^2(2-2s_j)(1-2s_j)y^{2-2s_j}e^{-\lambda_j t}$$

$$-\frac{1}{4\pi}\sum_{P\;\mathrm{cusp}}\int_{-\infty}^\infty y^2\frac{\partial^2}{\partial y^2}\Big(\delta_{P,\infty}y + |\phi_{P,\infty}(1/2+ir)|^2 y$$

$$+\delta_{P,\infty}\phi_{P,\infty}(1/2+ir)y^{1-2ir} + \delta_{P,\infty}\overline{\phi}_{P,\infty}(1/2+ir)y^{1+2ir}\Big)e^{-(r^2+1/4)t}\,dr$$

$$= \Delta_{\mathrm{hyp}} K^{\mathrm{cusp}}(t;z) - \sum_{0\le\lambda_j<1/4}|\alpha_{j,0}|^2(2-2s_j)(1-2s_j)y^{2-2s_j}e^{-\lambda_j t}$$

$$-\frac{1}{4\pi i}\int_{\mathrm{Re}(s)=1/2}\Big(\phi_{\infty,\infty}(s)(2-2s)(1-2s)y^{2-2s}$$

$$+\phi_{\infty,\infty}(1-s)2s(2s-1)y^{2s}\Big)e^{-s(1-s)t}\,ds.$$

Next, we integrate against t to get

$$\int_0^\infty \Delta_{\mathrm{hyp}} K(t;z)\,dt = \int_0^\infty \Delta_{\mathrm{hyp}} K^{\mathrm{cusp}}(t;z)\,dt$$

$$-\sum_{0\le\lambda_j<1/4}|\alpha_{j,0}|^2\frac{(2-2s_j)(1-2s_j)}{\lambda_j}y^{2-2s_j}$$

$$-\frac{1}{4\pi i}\int_{\mathrm{Re}(s)=1/2}\Big(\phi_{\infty,\infty}(s)(2-2s)(1-2s)y^{2-2s}$$

$$+\phi_{\infty,\infty}(1-s)2s(2s-1)y^{2s}\Big)\frac{ds}{s(1-s)}$$

$$= \int_0^\infty \Delta_{\mathrm{hyp}} K^{\mathrm{cusp}}(t;z)\,dt - \sum_{0\le\lambda_j<1/4}|\alpha_{j,0}|^2\frac{(2-2s_j)(1-2s_j)}{\lambda_j}y^{2-2s_j}$$

$$-\frac{4}{4\pi i}\int_{\mathrm{Re}(s)=1/2}\phi_{\infty,\infty}(s)\frac{1-2s}{s}y^{2-2s}\,ds.$$

Now we use the residue theorem to evaluate the last integral (be aware of the orientation).

$$-\frac{4}{4\pi i}\int_{\mathrm{Re}(s)=1/2}\phi_{\infty,\infty}(s)\frac{1-2s}{s}y^{2-2s}\,ds$$

$$=-\sum_{\text{residues }s_j}(-2)\mathrm{Res}_{s=s_j}(\phi_{\infty,\infty}(s))\frac{1-2s_j}{s_j}y^{2-2s_j}$$

$$+2\left(-\frac{1}{2\pi i}\right)\int_{\mathrm{Re}(s)=a}\phi_{\infty,\infty}(s)\frac{1-2s}{s}y^{2-2s}\,ds;$$

here $a>1$. It is known that the residues of $\phi_{\infty,\infty}$ occur at $s=1$ with residue $1/\mathrm{vol}_{\mathrm{hyp}}(X)$ and at $s=s_j$ such that $0<\lambda_j=s_j(1-s_j)<1/4$ with residue $|\alpha_{j,0}|^2$ (see [20], p. 652). Therefore, we get

$$-\frac{4}{4\pi i}\int_{\mathrm{Re}(s)=1/2}\phi_{\infty,\infty}(s)\frac{1-2s}{s}y^{2-2s}\,ds$$

$$=-\frac{2}{\mathrm{vol}_{\mathrm{hyp}}(X)}+\sum_{0<\lambda_j<1/4}|\alpha_{j,0}|^2\frac{(2-2s_j)(1-2s_j)}{\lambda_j}y^{2-2s_j}$$

$$+\frac{2}{2\pi i}\int_{\mathrm{Re}(s)=a}\phi_{\infty,\infty}(s)\frac{2s-1}{s}y^{2-2s}\,ds.$$

We are left to determine the latter integral. By substituting formula (7) for $\phi_{\infty,\infty}$ and using the functional equation for the Γ-function, we first compute

$$\frac{1}{2\pi i}\int_{\mathrm{Re}(s)=a}\phi_{\infty,\infty}(s)\frac{2s-1}{s}y^{2-2s}\,ds$$

$$=\sum_{n=1}^{\infty}2\sqrt{\pi}a_ny^2\cdot\frac{1}{2\pi i}\int_{\mathrm{Re}(s)=a}\frac{\Gamma(s+1/2)}{\Gamma(s+1)}\left(\frac{1}{(\mu_ny)^2}\right)^s\,ds$$

$$=\sum_{n=1}^{\infty}2a_ny^2\cdot\frac{1}{2\pi i}\int_{\mathrm{Re}(s)=a}\sqrt{\pi}\frac{\Gamma(s+1/2)}{\Gamma(s+1)}e^{st_n}\,ds,$$

where $t_n=-\log\left((\mu_ny)^2\right)$. Recalling formula (10.5) of [19], p. 307, namely

$$\frac{1}{2\pi i}\int_{\mathrm{Re}(s)=a}\sqrt{\pi}\frac{\Gamma(s+1/2)}{\Gamma(s+1)}e^{st}\,ds=\begin{cases}\dfrac{1}{\sqrt{e^t-1}}, & t>0,\\[2mm]0, & t<0,\end{cases}$$

we obtain

$$\frac{1}{2\pi i}\int_{\mathrm{Re}(s)=a}\phi_{\infty,\infty}(s)\frac{2s-1}{s}y^{2-2s}\,ds = \sum_{t_n>0}\frac{2a_n y^2}{\sqrt{e^{t_n}-1}} = \sum_{\mu_n<1/y}\frac{2a_n\mu_n y^3}{\sqrt{1-(\mu_n y)^2}}.$$

Summing up, we get

$$\int_0^\infty \Delta_{\mathrm{hyp}}K(t;z)\,dt = \int_0^\infty \Delta_{\mathrm{hyp}}K^{\mathrm{cusp}}(t;z)\,dt - \frac{2}{\mathrm{vol}_{\mathrm{hyp}}(X)} + \sum_{\mu_n<1/y}\frac{4a_n\mu_n y^3}{\sqrt{1-(\mu_n y)^2}}.$$

The claim now follows by observing that

$$\left(1+\frac{p}{2g}\right)\mu_{\mathrm{shyp}}(z) - \frac{1}{2g}\cdot\frac{2}{\mathrm{vol}_{\mathrm{hyp}}(X)}\mu_{\mathrm{hyp}}(z) = \frac{1}{4\pi g}\mu_{\mathrm{hyp}}(z).$$

This completes the proof of the proposition. $\qquad\qquad\qquad\qquad\qquad\qquad$ $\square$

Remark 3.4. By our definition, the partial heat kernel $K_\infty(t;z)$ is given by

$$K_\infty(t;z) = \sum_{n=-\infty}^{\infty} K_{\mathbb{H}}(t;z,z+nb).$$

Recalling the formula for the hyperbolic distance $d_{\mathrm{hyp}}(z,w)$, namely (see [3], p. 130)

$$\cosh\left(d_{\mathrm{hyp}}(z,w)\right) = 1 + \frac{|z-w|^2}{2\mathrm{Im}(z)\mathrm{Im}(w)},$$

which specializes to

$$\cosh\left(d_{\mathrm{hyp}}(z,z+nb)\right) = 1 + \frac{(nb)^2}{2y^2},$$

shows that the function $K_{\mathbb{H}}(t;z,z+nb)$ is independent of x, and hence can be represented in the form

$$K_{\mathbb{H}}(t;z,z+nb) = f_t\left(\frac{b}{\sqrt{2}y}n\right) \qquad\qquad (27)$$

with $f_t(w) = K_{\mathbb{H}}(t;\cosh^{-1}(1+w^2))$. Therefore, we can write

$$K_\infty(t;z) = \sum_{n=-\infty}^{\infty} f_t\left(\frac{b}{\sqrt{2}y}n\right). \qquad\qquad (28)$$

By the general Poisson formula, we then have

$$\sum_{n=-\infty}^{\infty} f_t\left(\frac{b}{\sqrt{2}y}n\right) = \frac{\sqrt{2}y}{b} \sum_{n=-\infty}^{\infty} \widehat{f}_t\left(\frac{2\pi\sqrt{2}y}{b}n\right),$$

where $\widehat{f}_t(v)$ denotes the Fourier transform of $f_t(w)$ given by

$$\widehat{f}_t(v) = \int_{-\infty}^{\infty} f_t(w)e^{-iwv}dw.$$

Summarizing we arrive at

$$K_\infty(t;z) = \frac{\sqrt{2}y}{b}\widehat{f}_t(0) + \frac{2\sqrt{2}y}{b}\sum_{n=1}^{\infty}\widehat{f}_t\left(\frac{2\pi\sqrt{2}y}{b}n\right). \qquad (29)$$

Definition 3.5. With the above notations, we set

$$K_\infty^{\mathrm{cusp}}(t;z) = K_\infty(t;z) - \frac{\sqrt{2}y}{b}\widehat{f}_t(0),$$

$$K_0^{\mathrm{cusp}}(t;z) = K^{\mathrm{cusp}}(t;z) - K_\infty^{\mathrm{cusp}}(t;z).$$

Lemma 3.6. *For the Fourier transform $\widehat{f}_t$ of f_t, we have the formula*

$$\widehat{f}_t(v) = \frac{\sqrt{2}}{\pi^2}\int_0^{\infty} r\sinh(\pi r)e^{-(r^2+1/4)t}K_{ir}^2(v/\sqrt{2})\,dr.$$

Proof. Using the explicit formula for the heat kernel on the upper half-plane (see [5], p. 246), we have

$$K_{\mathbb{H}}(t;z,w) = \frac{1}{2\pi}\int_0^{\infty} r\tanh(\pi r)e^{-(r^2+1/4)t}P_{-1/2+ir}\big(\cosh(d_{\mathrm{hyp}}(z,w))\big)\,dr,$$

from which we get

$$f_t(w) = \frac{1}{2\pi}\int_0^{\infty} r\tanh(\pi r)e^{-(r^2+1/4)t}P_{-1/2+ir}(1+w^2)\,dr. \qquad (30)$$

Taking into account that $f_t(w)$ is an even function, the Fourier transform $\widehat{f}_t$ of f_t can be written in the form

$$\widehat{f}_t(v) = \int_{-\infty}^{\infty} f_t(w)e^{-iwv}dw = 2\int_0^{\infty} f_t(w)\cos(wv)\,dw.$$

By means of formula 7.162 (5) of [7], p. 807, the proof of the lemma can now be easily completed.   □

Lemma 3.7. *The function $K_\infty^{\mathrm{cusp}}(t;z)$ decays exponentially as y tends to infinity.*

Proof. From Lemma 3.6, we note that the function $\widehat{f}_t(v)$ decays exponentially as v tends to infinity. From this, we immediately conclude that $K_\infty^{\mathrm{cusp}}(t;z)$ decays exponentially as y tends to infinity.   □

Lemma 3.8. *With the above notations, we have*

$$\int_0^\infty \Delta_{\mathrm{hyp}} K_\infty(t;z)\,dt = \frac{1}{2\pi}\left(\frac{2\pi y/b}{\sinh(2\pi y/b)}\right)^2 - \frac{1}{2\pi}. \tag{31}$$

Proof. First, we recall for $z, w \in \mathbb{H}$, $z \neq w$, the relation

$$\int_0^\infty K_{\mathbb{H}}(t;z,w)\,dt = -\frac{1}{4\pi}\log\left(\left|\frac{z-w}{z-\overline{w}}\right|^2\right).$$

Substituting $w = \gamma(z)$, summing over $\gamma \in \Gamma_\infty$, $\gamma \neq \mathrm{id}$, and applying Δ_{hyp}, then yields the formula

$$\int_0^\infty \Delta_{\mathrm{hyp}} K_\infty(t;z)\,dt = -\frac{1}{4\pi}\sum_{\substack{n=-\infty\\n\neq 0}}^\infty \Delta_{\mathrm{hyp}}\log\left(\left|\frac{z-(z+nb)}{z-(\overline{z}+nb)}\right|^2\right)$$

$$= -\frac{2y^2}{\pi}\sum_{\substack{n=-\infty\\n\neq 0}}^\infty \frac{(nb)^2-4y^2}{((nb)^2+4y^2)^2} = -\frac{2y^2}{\pi b^2}\sum_{\substack{n=-\infty\\n\neq 0}}^\infty \frac{n^2-(2y/b)^2}{(n^2+(2y/b)^2)^2}.$$

Applying now formula 1.421 (5) of [7], p. 36, namely

$$\sum_{n=-\infty}^\infty \frac{n^2-w^2}{(n^2+w^2)^2} = -\left(\frac{\pi}{\sinh(\pi w)}\right)^2,$$

with $w = 2y/b$, immediately completes the proof of the lemma.   □

Theorem 3.9. *We set*

$$\Phi(y) = \left(\frac{2\pi y/b}{\sinh(2\pi y/b)}\right)^2.$$

With the above notations, we then have the fundamental identity

$$\mu_{\text{can}}(z) = \frac{1}{2g} \int_0^\infty \Delta_{\text{hyp}} K_0^{\text{cusp}}(t;z)\, dt\, \mu_{\text{hyp}}(z) + \frac{1}{4\pi g} \Phi(y)\mu_{\text{hyp}}(z)$$

$$+ \frac{1}{g} \sum_{\mu_n < 1/y} \frac{2a_n\mu_n y^3}{\sqrt{1-(\mu_n y)^2}} \mu_{\text{hyp}}(z). \tag{32}$$

Proof. The proof consists in combining Proposition 3.3 with Lemma 3.8 together with the observation that

$$\Delta_{\text{hyp}} K^{\text{cusp}}(t;z) = \Delta_{\text{hyp}}\big(K_0^{\text{cusp}}(t;z) + K_\infty^{\text{cusp}}(t;z)\big)$$

$$= \Delta_{\text{hyp}}\big(K_0^{\text{cusp}}(t;z) + K_\infty(t;z)\big),$$

since $\Delta_{\text{hyp}}\big(y\widehat{f}_t(0)\big) = 0$. $\square$

4 Preliminary Computations

We will multiply the fundamental identity (32) of Theorem 3.9 with the function

$$h(s,y) = \frac{2g}{b}\pi^{-s}\Gamma(s)\zeta(2s)y^s \tag{33}$$

and integrate the resulting form along x and y. In this section, we first calculate the integrals involving the form μ_{can}, the function Φ, and the sum over the μ_n's, respectively. In the second part of the section, we treat the term involving K_0^{cusp} partly; this computation will be completed in the next section.

Lemma 4.1. *With the above notations, we have*

$$\int_0^\infty \int_0^b h(s,y)\mu_{\text{can}}(z) = \sum_{j=1}^g \widetilde{L}\left(s, f_j \otimes \overline{f}_j\right). \tag{34}$$

Proof. The proof is elementary, so we omit further details. $\square$

Lemma 4.2. *With the above notations, we have*

$$\frac{1}{4\pi g} \int_0^\infty \int_0^b h(s,y)\Phi(y)\mu_{\text{hyp}}(z) = 4\pi\zeta(s)b^{s-1}G_\infty(s). \tag{35}$$

Proof. We start with the following observation. By differentiating the relation

$$\frac{1}{1-e^{-2w}} = \sum_{n=0}^{\infty} e^{-2nw}$$

we get

$$\frac{e^{-2w}}{(1-e^{-2w})^2} = \sum_{n=1}^{\infty} n e^{-2nw},$$

which gives

$$\frac{1}{\sinh^2(w)} = \frac{4}{(e^w - e^{-w})^2} = \frac{4e^{-2w}}{(1-e^{-2w})^2} = 4\sum_{n=1}^{\infty} n e^{-2nw}.$$

We now turn to the proof of the lemma. We compute

$$\frac{1}{4\pi g} \int_0^\infty \int_0^b h(s,y)\Phi(y)\mu_{\mathrm{hyp}}(z) = \frac{\pi^{-s}\Gamma(s)\zeta(2s)}{2\pi} \int_0^\infty y^s \Phi(y)\frac{dy}{y^2}$$

$$= \frac{\pi^{-s}\Gamma(s)\zeta(2s)}{2\pi} \cdot \frac{(2\pi)^2}{b^2} \int_0^\infty \frac{y^s}{\sinh^2(2\pi y/b)}dy$$

$$= \frac{\pi^{-s}\Gamma(s)\zeta(2s)}{2\pi} \cdot \frac{(2\pi)^2}{b^2} \int_0^\infty 4y^{s+1} \sum_{n=1}^{\infty} n e^{-4\pi ny/b}\frac{dy}{y}$$

$$= 2^3 \pi^{-s+1}\Gamma(s)\zeta(2s)b^{-2} \sum_{n=1}^{\infty} n \int_0^\infty y^{s+1} e^{-4\pi ny/b}\frac{dy}{y}$$

$$= 2^3 \pi^{-s+1}\Gamma(s)\zeta(2s)b^{-2}\Gamma(s+1) \sum_{n=1}^{\infty} \frac{n}{(4\pi n/b)^{s+1}}$$

$$= 2^{-2s+1} \pi^{-2s}\Gamma(s)\Gamma(s+1)\zeta(s)\zeta(2s)b^{s-1}.$$

The claim now follows using the definition of the function $G_\infty(s)$.   □

Lemma 4.3. *With the above notations, we have*

$$\frac{1}{g} \int_0^\infty \int_0^b h(s,y) \sum_{\mu_n < 1/y} \frac{a_n \mu_n y}{\sqrt{1-(\mu_n y)^2}}dxdy$$

$$= \pi^{-s}\frac{s}{s+1}\Gamma(s)\zeta(2s)\phi_{\infty,\infty}\left(\frac{s+1}{2}\right).$$ (36)

Proof. Using the *B*-function, we compute

$$\frac{1}{g}\int_0^\infty \int_0^b h(s,y) \sum_{\mu_n<1/y} \frac{a_n\mu_n y}{\sqrt{1-(\mu_n y)^2}}\,\mathrm{d}x\mathrm{d}y$$

$$= 2\pi^{-s}\Gamma(s)\zeta(2s)\int_0^\infty \sum_{y<1/\mu_n} \frac{a_n\mu_n y^{s+1}}{\sqrt{1-(\mu_n y)^2}}\,\mathrm{d}y$$

$$= 2\pi^{-s}\Gamma(s)\zeta(2s)\sum_{n=1}^\infty \frac{a_n}{\mu_n^{s+1}} \int_0^1 \frac{w^{s+1}}{\sqrt{1-w^2}}\,\mathrm{d}w$$

$$= \pi^{-s}\Gamma(s)\zeta(2s)B\left(\frac{s}{2}+1,\frac{1}{2}\right) \sum_{n=1}^\infty \frac{a_n}{\mu_n^{s+1}}$$

$$= \pi^{-s}\Gamma(s)\zeta(2s)\frac{\Gamma(s/2+1)\Gamma(1/2)}{\Gamma\big((s+3)/2\big)} \sum_{n=1}^\infty \frac{a_n}{\mu_n^{2\frac{s+1}{2}}}$$

$$= \pi^{-s}\Gamma(s)\zeta(2s)\frac{s/2\,\Gamma\big((s+1)/2-1/2\big)\sqrt{\pi}}{(s+1)/2\,\Gamma\big((s+1)/2\big)} \sum_{n=1}^\infty \frac{a_n}{\mu_n^{2\frac{s+1}{2}}}$$

$$= \pi^{-s}\frac{s}{s+1}\Gamma(s)\zeta(2s)\phi_{\infty,\infty}\left(\frac{s+1}{2}\right). \qquad\qquad \square$$

Remark 4.4. For $\varepsilon > 0$, we can write

$$\frac{1}{2g}\int_0^\infty \int_0^b \int_0^\infty h(s,y)\Delta_{\mathrm{hyp}}K_0^{\mathrm{cusp}}(t;z)\,\mathrm{d}t\mathrm{d}x\frac{\mathrm{d}y}{y^2}$$

$$= \frac{1}{2g}\int_\varepsilon^\infty \int_0^\infty \int_0^b h(s,y)\Delta_{\mathrm{hyp}}K_0^{\mathrm{cusp}}(t;z)\mathrm{d}x\frac{\mathrm{d}y}{y^2}\mathrm{d}t + o(1)$$

as $\varepsilon \to 0$. Using now the specific form of the hyperbolic Laplacian, we integrate by parts in each real variable x and y. Since the integrand is invariant under $x \mapsto x+b$, the terms involving derivatives with respect to x will vanish. What remains to be done is the integration by parts with respect to y. Substituting

$$K_0^{\mathrm{cusp}}(t;z) = K^{\mathrm{cusp}}(t;z) - K_\infty^{\mathrm{cusp}}(t;z),$$

we arrive in this way at the formula

$$\frac{1}{2g} \int_0^\infty \int_0^b \int_0^\infty h(s, y) \Delta_{\text{hyp}} K_0^{\text{cusp}}(t; z) \, dt \, dx \, \frac{dy}{y^2}$$

$$= \frac{s(1-s)}{2g} \lim_{\varepsilon \to 0} \left[\int_\varepsilon^\infty \int_0^\infty \int_0^b h(s, y) K^{\text{cusp}}(t; z) \, dx \, \frac{dy}{y^2} \, dt \right.$$

$$\left. - \int_\varepsilon^\infty \int_0^\infty \int_0^b h(s, y) K_\infty^{\text{cusp}}(t; z) \, dx \, \frac{dy}{y^2} \, dt \right]. \quad (37)$$

We point out that for the right-hand side of formula (37) the individual triple integrals over $h(s, y) K^{\text{cusp}}(t; z)$ and $h(s, y) K_\infty^{\text{cusp}}(t; z)$ do not exist for $\varepsilon = 0$, which justifies the need to introduce the parameter ε. For further discussion of this point, see also Proposition 5.5 below.

Lemma 4.5. *With the above notations, we have*

$$\frac{1}{2g} \int_0^\infty \int_0^b h(s, y) \left(|\varphi_j(z)|^2 - |\alpha_{j,0}(y)|^2 \right) dx \frac{dy}{y^2} = \frac{\cosh(\pi r_j)}{2s(1-s)} \widetilde{L}\left(s, \varphi_j \otimes \overline{\varphi}_j\right).$$

$$(38)$$

Proof. We compute

$$\frac{1}{2g} \int_0^\infty \int_0^b h(s, y) \left(|\varphi_j(z)|^2 - |\alpha_{j,0}(y)|^2 \right) dx \frac{dy}{y^2}$$

$$= \frac{\pi^{-s}}{b} \Gamma(s) \zeta(2s) \int_0^\infty \int_0^b y^{s-2} \left(|\varphi_j(z)|^2 - |\alpha_{j,0}(y)|^2 \right) dx \, dy$$

$$= \frac{\pi^{-s}}{b} \Gamma(s) \zeta(2s) \int_0^\infty \int_0^b y^{s-2} \left[\sum_{n,m \neq 0} \alpha_{j,n} \overline{\alpha}_{j,m} W_{s_j}\left(\frac{nz}{b}\right) \overline{W}_{s_j}\left(\frac{mz}{b}\right) \right.$$

$$\left. + \overline{\alpha}_{j,0}(y) \sum_{n \neq 0} \alpha_{j,n} W_{s_j}\left(\frac{nz}{b}\right) + \alpha_{j,0}(y) \sum_{m \neq 0} \overline{\alpha}_{j,m} \overline{W}_{s_j}\left(\frac{mz}{b}\right) \right] dx \, dy$$

$$= \pi^{-s} \Gamma(s) \zeta(2s) \int_0^\infty y^{s-2} \sum_{n \neq 0} |\alpha_{j,n}|^2 \left| W_{s_j}\left(\frac{nz}{b}\right) \right|^2 dy$$

$$= \pi^{-s} \Gamma(s) \zeta(2s) \cosh(\pi r_j) \int_0^\infty y^{s-2} \sum_{n \neq 0} |\alpha_{j,n}|^2 \left(\frac{4|n|y}{b}\right) K_{ir_j}^2\left(\frac{2\pi|n|y}{b}\right) dy$$

$$= 4\pi^{-s}\Gamma(s)\zeta(2s)\cosh(\pi r_j)\sum_{n\neq 0}|\alpha_{j,n}|^2\left(\frac{|n|}{b}\right)\int_0^\infty y^s K_{ir_j}\left(\frac{2\pi|n|y}{b}\right)$$

$$\times K_{ir_j}\left(\frac{2\pi|n|y}{b}\right)\frac{dy}{y}.$$

With the change of variables

$$u = \frac{2\pi|n|y}{b},$$

we then obtain (see [11], p. 205)

$$\frac{1}{2g}\int_0^\infty\int_0^b h(s,y)\left(|\varphi_j(z)|^2 - |\alpha_{j,0}(y)|^2\right)dx\frac{dy}{y^2}$$

$$= 4\pi^{-s}\Gamma(s)\zeta(2s)\cosh(\pi r_j)\sum_{n\neq 0}|\alpha_{j,n}|^2\left(\frac{|n|}{b}\right)\int_0^\infty u^s K_{ir_j}(u)$$

$$\times K_{ir_j}(u)\left(\frac{2\pi|n|}{b}\right)^{-s}\frac{du}{u}$$

$$= 4\pi^{-s}(2\pi)^{-s}\Gamma(s)\zeta(2s)\cosh(\pi r_j)\sum_{n\neq 0}|\alpha_{j,n}|^2\left(\frac{|n|}{b}\right)\left(\frac{|n|}{b}\right)^{-s}$$

$$\times\int_0^\infty u^s K_{ir_j}(u)K_{ir_j}(u)\frac{du}{u}$$

$$= 4\pi^{-s}(2\pi)^{-s}\Gamma(s)\zeta(2s)\cosh(\pi r_j)\sum_{n\neq 0}|\alpha_{j,n}|^2\left(\frac{|n|}{b}\right)\left(\frac{|n|}{b}\right)^{-s}$$

$$\times\frac{2^{s-3}}{\Gamma(s)}\Gamma\left(\frac{s}{2}\right)\Gamma\left(\frac{s}{2}\right)\Gamma\left(\frac{s}{2}+ir_j\right)\Gamma\left(\frac{s}{2}-ir_j\right)$$

$$= \frac{2^{2-s+s-3}G_{r_j}(s)\cosh(\pi r_j)}{s(1-s)}\sum_{n\neq 0}\frac{|\alpha_{j,n}|^2}{(|n|/b)^{s-1}} = \frac{\cosh(\pi r_j)}{2s(1-s)}\widetilde{L}(s,\varphi_j\otimes\overline{\varphi}_j). \quad\square$$

Lemma 4.6. *With the above notations, we have*

$$\frac{1}{2g}\int_0^\infty\int_0^b h(s,y)\left(|E_{P,1/2+ir}(z)|^2 - |\alpha_{P,1/2+ir,0}(y)|^2\right)dx\frac{dy}{y^2}$$

$$= \frac{\cosh(\pi r)}{2s(1-s)}\widetilde{L}(s,E_{P,1/2+ir}\otimes\overline{E}_{P,1/2+ir}).$$

Proof. The proof runs along the same lines as the proof of Lemma 4.5. $\square$

Proposition 4.7. *With the above notations, we have for any $\varepsilon > 0$*

$$\frac{s(1-s)}{2g} \int_\varepsilon^\infty \int_0^\infty \int_0^b h(s,y) K^{\mathrm{cusp}}(t;z)\, dx\, \frac{dy}{y^2}\, dt$$

$$= \sum_{\lambda_j > 0} \frac{\cosh(\pi r_j) e^{-\lambda_j \varepsilon}}{2\lambda_j} \widetilde{L}(s, \varphi_j \otimes \overline{\varphi}_j)$$

$$+ \frac{1}{8\pi} \sum_{P \ \mathrm{cusp}} \int_{-\infty}^\infty \frac{\cosh(\pi r) e^{-(r^2+1/4)\varepsilon}}{r^2 + 1/4} \widetilde{L}(s, E_{P,1/2+ir} \otimes \overline{E}_{P,1/2+ir})\, dr. \quad (39)$$

Proof. Recall that

$$K^{\mathrm{cusp}}(t;z) = K(t;z) - \sum_{0 \le \lambda_j < 1/4} |\alpha_{j,0}(y)|^2 e^{-\lambda_j t}$$

$$- \frac{1}{4\pi} \sum_{P \ \mathrm{cusp}} \int_{-\infty}^\infty |\alpha_{P,1/2+ir,0}(y)|^2 e^{-(r^2+1/4)t}\, dr$$

$$= \sum_{\lambda_j > 0} \left(|\varphi_j(z)|^2 - |\alpha_{j,0}(y)|^2 \right) e^{-\lambda_j t}$$

$$+ \frac{1}{4\pi} \sum_{P \ \mathrm{cusp}} \int_{-\infty}^\infty \left(|E_{P,1/2+ir}(z)|^2 - |\alpha_{P,1/2+ir,0}(y)|^2 \right) e^{-(r^2+1/4)t}\, dr.$$

$$(40)$$

By multiplying (38) by $e^{-\lambda_j t}$, adding over all positive eigenvalues λ_j, and integrating along t from ε to ∞, we get

$$\frac{1}{2g} \sum_{\lambda_j > 0} \int_\varepsilon^\infty \int_0^\infty \int_0^b h(s,y) \left(|\varphi_j(z)|^2 - |\alpha_{j,0}(y)|^2 \right) e^{-\lambda_j t}\, dx\, \frac{dy}{y^2}\, dt$$

$$= \sum_{\lambda_j > 0} \int_\varepsilon^\infty \frac{\cosh(\pi r_j)}{2s(1-s)} \widetilde{L}(s, \varphi_j \otimes \overline{\varphi}_j) e^{-\lambda_j t}\, dt$$

$$= \frac{1}{s(1-s)} \sum_{\lambda_j > 0} \frac{\cosh(\pi r_j) e^{-\lambda_j \varepsilon}}{2\lambda_j} \widetilde{L}(s, \varphi_j \otimes \overline{\varphi}_j). \quad (41)$$

Using Lemma 4.6, we analogously find

$$\frac{1}{4\pi}\frac{1}{2g}\sum_{P \text{ cusp}}\int_{-\infty}^{\infty}\int_{\varepsilon}^{\infty}\int_{0}^{\infty}\int_{0}^{b} h(s,y)\left(|E_{P,1/2+ir}(z)|^2 - |\alpha_{P,1/2+ir,0}(y)|^2\right)$$

$$\times e^{-(r^2+1/4)t}\,dx\,\frac{dy}{y^2}\,dt\,dr$$

$$= \frac{1}{4\pi}\sum_{P \text{ cusp}}\int_{-\infty}^{\infty}\int_{\varepsilon}^{\infty}\frac{\cosh(\pi r)}{2s(1-s)}\widetilde{L}(s, E_{P,1/2+ir}\otimes\overline{E}_{P,1/2+ir})e^{-(r^2+1/4)t}\,dt\,dr$$

$$= \frac{1}{8\pi}\frac{1}{s(1-s)}\sum_{P \text{ cusp}}\int_{-\infty}^{\infty}\frac{\cosh(\pi r)e^{-(r^2+1/4)\varepsilon}}{r^2+1/4}\widetilde{L}(s, E_{P,1/2+ir}\otimes\overline{E}_{P,1/2+ir})\,dr.$$

$$(42)$$

By combining (41) and (42) with (40), and multiplying by $s(1-s)$, we complete the proof of the proposition.   □

5 The L-Function Relation

As stated before, our computations amount to computing the integral of the identity in Theorem 3.9 when multiplied by $h(s,y)$. As stated in Remark 4.4, we write

$$K_0^{\text{cusp}}(t;z) = K^{\text{cusp}}(t;z) - K_\infty^{\text{cusp}}(t;z).$$

The computations in the previous section allow us to compute the integral involving the term $K^{\text{cusp}}(t;z)$. In this section, we begin by computing the integral involving $K_\infty^{\text{cusp}}(t;z)$, after which we complete the proof of our main theorem, which we state in Theorem 5.4. To conclude this section, we show the necessity of introducing the parameter $\varepsilon > 0$, as stated in Remark 4.4, by computing the asymptotic behavior of the integral arising from the $K_\infty^{\text{cusp}}(t;z)$-term from Theorem 3.9. This computation is given in Proposition 5.5.

Lemma 5.1. *With the above notations, we have*

$$\frac{1}{2g}\int_0^\infty\int_0^b h(s,y)K_\infty^{\text{cusp}}(t;z)\,dx\,\frac{dy}{y^2}$$

$$= 2^{3/2(-s+1)}\pi^{-2s}\Gamma(s)\zeta(s)\zeta(2s)b^{s-1}\mathcal{M}(\widehat{f}_t)(s),$$ (43)

where $\mathcal{M}(\widehat{f}_t)$ is the Mellin transform of the function $\widehat{f}_t$ defined in Remark 3.4 given by

$$\mathcal{M}(\widehat{f}_t)(s) = \int_0^\infty v^s \widehat{f}_t(v) \frac{dv}{v}. \tag{44}$$

Proof. By Remark 3.4 and Definition 3.5, we have

$$\frac{1}{2g} \int_0^\infty \int_0^b h(s,y) K_\infty^{\mathrm{cusp}}(t;z) \, dx \frac{dy}{y^2}$$

$$= \frac{1}{2g} \int_0^\infty \int_0^b h(s,y) \frac{2\sqrt{2}y}{b} \sum_{n=1}^\infty \widehat{f}_t\left(\frac{2\pi\sqrt{2}y}{b} n\right) dx \frac{dy}{y^2}$$

$$= 2\sqrt{2}\pi^{-s} \Gamma(s) \zeta(2s) b^{-1} \int_0^\infty y^s \sum_{n=1}^\infty \widehat{f}_t\left(\frac{2\pi\sqrt{2}y}{b} n\right) \frac{dy}{y}$$

$$= 2\sqrt{2}\pi^{-s} \Gamma(s) \zeta(2s) b^{-1} \sum_{n=1}^\infty \int_0^\infty y^s \widehat{f}_t\left(\frac{2\pi\sqrt{2}y}{b} n\right) \frac{dy}{y}.$$

By the change of variables

$$v = \frac{2\pi\sqrt{2}y}{b} n,$$

we find

$$\frac{1}{2g} \int_0^\infty \int_0^b h(s,y) K_\infty^{\mathrm{cusp}}(t;z) \, dx \frac{dy}{y^2}$$

$$= 2\sqrt{2}\pi^{-s} \Gamma(s) \zeta(2s) b^{-1} \sum_{n=1}^\infty \frac{b^s}{(2\pi\sqrt{2}n)^s} \int_0^\infty v^s \widehat{f}_t(v) \frac{dv}{v}$$

$$= 2^{3/2(-s+1)} \pi^{-2s} \Gamma(s) \zeta(s) \zeta(2s) b^{s-1} \mathcal{M}(\widehat{f}_t)(s),$$

which completes the proof. $\square$

Lemma 5.2. *The Mellin transform* $\mathcal{M}(\widehat{f}_t)$ *of the function* $\widehat{f}_t$ *is given by*

$$\mathcal{M}(\widehat{f}_t)(s)$$

$$= \frac{2^{3s/2-5/2}}{\pi^2} \frac{\Gamma^2(s/2)}{\Gamma(s)} \int_0^\infty r \sinh(\pi r) e^{-(r^2+1/4)t} \Gamma(s/2+ir) \Gamma(s/2-ir) \, dr.$$

Proof. By Lemma 3.6, we have

$$\widehat{f}_t(v) = \frac{\sqrt{2}}{\pi^2} \int_0^\infty r \sinh(\pi r) e^{-(r^2+1/4)t} K_{ir}^2(v/\sqrt{2}) \, dr.$$

This gives

$$\mathcal{M}(\widehat{f}_t)(s) = \int_0^\infty v^s \widehat{f}_t(v) \frac{dv}{v}$$

$$= \int_0^\infty v^s \left(\frac{\sqrt{2}}{\pi^2} \int_0^\infty r \sinh(\pi r) e^{-(r^2+1/4)t} K_{ir}^2(v/\sqrt{2}) \, dr \right) \frac{dv}{v}$$

$$= \frac{\sqrt{2}}{\pi^2} \int_0^\infty r \sinh(\pi r) e^{-(r^2+1/4)t} \left(\int_0^\infty v^s K_{ir}^2(v/\sqrt{2}) \frac{dv}{v} \right) dr.$$

From [11], p. 205, we find

$$\int_0^\infty v^s K_{ir}^2(v/\sqrt{2}) \frac{dv}{v} = 2^{3s/2-3} \Gamma^2(s/2) \frac{\Gamma(s/2+ir)\Gamma(s/2-ir)}{\Gamma(s)}.$$

Summing up, we get

$$\mathcal{M}(\widehat{f}_t)(s) = \frac{2^{3s/2-5/2}}{\pi^2} \frac{\Gamma^2(s/2)}{\Gamma(s)} \int_0^\infty r \sinh(\pi r) e^{-(r^2+1/4)t} \Gamma(s/2+ir)\Gamma(s/2-ir) \, dr,$$

which is the claimed formula.   □

Proposition 5.3. *With the above notations, we have for any $\varepsilon > 0$*

$$\frac{s(1-s)}{2g} \int_\varepsilon^\infty \int_0^\infty \int_0^b h(s,y) K_\infty^{\mathrm{cusp}}(t;z) \, dx \frac{dy}{y^2} dt$$

$$= \frac{\zeta(s)b^{s-1}}{2\pi^2} \int_0^\infty \frac{r \sinh(\pi r) e^{-(r^2+1/4)\varepsilon}}{r^2+1/4} G_r(s) \, dr.$$

Proof. Using Lemma 5.1, we compute for the inner double integral

$$\frac{1}{2g} \int_0^\infty \int_0^b h(s,y) K_\infty^{\mathrm{cusp}}(t;z) \, dx \frac{dy}{y^2}$$

$$= 2^{3/2(-s+1)} \pi^{-2s} \Gamma(s)\zeta(s)\zeta(2s)b^{s-1} \mathcal{M}(\widehat{f}_t)(s)$$

$$= 2^{-1}\pi^{-2}\pi^{-2s} \Gamma^2(s/2)\zeta(s)\zeta(2s)b^{s-1}$$

$$\times \int_0^\infty r \sinh(\pi r) e^{-(r^2+1/4)t} \Gamma(s/2+ir)\Gamma(s/2-ir) \, dr.$$

The claim now follows using the definition of the function $G_r(s)$ and integrating along t from ε to ∞. $\square$

Theorem 5.4. *With the above notations, we define for any $\varepsilon > 0$ and $s \in \mathbb{C}$ with $\mathrm{Re}(s) > 1$, the Θ-function*

$$\Theta_\varepsilon(s) = \sum_{\lambda_j > 0} \frac{\cosh(\pi r_j)e^{-\lambda_j \varepsilon}}{2\lambda_j} \widetilde{L}(s, \varphi_j \otimes \overline{\varphi}_j)$$

$$+ \frac{1}{8\pi} \sum_{P \text{ cusp}} \int_{-\infty}^{\infty} \frac{\cosh(\pi r)e^{-(r^2+1/4)\varepsilon}}{r^2 + 1/4} \widetilde{L}(s, E_{P,1/2+ir} \otimes \overline{E}_{P,1/2+ir}) \, dr$$

and the universal function

$$F_\varepsilon(s) = \frac{\zeta(s)b^{s-1}}{2\pi^2} \int_0^\infty \frac{r \sinh(\pi r)e^{-(r^2+1/4)\varepsilon}}{r^2 + 1/4} G_r(s) \, dr. \tag{45}$$

Then we have the relation

$$\lim_{\varepsilon \to 0} \left(\Theta_\varepsilon(s) - F_\varepsilon(s) \right)$$

$$= \sum_{j=1}^{g} \widetilde{L}(s, f_j \otimes \overline{f}_j) - 4\pi\zeta(s)b^{s-1}G_\infty(s) - \pi^{-s}\frac{2s}{s+1}\Gamma(s)\zeta(2s)\phi_{\infty,\infty}\left(\frac{s+1}{2}\right).$$

Proof. The proof follows immediately from Lemma 4.1, Lemma 4.2, and Lemma 4.3, as well as Proposition 4.7 and Proposition 5.3 in conjunction with Remark 4.4. $\square$

Proposition 5.5. *With the above notations, we have the following asymptotics for the universal function (45) for $s \in \mathbb{R}$, $s > 1$,*

$$\lim_{\varepsilon \to 0} \left(\varepsilon^{\frac{s-1}{2}} F_\varepsilon(s) \right) = \frac{\zeta(s)b^{s-1}}{4\pi} \frac{G_{i/2}(s)}{\Gamma(s/2 + 1/2)}.$$

Proof. Substituting $v = \sqrt{\varepsilon}r$, we get

$$F_\varepsilon(s) = \frac{\zeta(s)b^{s-1}}{2\pi^2} e^{-\varepsilon/4} \int_0^\infty \frac{v \sinh(\pi v/\sqrt{\varepsilon})e^{-v^2}}{v^2 + \varepsilon/4} G_{v/\sqrt{\varepsilon}}(s) \, dv.$$

Now, recall the formula

$$\lim_{\varepsilon \to 0} \left(e^{-\frac{\pi v}{\sqrt{\varepsilon}}} \sinh\left(\frac{\pi v}{\sqrt{\varepsilon}}\right) \right) = \frac{1}{2}, \tag{46}$$

and, using Stirling's formula, the asymptotics

$$\lim_{|y|\to\infty}\left(|\Gamma(x+iy)|e^{\frac{\pi|y|}{2}}|y|^{\frac{1}{2}-x}\right)=\sqrt{2\pi} \tag{47}$$

for fixed $x\in\mathbb{R}$ (see formula (6) of [6], p. 47). Writing

$$\sinh\left(\frac{\pi v}{\sqrt{\varepsilon}}\right)\left|\Gamma\left(\frac{s}{2}+i\frac{v}{\sqrt{\varepsilon}}\right)\right|\left|\Gamma\left(\frac{s}{2}-i\frac{v}{\sqrt{\varepsilon}}\right)\right|\left(\frac{v}{\sqrt{\varepsilon}}\right)^{1-s}$$

$$=e^{-\frac{\pi v}{\sqrt{\varepsilon}}}\sinh\left(\frac{\pi v}{\sqrt{\varepsilon}}\right)\left|\Gamma\left(\frac{s}{2}+i\frac{v}{\sqrt{\varepsilon}}\right)\right|e^{\frac{\pi v}{2\sqrt{\varepsilon}}}\left(\frac{v}{\sqrt{\varepsilon}}\right)^{\frac{1-s}{2}}\left|\Gamma\left(\frac{s}{2}-i\frac{v}{\sqrt{\varepsilon}}\right)\right|$$

$$\times e^{\frac{\pi v}{2\sqrt{\varepsilon}}}\left(\frac{v}{\sqrt{\varepsilon}}\right)^{\frac{1-s}{2}},$$

we obtain, using (46) and (47),

$$\lim_{\varepsilon\to 0}\left(\varepsilon^{\frac{s-1}{2}}\sinh\left(\frac{\pi v}{\sqrt{\varepsilon}}\right)\left|\Gamma\left(\frac{s}{2}+i\frac{v}{\sqrt{\varepsilon}}\right)\right|\left|\Gamma\left(\frac{s}{2}-i\frac{v}{\sqrt{\varepsilon}}\right)\right|\right)=\pi v^{s-1}. \tag{48}$$

We have

$$G_{v/\sqrt{\varepsilon}}(s)=H(s)\Gamma\left(\frac{s}{2}+i\frac{v}{\sqrt{\varepsilon}}\right)\Gamma\left(\frac{s}{2}-i\frac{v}{\sqrt{\varepsilon}}\right)$$

$$=H(s)\left|\Gamma\left(\frac{s}{2}+i\frac{v}{\sqrt{\varepsilon}}\right)\right|\left|\Gamma\left(\frac{s}{2}-i\frac{v}{\sqrt{\varepsilon}}\right)\right|$$

with

$$H(s)=s(1-s)\pi^{-2s}\Gamma^2\left(\frac{s}{2}\right)\zeta(2s).$$

From (48), we then find

$$\lim_{\varepsilon\to 0}\left(\varepsilon^{\frac{s-1}{2}}\sinh\left(\frac{\pi v}{\sqrt{\varepsilon}}\right)G_{v/\sqrt{\varepsilon}}(s)\right)=\pi v^{s-1}H(s),$$

from which we derive

$$\lim_{\varepsilon\to 0}\left(\varepsilon^{\frac{s-1}{2}}F_{\varepsilon}(s)\right)=\frac{\zeta(s)b^{s-1}}{2\pi^2}\lim_{\varepsilon\to 0}\left(\varepsilon^{\frac{s-1}{2}}e^{-\varepsilon/4}\int_0^\infty\frac{v\sinh(\pi v/\sqrt{\varepsilon})e^{-v^2}}{v^2+\varepsilon/4}G_{v/\sqrt{\varepsilon}}(s)dv\right)$$

$$=\frac{\zeta(s)b^{s-1}}{2\pi^2}\int_0^\infty\lim_{\varepsilon\to 0}\left(\varepsilon^{\frac{s-1}{2}}\sinh\left(\frac{\pi v}{\sqrt{\varepsilon}}\right)G_{v/\sqrt{\varepsilon}}(s)\right)e^{-v^2}\frac{dv}{v}$$

$$=\frac{H(s)\zeta(s)b^{s-1}}{2\pi}\int_0^\infty e^{-v^2}v^{s-1}\frac{dv}{v}.$$

Using the substitution $w = v^2$, the remaining integral simplifies to

$$\int_0^\infty e^{-v^2} v^{s-1} \frac{dv}{v} = \frac{1}{2} \int_0^\infty e^{-w} w^{\frac{s-1}{2}} \frac{dw}{w} = \frac{1}{2} \Gamma\left(\frac{s}{2} - \frac{1}{2}\right).$$

Summing up, we get

$$\lim_{\varepsilon \to 0} \left(\varepsilon^{\frac{s-1}{2}} F_\varepsilon(s)\right) = \frac{\zeta(s) b^{s-1}}{4\pi} \frac{G_{i/2}(s)}{\Gamma(s/2 + 1/2)},$$

which is the claimed formula.   □

Acknowledgements The results of this article were presented by the first named author during the MPIM/HIM Program in Representation Theory, Complex Analysis, and Integral Geometry. We thank the organizers Gindikin and Krötz for the opportunity to participate in this Program. Both authors thank the referee for the suggestions, which helped to improve the manuscript.

The first author acknowledges support from grants from the NSF and PSC-CUNY. The second author acknowledges support from the DFG Graduate School *Berlin Mathematical School* and from the DFG Research Training Group *Arithmetic and Geometry*.

References

[1] W. Abikoff: *Degenerating families of Riemann surfaces.* Ann. of Math. (2) **105** (1977), 29–44.

[2] A. Aryasomayajula: Ph.D. Dissertation, Humboldt-Universität zu Berlin, Institut für Mathematik, in preparation.

[3] A.F. Beardon: *The geometry of discrete groups.* Graduate Texts in Mathematics 91. Springer-Verlag, New York, 1995.

[4] D. Bump: *The Rankin-Selberg Method: A Survey.* In: Number Theory, Trace Formulas, and Discrete Groups. K.E. Aubert, E. Bombieri, and D. Goldfeld (eds.). Academic Press, Boston, MA, 1989, 49–109.

[5] I. Chavel: *Eigenvalues in Riemannian geometry.* Pure and Applied Mathematics 115. Academic Press, Orlando, FL, 1984.

[6] A. Erdélyi, W. Magnus, F. Oberhettinger, F.G. Tricomi: *Higher Transcendental Functions.* Vol. I. McGraw-Hill Book Company, Inc., New York-Toronto-London, 1953.

[7] I. Gradshteyn, I. Ryzhik: *Tables of Integrals, Series, and Products.* Academic Press, 1981.

[8] D. Hejhal: *The Selberg trace formula for congruence subgroups.* Bull. Amer. Math. Soc. **81** (1975), 752–755.

[9] J. Huntley, J. Jorgenson, R. Lundelius: *Continuity of small eigenfunctions on degenerating Riemann surfaces with hyperbolic cusps.* Bol. Soc. Mat. Mexicana **1** (1995), 119–125.

[10] M. Huxley: *Scattering matrices for congruence subgroups.* In: Modular forms. Durham, 1983, 141–156.

[11] H. Iwaniec: *Spectral methods of automorphic forms.* Graduate Studies in Mathematics 53. Amer. Math. Soc., Providence, RI, 2002.

[12] J. Jorgenson: *Asymptotic behavior of Faltings's delta function.* Duke Math J. **61** (1990), 221–254.

[13] J. Jorgenson, J. Kramer: *Expressing Arakelov invariants using hyperbolic heat kernels.* In: The Ubiquitous Heat Kernel. J. Jorgenson and L. Walling (eds.). AMS Contemp. Math. 398, Providence, RI, 2006, 295–309.

[14] J. Jorgenson, J. Kramer: *Non-completeness of the Arakelov-induced metric on moduli space of curves.* Manuscripta Math. **119** (2006), 453–463.

[15] J. Jorgenson, J. Kramer: *Bounds on canonical Green's functions.* Compos. Math. **142** (2006), 679–700.

[16] J. Jorgenson, J. Kramer: *Bounds on Faltings's delta function through covers.* Ann. of Math. (2) **170** (2009), 1–42.

[17] J. Jorgenson, R. Lundelius: *Convergence of the normalized spectral function on degenerating hyperbolic Riemann surfaces of finite volume.* J. Func. Anal. **149** (1997), 25–57.

[18] J. Jorgenson, R. Lundelius: *A regularized heat trace for hyperbolic Riemann surfaces of finite volume.* Commun. Math. Helv. **72** (1997), 636–659.

[19] F. Oberhettinger, L. Badii: *Tables of Laplace Transforms.* Springer-Verlag, New York-Heidelberg-Berlin, 1973.

[20] A. Selberg: *Collected Papers, Volume I.* Springer-Verlag, New York-Heidelberg-Berlin, 1989.

[21] S. Wolpert: *The hyperbolic metric and the geometry of the universal curve.* J. Diff. Geom. **31** (1990), 417–472.

Orthogonal Period of a $GL_3(\mathbb{Z})$ Eisenstein Series

Gautam Chinta and Omer Offen

Abstract We provide an explicit formula for the period integral of the unramified Eisenstein series on $GL_3(\mathbb{A}_\mathbb{Q})$ over the orthogonal subgroup associated with the identity matrix. The formula expresses the period integral as a finite sum of products of double Dirichlet series that are Fourier coefficients of Eisenstein series on the metaplectic double cover of GL_3.

Keywords Eisenstein series • Metaplectic group • Multiple Dirichlet series

Mathematics Subject Classification (2010): 11F30, 11F37,11M36

1 Introduction

Let F be a number field, G a connected reductive group defined over F, and H a reductive F-subgroup of G. The period integral $P^H(\phi)$ of a cuspidal automorphic form on $G(\mathbb{A}_F)$ is defined by the absolutely convergent integral (cf. [AGR93, Proposition 1])

$$P^H(\phi) = \int_{(H(F)\backslash(H(\mathbb{A}_F)\cap G(\mathbb{A}_F)^1)} \phi(h)\, dh,$$

G. Chinta (✉)
Department of Mathematics, The City College of CUNY, New York, NY 10031, USA
e-mail: chinta@sci.ccny.cuny.edu

O. Offen
Mathematics Department, Technion - Israel Institute of Technology, Haifa, 32000 Israel
e-mail: offen@tx.technion.ac.il

B. Krötz et al. (eds.), *Representation Theory, Complex Analysis, and Integral Geometry*,
DOI 10.1007/978-0-8176-4817-6_3, © Springer Science+Business Media, LLC 2012

where $G(\mathbb{A}_F)^1$ is the intersection of $\ker |\chi(\cdot)|_{\mathbb{A}_F}$ for all rational characters χ of G. For more general automorphic forms, the period integral $P^H(\phi)$ fails to converge but in many cases it is known how to regularize it [LR03]. Case study indicates that the value $P^H(\phi)$, when not zero, carries interesting arithmetic information.

Roughly speaking, in cases of local multiplicity one, i.e. when at every place v of F the space of H_v-invariant linear forms of an irreducible representation of G_v is one dimensional, the period integral P^H on an irreducible automorphic representation $\pi = \otimes_v \pi_v$ factorizes as a tensor product $P^H = \otimes_v P_v$ of H_v-invariant linear forms on π_v. This indicates a relation between $P^H(\phi)$ and automorphic L-functions. For example, the setting were $H = GL_n$ over F, E/F is a quadratic extension and G is the restriction of scalars from E to F of GL_n over E, is an example where local multiplicity one holds. In this case, the nonvanishing of the period $P^H(\phi)$ of a cusp form depicts the existence of a pole at $s = 1$ of the associated Asai L-function (cf. [Fli88, Sect. 1, Theorem]) and the (regularized) period $P^H(E(\varphi, \lambda))$ of an Eisenstein series is related to special values of the Asai L-function (cf. [JLR99, Theorems 23 and 36]).

Remarkably, the period integral P^H is sometimes factorizable even though local multiplicity one fails. Consider now the case where G is defined as in the previous example, but H is the quasi split unitary group with respect to E/F. For cuspidal representations, nonvanishing of P^H characterizes the image of quadratic base change from $G' = GL_n$ over F to G (cf. [Jac05] and [Jac10]). Furthermore, although for "most" irreducible representations of G_v, the space of H_v-invariant linear forms has dimension 2^{n-1}, on a cuspidal representation the period P^H is factorizable (cf. [Jac01]). This factorization is best understood through the *relative trace formula* (RTF) of Jacquet. Roughly speaking, the RTF is a distribution on $G(\mathbb{A}_F)$ with a spectral expansion ranging over the H-*distinguished* spectrum, i.e. the part of the automorphic spectrum of $G(\mathbb{A}_F)$, where P^H is nonvanishing. In the case at hand the RTF for (G, H) is compared with the Kuznetsov trace formula for $G' = GL_n$ over F. If π is a cuspidal representation of $G(\mathbb{A}_F)$ and it is the base change of π', a cuspidal representation of $G'(\mathbb{A}_F)$, then the contribution of π to the RTF is compared with the contribution of π' to the Kuznetsov trace formula. The multiplicity one of Whittaker functionals for G' allows the factorization of the contribution of π', hence that of the contribution of π and finally of P^H on π. The value $P^H(\phi)$ (or rather its absolute value squared) for a cusp form is related to special values of Rankin–Selberg L-functions (cf. [LO07]). Essential to the factorization of P^H in this case is the fact that (up to a quadratic twist) π' base-changing to π is unique. In some sense, the local factors π'_v of π' pick a one-dimensional subspace of H_v-invariant linear forms on π_v and with the appropriate normalization, these give the local factors of P^H. For π an Eisenstein automorphic representation in the image of base change π' is no longer unique (but the base-change fiber is finite). This is the reason that the (regularized) period $P^H(E(\varphi, \lambda))$ of an Eisenstein series can be expressed as a finite sum of factorizable linear forms. In effect, this was carried out using a *stabilization* process (stabilizing the open double cosets in $P \backslash G / H$ over the algebraic closure of F where the Eisenstein series is induced from the parabolic subgroup P) for Eisenstein series induced from the Borel subgroup (cf. [LR00] for $n = 3$ and [Off07] for general n) and is work in progress for more general Eisenstein series.

Consider now the case where $G = GL_n$ over F and H is an orthogonal subgroup. Using his RTF formalism and evidence from the $n = 2$ case, Jacquet suggests that in this setting *the role of G'* is played by the metaplectic double cover of G [Jac91]. For this G' local multiplicity one of Whittaker functionals fails. This leads us to expect that the period integral $P^H(\phi)$ of a cusp form is not factorizable. In the last paragraph of [Jac91], Jacquet remarks that it is natural to conjecture that the period is related to Whittaker–Fourier coefficients of a form on G' related to ϕ under the metaplectic correspondence [FK86]. Nevertheless, to date, the arithmetic interpretation of the period at hand is a mystery. Even precise conjectures are yet to be made.

This brings us, finally, to the subject matter of this note. Often, studying the period integral of an Eisenstein series is more approachable than that of a cusp form and yet may help to predict expectations for the cuspidal case (this was the case for $G = GL_2$ and H an anisotropic torus, where the classical formula of Maass for the period of an Eisenstein series in terms of the zeta function of an imaginary quadratic field significantly predates the analogous formula of Waldspurger for the absolute value squared of the period of a cusp form, cf. [Wal80, Wal81]). In this work, we provide a very explicit formula for the period integral $P^H(E(\varphi, \lambda))$ in the special case that $n = 3$, H is the orthogonal group associated with the identity matrix and $E(\varphi, \lambda)$ is the unramified Eisenstein series induced from the Borel subgroup. The formula we obtain expresses the period integral as a *finite* sum of products of certain double Dirichlet series. This formula, given in Theorem 6.1, is our main result. The double Dirichlet series that appear are related to the Fourier coefficients of Eisenstein series on $G'(\mathbb{A}_F)$ (cf. [BBFH07]). This fits perfectly into Jacquet's formalism and it is our hope that the formula in this very special case can shed a light on the arithmetic information carried by orthogonal periods in the general context.

We conclude this introduction with a description of the computation of Maass alluded to above. Let $E(z, s)$ be the real analytic Eisenstein series on $SL_2(\mathbb{Z})$. A classical result of Maass relates a weighted sum of $E(z, s)$ over CM points of discriminant $d < 0$ with the ζ function of the imaginary extension $\mathbb{Q}(\sqrt{d})$. This can be reinterpreted as relating an orthogonal period of the Eisenstein series with a Fourier coefficient of a half-intgeral weight automorphic form. Indeed, the ζ function of $\mathbb{Q}(\sqrt{d})$ shows up in the Fourier expansion of a half-integral weight Eisenstein series.

Let $z = x + iy$ with $x, y \in \mathbb{R}, y > 0$ be an element of the complex upper halfplane. Let Γ_∞ be the subgroup of $SL_2(\mathbb{Z})$ consisting of matrices of the form $\begin{pmatrix} \pm 1 & * \\ 0 & \pm 1 \end{pmatrix}$. The weight zero real analytic Eisenstein series for $SL_2(\mathbb{Z})$ is defined by the absolutely convergent series

$$E(z, s) = \sum_{\gamma = \begin{pmatrix} * & * \\ c & d \end{pmatrix} \in \Gamma_\infty \backslash SL_2(\mathbb{Z})} \text{Im}(\gamma z)^s \qquad (1.1)$$

for $s \in \mathbb{C}$ with $\mathrm{Re}(s) > 1$ and by analytic continuation for $s \in \mathbb{C}, s \neq 1$. Similarly, the Eisenstein series of weight $\frac{1}{2}$ for $\Gamma_0(4)$ is defined by

$$\tilde{E}(z, s) = \sum_{\gamma = \left(\begin{smallmatrix} * & * \\ c & d \end{smallmatrix}\right) \in (\Gamma_\infty \cap \Gamma_0(4)) \backslash \Gamma_0(4)} \epsilon_d^{-1}\left(\frac{c}{d}\right) \frac{\mathrm{Im}(\gamma z)^s}{\sqrt{cz + d}}, \tag{1.2}$$

where

$$\epsilon_d = \begin{cases} 1 & \text{if } d \equiv 1 \pmod 4 \\ i & \text{if } d \equiv 3 \pmod 4. \end{cases}$$

The Fourier expansion of the half integral weight Eisenstein series was first computed by Maass [Maa38]. To describe the expansion, first define

$$K_m(s, y) = \int_{-\infty}^{\infty} \frac{e^{2\pi i m x}}{(x^2 + y^2)^s (x + iy)^{1/2}} dx.$$

Then

$$\tilde{E}(z, s) = y^s + c_0(s) y^s \frac{\zeta(4s - 1)}{\zeta(4s)} + \sum_{m \neq 0} b_m(s) K_m(s, y) e^{2\pi i m x}, \tag{1.3}$$

where, for m squarefree,

$$b_m(s) = c_m(s) \frac{L(2s, \chi_m)}{\zeta(4s + 1)}. \tag{1.4}$$

In the above equations, $c_m(s)$ is a quotient of Dirichlet polynomials in 2^{-s} and χ_m is the real primitive character corresponding to the extension $\mathbb{Q}(\sqrt{m})/\mathbb{Q}$. See Propositions 1.3 and 1.4 of Goldfeld–Hoffstein [GH85] for precise formulas.

On the other hand, quadratic Dirichlet L-functions also arise as sums of the nonmetaplectic Eisenstein series over CM points. Let $z = x + iy$ in the upper half plane be an element of an imaginary quadratic field K of discriminant d_K. Let A be the ideal class in the ring of integers of K corresponding to $\mathbb{Z} + z\mathbb{Z}$. Let $q(m, n)$ be the binary quadratic form

$$q(m, n) = \frac{\sqrt{|d_K|}}{2 \, \mathrm{Im}(z)} \mathrm{N}(mz + n) = \frac{\sqrt{|d_K|}}{2 \, \mathrm{Im} z} |mz + n|^2$$

and ζ_q the Epstein zeta function

$$\zeta_q(s) = \sum_{\substack{m, n \in \mathbb{Z} \\ (m, n) \neq (0, 0)}} \frac{1}{q(m, n)^s}. \tag{1.5}$$

Then

$$\zeta_K(s, A^{-1}) = \frac{1}{w_K} \zeta_q(s), \tag{1.6}$$

where w_K is the number of roots of unity in K. These zeta functions can be expressed in terms of the nonmetaplectic Eisenstein series:

$$\zeta_K(s, A^{-1}) = \frac{1}{w_K} \zeta_q(s) = \frac{2^{1+s}}{w_K |d_K|^{s/2}} \zeta(2s) E(z, s) \tag{1.7}$$

By virtue of the bijective correspondences between ideal classes in the ring of integers of K, binary quadratic forms and CM points in the upper halfplane, we arrive at the identity

$$\zeta_K(s) = \frac{1}{w_K} \sum_q \zeta_q(s) = \frac{2^{1+s}}{w_K |d_K|^{s/2}} \zeta(2s) \sum_z E(z, s), \tag{1.8}$$

where the sum in the middle is over equivalence classes of integral binary quadratic forms of discriminant d_K and the rightmost sum is over $SL_2(\mathbb{Z})$ inequivalent CM points of discriminant d_K. Writing the zeta function of K as $\zeta_K(s) = \zeta(s)L(s, \chi_{d_K})$ gives the relation between the Fourier coefficients $b_{d_K}(s)$ of metaplectic Eisenstein series (see (1.4)) and sums of nonmetaplectic Eisenstein series over CM points. The sum over CM points is actually a finite sum of orthogonal periods. In this setting, an orthogonal period of an $SL_2(\mathbb{Z})$ invariant function on the upper halfplane is a sum over a subset of CM points corresponding to a fixed genus class of binary quadratic forms. Therefore, the precise relation between an orthogonal period of the nonmetaplectic Eisenstein series and Fourier coefficients of the metaplectic Eisenstein series is slightly more complicated. See Sect. 4 (in particular Proposition 4.1) for an expression for the orthogonal period in terms of a finite sum of quadratic Dirichlet L-functions.

2 Adelic Versus Classical Periods

Let $G = GL_n$ over $\mathbb{Q}$ and let $X = \{g \in G : {}^t g = g\}$ be the algebraic subset of symmetric matrices. Let $K = \prod_v K_v$ be the standard maximal compact subgroup of $G(\mathbb{A}_{\mathbb{Q}})$, where the product is over all places v of $\mathbb{Q}$, $K_p = G(\mathbb{Z}_p)$ for every prime number p and $K_\infty = O(n) = \{g \in G(\mathbb{R}) : g {}^t g = I_n\}$.

2.1 The Genus Class

For $x, y \in X(\mathbb{Q})$, we say that x and y are in the same *class* and write $x \sim y$ if there exists $g \in G(\mathbb{Z})$ such that $y = g x {}^t g$ and we say that x and y are in the same *genus class* and write $x \approx y$ if for every place v of $\mathbb{Q}$ there exists $g \in K_v$ such that $y = g\,x\,{}^t g$. Of course classes refine genus classes. If $x \in X(\mathbb{Q})$ is positive definite, it is well known that there are finitely many classes in the genus class of x.

2.2 An Anisotropic Orthogonal Period as a Sum Over the Genus

Fix once and for all $x \in X(\mathbb{Q})$ positive definite and let

$$H = \{g \in G : g\,x\,{}^t g = x\}$$

be the orthogonal group associated with x. Thus, H is anisotropic and the orthogonal period integral

$$P^H(\phi) = \int_{H(\mathbb{Q})\backslash H(\mathbb{A}_{\mathbb{Q}})} \phi(h)\,\mathrm{d}h$$

is well defined and absolutely convergent for any say continuous function ϕ on $H(\mathbb{Q})\backslash H(\mathbb{A}_{\mathbb{Q}})$.

Note that the imbedding of $G(\mathbb{R})$ in $G(\mathbb{A}_{\mathbb{Q}})$ in the "real coordinate" defines a bijection $G(\mathbb{Z})\backslash G(\mathbb{R})/K_\infty \simeq G(\mathbb{Q})\backslash G(\mathbb{A}_{\mathbb{Q}})/K$. Furthermore, the map $g \mapsto g\,{}^t g$ defines a bijection from $G(\mathbb{R})/K_\infty$ to the space $X^+(\mathbb{R})$ of positive definite symmetric matrices in $X(\mathbb{R})$. The resulting bijection

$$G(\mathbb{Q})\backslash G(\mathbb{A}_{\mathbb{Q}})/K \simeq G(\mathbb{Z})\backslash X^+(\mathbb{R}) \tag{2.1}$$

allows us to view any function $\phi(g)$ on $G(\mathbb{Q})\backslash G(\mathbb{A}_{\mathbb{Q}})/K$ as a function (still denoted by) $\phi(x)$ on $G(\mathbb{Z})\backslash X^+(\mathbb{R})$.

By [Bor63, Proposition 2.3], there is a natural bijection between the double coset space $H(\mathbb{Q})\backslash H(\mathbb{A}_{\mathbb{Q}})/(H(\mathbb{A}_{\mathbb{Q}}) \cap K)$ and the set $\{y \in X_{\mathbb{Q}} : y \approx x\}/\sim$ of classes in the genus class of x. Let $g_\infty \in G(\mathbb{R})$ be such that $x = g_\infty\,{}^t g_\infty$ and let $g_0 \in G(\mathbb{A}_{\mathbb{Q}})$ have g_∞ in the infinite place and the identity matrix at all finite places. As in [CO07, Lemma 2.1], it can be deduced that for any function ϕ on $G(\mathbb{Q})\backslash G(\mathbb{A}_{\mathbb{Q}})/K$ we have

$$\int_{H(\mathbb{Q})\backslash H(\mathbb{A}_{\mathbb{Q}})} \phi(h\,g_0)\,\mathrm{d}h$$

$$= \mathrm{vol}(H(\mathbb{A}_{\mathbb{Q}}) \cap g_0 K g_0^{-1}) \sum_{\{y \in X_{\mathbb{Q}}:y\approx x\}/\sim} \frac{\phi(y)}{\#\{g \in G(\mathbb{Z}) : g\,y\,{}^t g = y\}} \tag{2.2}$$

where ϕ on the left- and right-hand sides correspond via (2.1). In short, the anisotropic orthogonal period associated with x of an automorphic form ϕ equals a finite weighted sum of point evaluations of ϕ over classes in the genus class of x.

2.3 The Unramified Adelic Eisenstein Series as a Classical One

Let $B = AU$ be the Borel subgroup of upper triangular matrices in G, where A is the subgroup of diagonal matrices and U is the subgroup of upper triangular unipotent matrices. For $\lambda = (\lambda_1, \ldots, \lambda_n) \in \mathbb{C}^n$, let

$$\varphi_\lambda(\mathrm{diag}(a_1, \ldots, a_n) \, u \, k) = \prod_{i=1}^{n} |a_i|^{\lambda_i + \frac{n+1}{2} - i}$$

for $\mathrm{diag}(a_1, \ldots, a_n) \in A(\mathbb{A}_\mathbb{Q})$, $u \in U(\mathbb{A}_\mathbb{Q})$ and $k \in K$. The unramified Eisenstein series $\mathcal{E}(g, \lambda)$ induced from B is defined by the meromorphic continuation of the series

$$\mathcal{E}(g, \lambda) = \sum_{\gamma \in B(\mathbb{Q}) \backslash G(\mathbb{Q})} \varphi_\lambda(\gamma \, g).$$

Note that $\mathcal{E}(g, \lambda)$ is a function on $G(\mathbb{Q}) \backslash G(\mathbb{A}_\mathbb{Q}) / K$. With the identification (2.1), for $x \in X^+(\mathbb{R})$ we have

$$\mathcal{E}(x, \lambda) = \det x^{\frac{1}{2}(\lambda_1 + \frac{n-1}{2})} \sum_{\gamma \in B(\mathbb{Z}) \backslash G(\mathbb{Z})} \prod_{i=1}^{n-1} d_{n-i}(\delta \, x \, {}^t\delta)^{\frac{1}{2}(\lambda_{i+1} - \lambda_i - 1)}, \tag{2.3}$$

where $d_i(x)$ denotes the determinant of the lower right $i \times i$ block of x.

Assume now that $n = 3$. Arguing along the same lines as in [CO07, Sect. 4.2] we may write (2.3) as

$$\mathcal{E}(x, \lambda_1, \lambda_2, \lambda_3) = \frac{1}{4} \zeta(\lambda_2 - \lambda_3 + 1)^{-1} \zeta(\lambda_1 - \lambda_2 + 1)^{-1} (\det x)^{\frac{\lambda_2}{2}}$$

$$\times \sum_{\substack{0 \neq v, w \in \mathbb{Z}^3 \\ v \perp w}} Q_{x,1}(v)^{\frac{1}{2}(\lambda_3 - \lambda_2 - 1)} Q_{x,2}(w)^{\frac{1}{2}(\lambda_2 - \lambda_1 - 1)}, \tag{2.4}$$

where $Q_{x,1}$ (resp. $Q_{x,2}$) is the quadratic form on $V = \mathbb{R}^3$ defined on the row vector $v \in V$ by $v \mapsto vxv^t$ (resp. $v \mapsto vx^{-1}v^t$). The genus class of the identity matrix $x = I_3$ consists of a unique class. Let $Q = Q_{I_3,1} = Q_{I_3,2}$. Combining (2.2) and

(2.4) we see that when $x = I_3$ there exists a normalization of the Haar measure on $H(\mathbb{A}_{\mathbb{Q}})$ such that as a meromorphic function in $\lambda = (\lambda_1, \lambda_2, \lambda_3) \in \mathbb{C}^3$ we have

$$\int_{H(\mathbb{Q})\backslash H(\mathbb{A}_{\mathbb{Q}})} \mathcal{E}(h, \lambda) \, dh = 4\,\mathcal{E}(I_3, \lambda)$$

$$= \zeta(\lambda_2 - \lambda_3 + 1)^{-1}\,\zeta(\lambda_1 - \lambda_2 + 1)^{-1} \sum_{\substack{0 \neq v,w \in \mathbb{Z}^3 \\ v \perp w}} Q(v)^{\frac{1}{2}(\lambda_3 - \lambda_2 - 1)}\, Q(w)^{\frac{1}{2}(\lambda_2 - \lambda_1 - 1)}.$$

$$(2.5)$$

Introduce the new variables $s_2 = (\lambda_2 - \lambda_3 + 1)/2, s_1 = (\lambda_1 - \lambda_2 + 1)/2$ and write the right hand side of (2.5) as

$$E(I_3; s_1, s_2) := \zeta(2s_1)^{-1}\zeta(2s_2)^{-1} \sum_{\substack{0 \neq v,w \in \mathbb{Z}^3 \\ v \perp w}} Q(v)^{-s_2} Q(w)^{-s_1}. \qquad (2.6)$$

The rest of this work is devoted to the explicit computation of (2.6), which is given in Theorem 6.1.

3 The Double Dirichlet Series

We define the double Dirichlet series which arise in our evaluation of the $GL_3(\mathbb{Z})$ Eisenstein series at the identity. Let ψ_1, ψ_2 be two quadratic characters unramified away from 2. Then the double Dirichlet series $Z(s_1, s_2; \psi_1, \psi_2)$ is roughly of the form

$$\sum_d \frac{L(s_1, \chi_d)}{d^{s_2}}. \qquad (3.1)$$

More precisely,

$$Z(s_1, s_2; \psi_1, \psi_2) = \sum_{\substack{d_1, d_2 > 0 \\ \text{odd}}} \frac{\chi_{d_2'}(\hat{d}_1)}{d_1^{s_1} d_2^{s_2}} a(d_1, d_2)\psi_1(d_1)\psi_2(d_2), \qquad (3.2)$$

where

- $d_2' = (-1)^{(d_2-1)/2} d_2$ and $\chi_{d_2'}$ is the Kronecker symbol associated with the squarefree part of d_2'.
- $\hat{d}_1$ is the part of d_1 relatively prime to the squarefree part of d_2.

- The coefficients $a(d_1, d_2)$ are multiplicative in both entries and are defined on prime powers by

$$a(p^k, p^l) = \begin{cases} \min(p^{k/2}, p^{l/2}) & \text{if } \min(k, l) \text{ is even,} \\ 0 & \text{otherwise.} \end{cases} \tag{3.3}$$

It can be shown that the functions $Z(s_1, s_2; \psi_1, \psi_2)$ appear in the Whittaker expansion of the metaplectic Eisenstein series on the double cover of $GL_3(\mathbb{R})$, see e.g [BBFH07]. As such these functions have an analytic continuation to $s_1, s_2 \in \mathbb{C}$ and satisfy a group of 6 functional equations.

We conclude this section by relating the heuristic definition (3.1) to the precise definition (3.2).

Theorem 3.1. *Let* ψ_1, ψ_2 *be quadratic characters ramified only at 2. Then*

$$Z(s, w; \psi_1, \psi_2) = \zeta_2(2w)\zeta_2(2s + 2w - 1) \sum_{\substack{d_2 > 0, \text{ odd} \\ \text{sqfree}}} \frac{L_2(s, \chi_{d_2'}\psi_1)}{L_2(s + 2w, \chi_{d_2'}\psi_1)} \frac{\psi_2(d_2)}{d_2^w},$$

where $L_2(s, \chi)$ *denotes the Dirichlet L-function with the Euler factor at 2 removed.*

Proof. See [CFH05]. □

4 Genus Theory for Binary Quadratic Forms

Our description of the genus characters follows the presentation in Sect. 3 of Bosma and Stevenhagen, [BS96]. Let D be a negative discriminant. Write $D = df^2$ where d is a fundamental discriminant. We will assume f is odd. Let $\text{Cl}(D)$ be the group of $SL_2(\mathbb{Z})$ equivalence classes of primitive integral binary quadratic forms of discriminant D. We will denote the quadratic form $q(x, y) = ax^2 + bxy + cy^2$ by $[a, b, c]$. We call e a prime discriminant if $e = -4, 8, -8$ or $p' = (-1)^{(p-1)/2}p$ for an odd prime. Note that e is a fundamental discriminant. Write $D = D_1 D_2$ where D_1 is an even fundamental discriminant and D_2 is an odd discriminant. Let D_0 be D_1 times the product of the prime discriminants dividing D_2.

For each odd prime p dividing D we define a character $\chi^{(p)}$ on $\text{Cl}(D)$ by

$$\chi^{(p)}([a, b, c]) = \begin{cases} \chi_{p'}(a) & \text{if } (p, a) = 1 \\ \chi_{p'}(c) & \text{if } (p, c) = 1. \end{cases} \tag{4.1}$$

The primitivity of $[a, b, c]$ ensures that at least one of these two conditions will be satisfied. These characters generate a group $\mathcal{X}(D)$, called the group of genus class characters of $Cl(D)$. The order of $\mathcal{X}(D)$ is $2^{\omega(D)-1}$, where $\omega(D)$ is the number of distinct prime divisors of D. For each squarefree odd number e_1 dividing D we define the genus class character

$$\chi_{e_1',e_2'} = \prod_{p|e_1} \chi^{(p)},$$

where $e_1'e_2' = D_0$. Then as e_1 ranges over the squarefree positive odd divisors of D, $\chi_{e_1',e_2'}$ will range over all the genus character exactly once (i D is even) or twice (if D is odd).

Two forms q_1 and q_2 are in the same genus if and only if $\chi(q_1) = \chi(q_2)$ for all $\chi \in \mathcal{X}(D)$. As in Sect. 2, we denote this by $q_1 \approx q_2$.

Using the identification between primitive integral binary quadratic forms of discriminant D and invertible ideal classes in the order $\mathcal{O}_D = \mathbb{Z}[(D + \sqrt{D})/2]$, we may define the genus characters on the group $Pic(\mathcal{O}_D)$. This allows us to associate with a genus class character χ the L-function

$$L_{\mathcal{O}_D}(s, \chi) = \sum_{\mathfrak{a}} \frac{\chi(\mathfrak{a})}{\mathbf{N}(\mathfrak{a})^s},$$

where the sum is over all invertible ideals of $\mathcal{O}_D$. In terms of the Epstein zeta function, we have

$$L_{\mathcal{O}_D}(s, \chi) = \frac{1}{\#\mathcal{O}_D^\times} \sum_{q \in Cl(D)} \chi(q)\zeta_q(s). \tag{4.2}$$

Using the group of characters $\mathcal{X}(D)$, we may isolate individual genus classes on the right-hand side of (4.2).

Proposition 4.1. *Let q_0 be a fixed form in $Cl(D)$. Then*

$$\sum_{q \approx q_0} \zeta_q(s) = \frac{\#\mathcal{O}_D^\times}{2^{\omega(D)-1}} \sum_{\chi \in \mathcal{X}(D)} \chi(q_0) L_{\mathcal{O}_D}(s, \chi).$$

Finally, the following proposition shows how to write an L-function associated with a genus class character in terms of ordinary Dirichlet L-functions.

Proposition 4.2. *Let e_1, e_2 be fundamental discriminants and let $D = e_1 e_2 f^2$. Then*

$$L_{\mathcal{O}_D}(s, \chi_{e_1,e_2}) = L(s, \chi_{e_1})L(s, \chi_{e_2}) \prod_{p^k \| f} \mathcal{P}_k(p^{-s}, \chi_{e_1}(p), \chi_{e_2}(p)),$$

where $\mathcal{P}_k(p^{-s}, \chi_{e_1}(p), \chi_{e_2}(p))$ is a Dirichlet polynomial defined by the generating series

$$F(u, X; \alpha, \beta) = \sum_{k \geq 0} \mathcal{P}_k(u, \alpha, \beta) X^k = \frac{(1 - \alpha u X)(1 - \beta u X)}{(1 - X)(1 - pu^2 X)}. \qquad (4.3)$$

Proof. See Remark 3 of Kaneko, [Kan05]. Actually, Kaneko considers only zeta functions of orders, not genus character L-functions as in the proposition, but the ideas are similar.   □

5 The Gauss Map

Let $V = \mathbb{Q}^3$ equipped with the quadratic form Q, $Q(x, y, z) = x^2 + y^2 + z^2$. We also let Q denote the associated bilinear form on $V \times V$. Let $L = \mathbb{Z}^3$ and let $L[n]$ be the set of vectors in L such that $Q(v) = n$. Let L_0 be the set of primitive integral vectors and let $L_0[n] = L_0 \cap L[n]$. Let

$$D = \begin{cases} -4n & \text{if } n \equiv 1 \text{ or } 2 \pmod 4 \\ -n & \text{if } n \equiv 3 \pmod 4. \end{cases} \qquad (5.1)$$

(The case $n \equiv 0 \pmod 4$ will not occur in our computations below.)

We have a map from $L_0[n]$ to equivalence classes of primitive binary quadratic forms of discriminant D defined as follows. Let $v \in L_0[n]$. Let W be the orthogonal complement of v (with respect to Q) and let M be a maximal Q-integral sublattice in W. Explicitly, we take $M = L \cap W$ if $n \equiv 0, 1 \mod 4$ and $M = \frac{1}{2}L \cap W$ if $n \equiv 3 \mod 4$. Let u, w be an integral basis for M. The restriction of Q to the two-dimensional subspace W is a binary quadratic form, which we'll denote by q. With respect to an integral basis u, w of M, the Gram matrix of this restriction is

$$\begin{pmatrix} Q(u, u) & Q(u, v) \\ Q(u, v) & Q(v, v) \end{pmatrix}. \qquad (5.2)$$

We call the map $\mathcal{G} : L_0[n] \to \mathrm{Cl}(D)$ defined by $\mathcal{G}(v) = Q|_{v^\perp}$ the Gauss map. We now describe the image of this map more explicitly for fixed n.

We begin with three observations.

1. By the Hasse–Minkowski principle, if $q \in \mathrm{Cl}(D)$ is in the image of $\mathcal{G}$, then every form in the genus of q is also in the image.
2. If $q_1 \approx q_2$ are two forms in the image of $\mathcal{G}$, then by Siegel's mass formula, the fiber over both forms has the same cardinality.
3. If q_1 and q_2 are two forms in the image, then q_1 and q_2 are in the same genus.

These three facts follow because the ternary quadratic form Q is the only form in its genus. We refer the reader to Theorems 1 and 2 of the survey paper of Shimura [Shi06] for further details. More explicitly, we have the following theorem.

Theorem 5.1. *Let n be a positive integer which is not divisible by 4, D as in (5.1) above and let $q \in Cl(D)$ be a form in the image of $\mathcal{G}$. For any genus character χ_{e_1,e_2} of $Cl(D)$ with e_1 odd, we have*

$$\chi_{e_1,e_2}(q) = \begin{cases} \chi_{-8}(|e_1|) & \text{if } n \equiv 3 \pmod 4 \\ \chi_{-4}(|e_1|) & \text{if } n \equiv 1,2 \pmod 4. \end{cases}$$

Moreover,

$$\#\mathcal{G}^{-1}(\{q\}) = \frac{24 \cdot 2^{\omega(n)}}{\#\mathcal{O}_D^\times} = \begin{cases} 48/\#\mathcal{O}_D^\times \cdot 2^{\omega(D)-1} & \text{if } n \equiv 3 \pmod 4 \\ 24/\#\mathcal{O}_D^\times \cdot 2^{\omega(D)-1} & \text{if } n \equiv 1,2 \pmod 4. \end{cases}$$

This theorem was first proven by Gauss [Gau86]. We again refer the reader to [Shi06] for a more modern presentation.

6   Proof of the Main Theorem

We will evaluate the minimal parabolic $GL_3(\mathbb{Z})$ Eisenstein series at the identity matrix. We recall

$$\zeta(2s_1)\zeta(2s_2)E(I, s_1, s_2) = \sum_{\substack{0 \neq v \in L \\ 0 \neq w \in L \cap v^\perp}} Q(v)^{-s_2} Q(w)^{-s_1}. \qquad (6.1)$$

Our goal is the following theorem.

Theorem 6.1. *The Eisenstein series $E(I, s_1, s_2)$ can be expressed as a linear combination of products of the double Dirichlet series $Z(\psi_1, \psi_2) := Z(s_1, s_2; \psi_1, \psi_2)$, where ψ_1, ψ_2 range over the characters ramified only at 2. Explicitly,*

$$\zeta_2(2s_1)\zeta_2(2s_2)\zeta_2(2s_1 + 2s_2 - 1)E(I_3, s_1, s_2)/12$$

$$= Z(1, \chi_{-4})Z(\chi_{-4}, 1) + Z(1, 1)Z(\chi_{-4}, \chi_{-4})$$

$$+ 2^{-s_1} Z(1, \chi_{-8})Z(\chi_{-4}, 1) + 2^{-s_1} Z(1, \chi_{-8})Z(\chi_{-4}, \chi_{-4})$$

$$+ 2^{-s_2} Z(1, \chi_{-4}) Z(\chi_{-8}, 1) + 2^{-s_2} Z(1, \chi_{-4}) Z(\chi_8, 1)$$

$$+ 2^{-s_2} Z(1, 1) Z(\chi_{-8}, \chi_{-4}) - 2^{-s_2} Z(1, 1) Z(\chi_8, \chi_{-4})$$

$$+ 2^{-s_1 - s_2} Z(1, \chi_{-8}) Z(\chi_{-8}, 1) + 2^{-s_1 - s_2} Z(1, \chi_{-8}) Z(\chi_8, 1)$$

$$+ 2^{-s_1 - s_2} Z(1, \chi_8) Z(\chi_{-8}, \chi_{-4}) - 2^{-s_1 - s_2} Z(1, \chi_8) Z(\chi_8, \chi_{-4})$$

$$+ 2^{-s_2} Z(1, 1) Z(1, \chi_{-8}) - 2^{-s_2} Z(1, \chi_{-4}) Z(1, \chi_8). \tag{6.2}$$

In particular, according to (2.5) and (2.6), this expresses the orthogonal period $P^H(\mathcal{E}(\cdot, \lambda))$ in terms of the double Dirichlet series $Z(\psi_1, \psi_2)$.

Proof. Begin by breaking up the sum in (6.1) into congruence classes of $Q(v)$ mod 4. Because multiplication by 2 gives a bijection between $L(n)$ and $L(4n)$, we have

$$\sum_{\substack{0 \neq v \in L \\ Q(v) \equiv 0 \bmod 4 \\ 0 \neq w \in L \cap v^\perp}} Q(v)^{-s_2} Q(w)^{-s_1} = 4^{-s_2} \zeta(2s_1) \zeta(2s_2) E(I, s_1, s_2). \tag{6.3}$$

Therefore,

$$(1 - 4^{-s_2}) \zeta(2s_1) \zeta(2s_2) E(I, s_1, s_2) = \sum_{\substack{0 \neq v \in L \\ Q(v) \not\equiv 0 \bmod 4 \\ w \in L \cap v^\perp}} Q(v)^{-s_2} Q(w)^{-s_1}$$

$$= \zeta_2(2s_2) \sum_{\substack{v_0 \in L_0 \\ w \in L \cap v_0^\perp}} Q(v_0)^{-s_2} Q(w)^{-s_1}.$$

The second line follows after writing $v \in L$ as $c v_0$ with $v_0 \in L_0$ and c an odd positive integer. Note that we have dropped the condition $Q(v) \not\equiv 0 \bmod 4$ as it becomes redundant for $v_0 \in L_0$. Thus,

$$\zeta(2s_1) E(I, s_1, s_2)$$

$$= \left(\sum_{\substack{v \in L_0 \\ Q(v) \equiv 1 \bmod 4}} + \sum_{\substack{v \in L_0 \\ Q(v) \equiv 2 \bmod 4}} + \sum_{\substack{v \in L_0 \\ Q(v) \equiv 3 \bmod 4}} \right) Q(v)^{-s_2} Q(w)^{-s_1} \tag{6.4}$$

is equal to $S_1 + S_2 + S_3$, say. We treat each of these 3 sums separately. Begin with S_1:

$$S_1 = \sum_{\substack{n>0 \\ n\equiv 1 \bmod 4}} \frac{1}{n^{s_2}} \left(\sum_{v\in L_0[n]} \zeta_{\mathcal{G}(v)}(s_1) \right)$$

$$= \sum_{\substack{n>0 \\ n\equiv 1 \bmod 4}} \frac{1}{n^{s_2}} \left(\frac{24 \cdot 2^{\omega(n)}}{\#\mathcal{O}^{\times}_{-4n}} \sum_{q\sim q_{0,n}} \zeta_q(s_1) \right),$$

where $q_{0,n}$ is a form in $\mathrm{Cl}(-4n)$ satisfying $\chi_{e'_1,e'_2}(q_{0,n}) = \chi_{-4}(e_1)$ for all squarefree odd divisors e_1 of n. This follows from Theorem 5.1. Since $\omega(n) = \omega(-4n) - 1$, Proposition 4.1 now implies that

$$S_1 = 24 \sum_{\substack{n>0 \\ n\equiv 1 \bmod 4}} \frac{1}{n^{s_2}} \left(\sum_{\substack{e_1|n \\ \text{sqfree}}} \chi_{-4}(e_1) L_{\mathcal{O}_{-4n}}(s_1, \chi_{e'_1,e'_2}). \right) \tag{6.5}$$

As in Sect. 4, e'_2 is chosen to be the fundamental discriminant such that $e'_1 e'_2$ is equal to the product of the prime discriminants dividing $-4n$. Reintroduce the integers $n \equiv 3 \mod 4$ in (6.5):

$$S_1/12 = \sum_{\substack{n>0 \\ n\equiv 1 \bmod 4}} \frac{1 + \chi_{-4}(n)}{n^{s_2}} \left(\sum_{\substack{e_1|n \\ \text{sqfree}}} \chi_{-4}(e_1) L_{\mathcal{O}_{-4n}}(s_1, \chi_{e'_1,e'_2}) \right)$$

$$= \sum_{\substack{n>0 \\ \text{odd}}} \frac{1}{n^{s_2}} \left(\sum_{\substack{e_1|n \\ \text{sqfree}}} \chi_{-4}(e_1) L_{\mathcal{O}_{-4n}}(s_1, \chi_{e'_1,e'_2}) \right)$$

$$+ \sum_{\substack{n>0 \\ \text{odd}}} \frac{\chi_{-4}(n)}{n^{s_2}} \left(\sum_{\substack{e_1|n \\ \text{sqfree}}} \chi_{-4}(e_1) L_{\mathcal{O}_{-4n}}(s_1, \chi_{e'_1,e'_2}) \right). \tag{6.6}$$

Now write $n = e_1 e_2 f^2$ with e_1, e_2, f odd and reverse the order of summation in both sums in (6.6). For $\psi = 1$ or χ_{-4},

$$\sum_{\substack{n>0 \\ \text{odd}}} \frac{\psi(n)}{n^{s_2}} \left(\sum_{\substack{e_1 \mid n \\ \text{sqfree}}} \chi_{-4}(e_1) L_{\mathcal{O}_{-4n}}(s_1, \chi_{e_1', e_2'}) \right)$$

$$= \sum_{\substack{e_1, e_2>0 \\ \text{odd,sqfree}}} \frac{\psi(e_1 e_2)\chi_{-4}(e_1)}{(e_1 e_2)^{s_2}} \sum_{f>0,\text{odd}} \left(\frac{L_{\mathcal{O}_{-4e_1 e_2 f^2}}(s_1, \chi_{e_1', -4e_2'})}{f^{2s_2}} \right). \tag{6.7}$$

By virtue of Proposition 4.2, the inner sum in (6.7) is an Euler product, which may be explicitly evaluated as

$$\sum_{f>0,\text{odd}} \frac{L_{\mathcal{O}_{-4e_1 e_2 f^2}}(s_1, \chi_{e_1', -4e_2'})}{f^{2s_2}}$$

$$= L(s_1, \chi_{e_1'}) L(s_1, \chi_{-4e_2'}) \prod_{p \neq 2} \sum_{k=0}^{\infty} \frac{P_k(p^{-s_1}, \chi_{e_1'}(p), \chi_{-4e_2'}(p))}{p^{-2kw}}$$

$$= L(s_1, \chi_{e_1'}) L(s_1, \chi_{-4e_2'}) \frac{\zeta_2(2s_2)\zeta_2(2s_1 + 2s_2 - 1)}{L_2(s_1 + 2s_2, \chi_{e_1'}) L_2(s_1 + 2s_2, \chi_{-4e_2'})}. \tag{6.8}$$

Thus, (6.6) becomes

$$\zeta_2(2s_2)\zeta_2(2s_1 + 2s_2 - 1)$$

$$\times \sum_{\psi=1,\chi_{-4}} \left(\sum_{\substack{e_1>0 \\ \text{odd}}} \frac{\psi \chi_{-4}(e_1)}{e_1^{s_2}} \frac{L(s_1, \chi_{e_1'})}{L_2(s_1 + 2s_2, \chi_{e_1'})} \right) \left(\sum_{\substack{e_2>0 \\ \text{odd}}} \frac{\psi(e_2)}{e_2^{s_2}} \frac{L(s_1, \chi_{-4e_2'})}{L_2(s_1 + 2s_2, \chi_{-4e_2'})} \right).$$

$$\tag{6.9}$$

Comparing with Theorem 3.1, the second term in parentheses above is just

$$\frac{Z(s_1, s_2; \chi_{-4}, \psi)}{\zeta_2(2s_2)\zeta_2(2s_1 + 2s_2 - 1)}. \tag{6.10}$$

To write the first in terms of the double Dirichlet series of Sect. 3.1, we remove the Euler factor at 2 from the L function, which appear in the numerator:

$$L(s_1, \chi_{e_1'}) = L_2(s_1, \chi_{e_1'}) \left(1 + \frac{\chi_{e_1'}(2)}{2^{s_1}} \right) \left(1 - \frac{1}{4^{s_1}} \right)^{-1}.$$

Now $\chi_{e'_1}(2) = \chi_8(e)$, so the first term in parentheses in (6.9) is

$$\frac{(1 - \frac{1}{4^{s_1}})^{-1}}{\zeta_2(2s_2)\zeta_2(2s_1 + 2s_2 - 1)}[Z(s_1, s_2; 1, \psi\chi_{-4}) + 2^{-s_1} Z(s_1, s_2; 1, \psi\chi_{-8})]. \quad (6.11)$$

Putting (6.10),(6.11) into (6.9) completes our evaluation of S_1.

The evaluations of S_2 and S_3 are similar and will be omitted. We merely list the results below.

Proposition 6.2. *Abbreviate* $Z(s_1, s_2; \psi_1, \psi_2)$ *by* $Z(\psi_1, \psi_2)$. *Let*

$$S_i^* = \frac{S_i}{12}(1 - 4^{-s_1})\zeta_2(2s_2)\zeta_2(2s_1 + 2s_2 - 1)$$

for $i = 1, 2, 3$. *We have*

$$S_1^* = Z(1, \chi_{-4})Z(\chi_{-4}, 1) + Z(1, 1)Z(\chi_{-4}, \chi_{-4})$$
$$+ 2^{-s_1} Z(1, \chi_{-8})Z(\chi_{-4}, 1) + 2^{-s_1} Z(1, \chi_{-8})Z(\chi_{-4}, \chi_{-4})$$
$$2^{s_2} S_2^* = Z(1, \chi_{-4})Z(\chi_{-8}, 1) + Z(1, \chi_{-4})Z(\chi_8, 1)$$
$$+ Z(1, 1)Z(\chi_{-8}, \chi_{-4}) - Z(1, 1)Z(\chi_8, \chi_{-4})$$
$$+ 2^{-s_1} Z(1, \chi_{-8})Z(\chi_{-8}, 1) + 2^{-s_1} Z(1, \chi_{-8})Z(\chi_8, 1)$$
$$+ 2^{-s_1} Z(1, \chi_8)Z(\chi_{-8}, \chi_{-4}) - 2^{-s_1} Z(1, \chi_8)Z(\chi_8, \chi_{-4})$$
$$2^{s_2} S_3^* = Z(1, 1)Z(1, \chi_{-8}) - Z(1, \chi_{-4})Z(1, \chi_8)$$

Adding up $S_1 + S_2 + S_3$ completes the proof of the theorem □

7 Concluding Remarks

7.1 A Two-Variable Converse Theorem

Hamburger's converse theorem states that a Dirichlet series satisfying the same functional equation as the Riemann zeta function must be a constant multiple of the Riemann zeta function, [Ham21]. It is natural to ask for a two-variable analogue of this result. We formulate such an analogue here.

Conjecture 7.1. Let $D(s, w) = \sum_{m,n \geq 0} \frac{a(m,n)}{m^s n^w}$ be a double Dirichlet series in two complex variables, which is absolutely convergent for $\text{Re}(s), \text{Re}(w) > 1$. Define

$$D^*(s, w) = G(s, w)D(s, w),$$

where

$$G(s,w) = \zeta(2s)\zeta(2w)\zeta(2s+2w-1)\Gamma(s)\Gamma(w)\Gamma\left(s+w-\frac{1}{2}\right).$$

Suppose that:

1. $D^*(s,w)$ has a meromorphic continuation to $(s,w) \in \mathbb{C}^2$. Moreover,

$$(s-1)(w-1)\left(s+w-\frac{1}{2}\right)D^*(s,w)$$

   is entire, and for each fixed s, is bounded in each strip $a < \mathrm{Re}(w) < b$ of fixed width.
2. $D^*(s,w)$ is invariant under $(s,w) \mapsto \left(1-s, s+w-\frac{1}{2}\right)$ and $(s,w) \mapsto \left(s+w-\frac{1}{2}, 1-w\right)$
3. $D(s,w)$ satisfies the limits

$$\lim_{s\to\infty} D(s,w) = 24\frac{\zeta(s)}{\zeta(2s)}L(s,\chi_{-4}) \text{ and } \lim_{w\to\infty} D(s,w) = 24\frac{\zeta(w)}{\zeta(2w)}L(w,\chi_{-4})$$

Then $D(s,w) = E(I_3, s, w)$.

This conjecture would provide an alternate proof of our main result Theorem 6.1, since, with a little work, one can directly show that the double Dirichlet series on the right-hand side of (6.2) satisfies the same conditions as the $D(s,w)$ of the conjecture after multiplying by 12 and clearing the zeta factors. This would have the following arithmetic consequence. Whereas we proved the main identity using Gauss's result (Theorem 5.1) on the image of $\mathcal{G}$, a independent proof of the main identity will give a result almost as strong as Theorem 5.1. In particular, the conjecture would give a new proof of Gauss's result on the number of representations of an integer as a sum of 3 squares.

7.2 Siegel Modular Forms and Double Dirichlet Series

Let $r(m,n)$ be the number of pairs of vectors $v, w \in \mathbb{Z}^3$ such that $Q(v) = n$, $Q(w) = m$ and v is orthogonal to w. Comparing with (6.1), we see that the double Dirichlet series

$$D(s,w) = \sum_{n,m\geq 1} \frac{r(m,n)}{m^s n^w}$$

is equal to $\zeta(2s)\zeta(2w)E(I, s, w)$. From the theory of Eisenstein series, we know that $D(s, w)$ has a meromorphic continuation to $\mathbb{C}^2$ and satisfies a group of 6 functional equations. On the other hand, $r(m, n)$ are the diagonal Fourier coefficients of a Siegel modular theta series θ of genus 2. Thus, $D(s, w)$ can be obtained as an integral transform of θ. It is natural to ask if the analytic properties of $D(s, w)$ can be obtained from the automorphic properties of θ. If so, then presumably one can construct a double Dirichlet series with analytic continuation and functional equations by taking the same integral transform of any genus 2 Siegel modular form. We believe this warrants further investigation.

Acknowledgments This work originates from a suggestion of Professors D. Bump and A. Venkatesh. Professor S. Friedberg advised us on the comments of Sect. 7.2. We are grateful to all of them for their suggestions. This work was begun at the Hausdorff Research Institute for Mathematics during the special program on Representation Theory, Complex Analysis and Integral Geometry. We thank the Institute and the organizers of the program for inviting us to participate.

The first named author was supported by grants from the NSF and PSC-CUNY. The second named author was supported by THE ISRAEL SCIENCE FOUNDATION (grant No. 88/08).

References

[AGR93]   Avner Ash, David Ginzburg, and Steven Rallis. Vanishing periods of cusp forms over modular symbols. *Math. Ann.*, 296(4):709–723, 1993.

[BBFH07] B. Brubaker, D. Bump, S. Friedberg, and J. Hoffstein. Weyl group multiple Dirichlet series. III. Eisenstein series and twisted unstable A_r. *Ann. of Math. (2)*, 166(1): 293–316, 2007.

[Bor63]   Armand Borel. Some finiteness properties of adele groups over number fields. *Inst. Hautes Études Sci. Publ. Math.*, (16):5–30, 1963.

[BS96]   Wieb Bosma and Peter Stevenhagen. On the computation of quadratic 2-class groups. *J. Théor. Nombres Bordeaux*, 8(2):283–313, 1996.

[CFH05]   Gautam Chinta, Solomon Friedberg, and Jeffrey Hoffstein. Asymptotics for sums of twisted L-functions and applications. In *Automorphic representations, L-functions and applications: progress and prospects*, volume 11 of *Ohio State Univ. Math. Res. Inst. Publ.*, pages 75–94. de Gruyter, Berlin, 2005.

[CO07]   Gautam Chinta and Omer Offen. Unitary periods, Hermitian forms and points on flag varieties. *Math. Ann.*, 339(4):891–913, 2007.

[FK86]   Yuval Z. Flicker and David A. Kazhdan. Metaplectic correspondence. *Inst. Hautes Études Sci. Publ. Math.*, (64):53–110, 1986.

[Fli88]   Yuval Z. Flicker. Twisted tensors and Euler products. *Bull. Soc. Math. France*, 116(3):295–313, 1988.

[Gau86]   Carl Friedrich Gauss. *Disquisitiones arithmeticae*. Springer-Verlag, New York, 1986. Translated and with a preface by Arthur A. Clarke, Revised by William C. Waterhouse, Cornelius Greither and A. W. Grootendorst and with a preface by Waterhouse.

[GH85]   Dorian Goldfeld and Jeffrey Hoffstein. Eisenstein series of $\frac{1}{2}$-integral weight and the mean value of real Dirichlet L-series. *Invent. Math.*, 80(2):185–208, 1985.

[Ham21]   Hans Hamburger. Über die Riemannsche Funktionalgleichung der ξ-Funktion. *Math. Z.*, 11(3-4):224–245, 1921.

[Jac10]   Hervé Jacquet. Distinction by the quasi-split unitary group. *Israel J. Math.*, 178:269–324, 2010.

[Jac91]   Hervé Jacquet. Représentations distinguées pour le groupe orthogonal. *C. R. Acad. Sci. Paris Sér. I Math.*, 312(13):957–961, 1991.

[Jac01]   Hervé Jacquet. Factorization of period integrals. *J. Number Theory*, 87(1): 109–143, 2001.

[Jac05]   Hervé Jacquet. Kloosterman identities over a quadratic extension. II. *Ann. Sci. École Norm. Sup. (4)*, 38(4):609–669, 2005.

[JLR99]   Hervé Jacquet, Erez Lapid, and Jonathan Rogawski. Periods of automorphic forms. *J. Amer. Math. Soc.*, 12(1):173–240, 1999.

[Kan05]   Masanobu Kaneko. On the local factor of the zeta function of quadratic orders. In *Zeta functions, topology and quantum physics*, volume 14 of *Dev. Math.*, pages 75–79. Springer, New York, 2005.

[LO07]   Erez Lapid and Omer Offen. Compact unitary periods. *Compos. Math.*, 143(2): 323–338, 2007.

[LR00]   Erez Lapid and Jonathan Rogawski. Stabilization of periods of Eisenstein series and Bessel distributions on GL(3) relative to U(3). *Doc. Math.*, 5:317–350 (electronic), 2000.

[LR03]   Erez M. Lapid and Jonathan D. Rogawski. Periods of Eisenstein series: the Galois case. *Duke Math. J.*, 120(1):153–226, 2003.

[Maa38]   H. Maaß. Konstruktion ganzer Modulformen halbzahliger Dimension mit ϑ-Multiplikatoren in zwei Variablen. *Math. Z.*, 43(1):709–738, 1938.

[Off07]   Omer Offen. Stable relative Bessel distributions on GL(n) over a quadratic extension. *Amer. J. Math.*, 129(5):1183–1226, 2007.

[Shi06]   Goro Shimura. Quadratic Diophantine equations, the class number, and the mass formula. *Bull. Amer. Math. Soc. (N.S.)*, 43(3):285–304 (electronic), 2006.

[Wal80]   J.-L. Waldspurger. Correspondance de Shimura. *J. Math. Pures Appl. (9)*, 59(1): 1–132, 1980.

[Wal81]   J.-L. Waldspurger. Sur les coefficients de Fourier des formes modulaires de poids demi-entier. *J. Math. Pures Appl. (9)*, 60(4):375–484, 1981.

Regular Orbits of Symmetric Subgroups on Partial Flag Varieties

Dan Ciubotaru, Kyo Nishiyama, and Peter E. Trapa

Abstract We give a new parameterization of the orbits of a symmetric subgroup on a partial flag variety. The parameterization is in terms of Spaltenstein varieties and associated nilpotent orbits. We explain applications to enumerating special unipotent representations of real reductive groups, as well as (a portion of) the closure order on the set of nilpotent coadjoint orbits.

Keywords Partial flag varieties • Symmetric subgroups • Spaltenstein varieties • Unipotent representations

Mathematics Subject Classification (2010): 17B08

1 Introduction

The main result of this paper is a new parameterization of the orbits of a symmetric subgroup K on a partial flag variety $\mathcal{P}$. The parameterization is in terms of certain Spaltenstein varieties, on the one hand, and certain nilpotent orbits, on the other. One of our motivations, as explained below, is related to enumerating special unipotent representations of real reductive groups. Another motivation is understanding (a portion of) the closure order on the set of nilpotent coadjoint orbits.

D. Ciubotaru • P.E. Trapa (✉)
Department of Mathematics, University of Utah, Salt Lake City, UT 84112, USA
e-mail: ciubo@math.utah.edu; ptrapa@math.utah.edu

K. Nishiyama
Department of Mathematics and Physics, Aoyama Gaukin University,
Fuchinobe 5-10-1, Sagamihara 252-5258, Japan
e-mail: kyo@gem.aoyama.ac.jp

B. Krötz et al. (eds.), *Representation Theory, Complex Analysis, and Integral Geometry*,
DOI 10.1007/978-0-8176-4817-6_4, © Springer Science+Business Media, LLC 2012

In more detail, suppose G is a complex connected reductive algebraic group and let θ denote an involutive automorphism of G. Write K for the fixed points of θ, and $\mathcal{P}$ for a variety of parabolic subalgebras of a fixed type in $\mathfrak{g}$, the Lie algebra of G. Then K acts with finitely many orbits on $\mathcal{P}$, and these orbits may be parameterized in a number of ways (e.g. [M, RS, BH]), each of which may be viewed as a generalization of the classical Bruhat decomposition. (This latter decomposition arises if $G = G_1 \times G_1$, θ interchanges the two factors, and $\mathcal{P}$ is taken to be the full flag variety of (pairs of) Borel subalgebras.) We give our parameterization of $K \backslash \mathcal{P}$ in Corollary 2.14 and then turn to applications and examples in later sections.

As mentioned above, one of the applications we have in mind concerns the connection with nilpotent coadjoint orbits for K. To each orbit $Q = K \cdot \mathfrak{p}$ of parabolic subalgebras in $\mathcal{P}$, we obtain such a coadjoint orbit as follows. Let $\mathfrak{k}$ denote the Lie algebra of K, and consider

$$K \cdot \left[(\mathfrak{g}/\mathfrak{p})^* \cap (\mathfrak{g}/\mathfrak{k})^* \right] = K \cdot (\mathfrak{g}/(\mathfrak{p} + \mathfrak{k}))^* \subset \mathfrak{g}^*; \tag{1.1}$$

here and elsewhere we implicitly invoke the inclusion of $(\mathfrak{g}/\mathfrak{p})^*$ and $(\mathfrak{g}/\mathfrak{k})^*$ into $\mathfrak{g}^*$ and take the intersection there. Suppose for simplicity K is connected. Then the space in (1.1) is irreducible. It also consists of nilpotent elements and is K invariant. Since the number of nilpotent K orbits on $(\mathfrak{g}/\mathfrak{k})^*$ is finite [KR], the space must contain a unique dense K orbit, call it $\Phi_{\mathcal{P}}(Q)$. (It is easy to adapt this argument to yield the same conclusion if K is disconnected.) Thus, we obtain a natural map

$$\Phi = \Phi_{\mathcal{P}} : K \backslash \mathcal{P} \longrightarrow K \backslash \mathcal{N}_{\mathcal{P}}^\theta, \tag{1.2}$$

where $\mathcal{N}_{\mathcal{P}}^\theta$ denotes the cone of nilpotent elements in

$$\left[G \cdot (\mathfrak{g}/\mathfrak{p})^* \right] \cap (\mathfrak{g}/\mathfrak{k})^*. \tag{1.3}$$

In fact, the map $\Phi_{\mathcal{P}}$ is the starting point of our parameterization of $K \backslash \mathcal{P}$ in Sect. 2. For orientation, in the setting of the Bruhat decomposition mentioned above, the map may be interpretation as taking Weyl group elements to nilpotent coadjoint orbits. (Concretely, it amounts to taking an element w to the dense orbit in the G_1 saturation of the intersection of the nilradicals of two Borel subalgebras in relative position w.)

Just as the Bruhat order on a Weyl group is easier to understand than the classification and closure order on nilpotent orbits, the set of K orbits on $\mathcal{P}$ in some sense behaves more nicely than the set of K orbits on $\mathcal{N}_{\mathcal{P}}^\theta$. The former (and the closure order on it) can be described uniformly, for instance [RS]. This is not the case for $K \backslash \mathcal{N}_{\mathcal{P}}^\theta$, where any (known) classification involves at least some case-by-case analysis. So a natural question becomes: can one translate the uniform features of K orbits on $\mathcal{P}$ to the setting of K orbits on $\mathcal{N}_{\mathcal{P}}^\theta$ using $\Phi_{\mathcal{P}}$? This is the viewpoint we adopt in Sect. 2. In particular, one may ask the following: given a K orbit $\mathcal{O}_K$ in $\mathcal{N}_{\mathcal{P}}^\theta$, do these exist a canonical element Q of $K \backslash \mathcal{P}$ such that $\Phi_{\mathcal{P}}(Q) = \mathcal{O}_K$? If so, we would be able to embed the set of K orbits on $\mathcal{N}_{\mathcal{P}}^{\theta}$ into (the more uniformly behaved) set of K orbits on $\mathcal{P}$. One might optimistically hope to understand a parameterization of $K \backslash \mathcal{N}_{\mathcal{P}}^{\theta}$ (and understand its closure order) in this way.

The simplest way to produce affirmative answer to this last question is whether the fiber of $\Phi_{\mathcal{P}}$ over $\mathcal{O}_K$ consists of a single element Q. So it is desirable to have a formula for the cardinality of the fiber. Using ideas of Rossmann and Borho-MacPherson, we give such a formula in Proposition 2.10 in terms of certain Springer representations. The question of whether the fiber consists of a single element then becomes a multiplicity one question about certain Weyl group representations. We then turn to two natural questions:

(1) Can one find a natural class of orbits $\mathcal{O}_K$ for which the fiber $\Phi_{\mathcal{P}}^{-1}(\mathcal{O}_K)$ is indeed a singleton?
(2) If so, can one give an *effective* algorithm to determine the fiber? (This is clearly important if one really wants to use these ideas to try to classify K orbits on $\mathcal{N}_{\mathcal{P}}^{\theta}$ uniformly.)

We give affirmative answers to these questions in Proposition 3.7 and Remark 3.10 respectively. The class of K orbits we find are those $\mathcal{O}_K$ such that $\mathcal{O} = G \cdot \mathcal{O}_K$ is an *even* complex orbit; then $\Phi_{\mathcal{P}}^{-1}(\mathcal{O}_K)$ consists of a single element if $\mathcal{P}$ is taken to be the partial flag variety such that $T^*\mathcal{P}$ is a resolution of singularities of the closure of $\mathcal{O}$. (The corresponding K orbits on $\mathcal{P}$ are the regular orbits of the title.) Perhaps surprisingly the algorithm answering (2) relies on the Kazhdan–Lusztig–Vogan algorithm [V1] for computing the intersection homology groups (with coefficients) of K orbit closures on the full flag variety.

The setting of Sect. 3 may appear too restrictive to be of much practical value. But in Sect. 4 we recall that it is exactly the geometric setting of the Adams–Barbasch–Vogan definition of Arthur packets. More precisely, since the ground field is $\mathbb{C}$, θ arises as the complexification of a Cartan involution for a real form $G_{\mathbb{R}}$ of G. We show that the algorithm of Remark 3.10 gives an effective means to compute a distinguished constituent of each Arthur packet of integral special unipotent representations for $G_{\mathbb{R}}$. According to the Arthur conjectures, these representations should be unitary. This is a striking prediction (which is still open in general), since the constructions leading to their definition have nothing to do with unitarity.

Section 4 is highly technical unfortunately, but we have included it in the hope that it is perhaps more accessible than [ABV, Chap. 27] (upon which it is of course based). We have also included it for another reason, which is easy to understand from the current context. If it were possible to give affirmative answers to questions (1) and (2) above to a wider class of orbits than we consider in Sect. 3, then the ideas of Sect. 4 translate those answers into new conclusions about special unipotent representations of real reductive groups. In recent joint work with Barbasch, one of us (PT) has made progress in this direction. The precise formulation of these results involves a rather different set of ideas, and the details [BT] appear elsewhere.

Finally, in Sect. 5, we consider a number of examples illustrating some subtleties of the parameterization of Sect. 2.

2 Parametrizing $K \backslash \mathcal{P}$

The main result of this section is Corollary 2.14, which gives a parameterization of the K orbits on $\mathcal{P}$. As Propositions 2.10 and 2.15 show, the parameterization is closely related to Springer's Weyl group representations.

We begin with a discussion of the set $K \backslash \mathcal{B}$ of K orbits on $\mathcal{B}$, the full flag variety of Borel subalgebras in our fixed complex reductive Lie algebra $\mathfrak{g}$. Basic references for this material are [M] or [RS]. The set $K \backslash \mathcal{B}$ is partially ordered by the inclusion of orbit closures. It is generated by closure relations in codimension one. We will need to distinguish two kinds of such relations. To do so, we fix a base-point $\mathfrak{b}_{\mathrm{o}} \in \mathcal{B}$ and a Cartan $\mathfrak{h}_{\mathrm{o}}$ in $\mathfrak{b}_{\mathrm{o}}$. We write $\mathfrak{b}_{\mathrm{o}} = \mathfrak{h}_{\mathrm{o}} \oplus \mathfrak{n}_{\mathrm{o}}$ for the corresponding Levi decomposition, and let $\Delta^{+} = \Delta^{+}(\mathfrak{h}_{\mathrm{o}}, \mathfrak{n}_{\mathrm{o}})$ denote the roots of $\mathfrak{h}_{\mathrm{o}}$ in $\mathfrak{n}_{\mathrm{o}}$. For a simple root $\alpha \in \Delta^{+}$, let $\mathcal{P}_{\alpha}$ denote the set of parabolic subalgebras of type α, and write π_{α} for the projection $\mathcal{B} \to \mathcal{P}_{\alpha}$.

Fix K orbits Q and Q' on $\mathcal{B}$. If K is connected, then Q is irreducible, and hence so is $\pi_{\alpha}^{-1}(\pi_{\alpha}(Q))$. Thus, $\pi_{\alpha}^{-1}(\pi_{\alpha}(Q))$ contains a unique dense K orbit. In general, K need not be connected and Q need not be irreducible. But it is easy to see that the similar reasoning applies to conclude $\pi_{\alpha}^{-1}(\pi_{\alpha}(Q))$ always contains a dense K orbit. We write $Q \overset{\alpha}{\to} Q'$ if

$$\dim(Q') = \dim(Q) + 1$$

and

$$Q' \text{ is dense in } \pi_{\alpha}^{-1}(\pi_{\alpha}(Q)).$$

This implies that Q is codimension one in the closure of Q'. The relations $Q < Q'$ for $Q \overset{\alpha}{\to} Q'$ do not generate the full closure order, however. Instead, we must also consider a kind of saturation condition. More precisely, whenever a codimension one subdiagram of the form

$$
\begin{array}{ccc}
 & Q_1 & \\
\alpha \nearrow & & \\
Q_2 & & Q_3 \\
 & \searrow \quad \nearrow \alpha & \\
 & Q_4 &
\end{array}
\qquad (2.1)
$$

is encountered, we complete it to

$$
\begin{array}{ccc}
 & Q_1 & \\
\alpha \nearrow & & \nwarrow \\
Q_2 & & Q_3 \\
 & \searrow \quad \nearrow \alpha & \\
 & Q_4 &
\end{array}
\qquad (2.2)
$$

New edges added in this way are dashed in the diagrams below. Note that this operation must be applied recursively, and thus some of the edges in the original diagram (2.1) may be dashed as the recursion unfolds. Following the terminology of [RS, 5.1], we call the partially ordered set determined by the solid edges the weak closure order.

Now fix a variety of parabolic subalgebras $\mathcal{P}$ of an arbitrary fixed type and write $\pi_{\mathcal{P}}$ for the projection from $\mathcal{B}$ to $\mathcal{P}$. For definiteness fix $\mathfrak{p}_\circ \in \mathcal{P}$ containing $\mathfrak{b}_\circ$, and write $\mathfrak{p}_\circ = \mathfrak{l}_\circ \oplus \mathfrak{u}_\circ$ for the Levi decomposition such that $\mathfrak{h}_\circ \subset \mathfrak{l}_\circ$. Then $K \backslash \mathcal{P}$ may be parameterized from a knowledge of the weak closure on $K \backslash \mathcal{B}$ as follows. Consider the relation $Q \sim_{\mathcal{P}} Q'$ if $\pi_{\mathcal{P}}(Q) = \pi_{\mathcal{P}}(Q')$; this is generated by the relations $Q \sim Q'$ if $Q \overset{\alpha}{\to} Q'$ for α simple in $\Delta(\mathfrak{h}_\circ, \mathfrak{l}_\circ)$. Equivalence classes in $K \backslash \mathcal{B}$ clearly are in bijection with $K \backslash \mathcal{P}$. (See also the parameterization of [BH, Sect. 1], especially Proposition 4.) Fix an equivalence class C and fix a representative $Q \in C$. The same reasoning that shows that $\pi_\alpha^{-1}(\pi_\alpha(Q))$ contains a unique dense K orbit also shows that

$$\pi_{\mathcal{P}}^{-1}(\pi_{\mathcal{P}}(Q))$$

contains a unique dense K orbit $Q_C \in K \backslash \mathcal{B}$. In other words, Q_C is the unique largest dimensional orbit among the elements in C. In fact, Q_C is characterized among the elements of C by the condition

$$\dim \pi_\alpha^{-1}(\pi_\alpha(Q_C)) = \dim(Q_C) \tag{2.3}$$

for all α simple in $\Delta(\mathfrak{h}_\circ, \mathfrak{l}_\circ)$. It follows that the full closure order on $K \backslash \mathcal{P}$ is simply the restriction of the full closure order on $K \backslash \mathcal{B}$ to the subset of all maximal-dimensional representatives of the form Q_C. By restricting only the weak closure order, we may speak of the weak closure order on $K \backslash \mathcal{P}$.

We next place the map $\Phi_{\mathcal{P}}$ of (1.2) in a more natural context. Consider the cotangent bundle $T^*\mathcal{P} \subset \mathcal{P} \times \mathfrak{g}^*$. It consists of pairs $(\mathfrak{p}, \xi)$ with

$$\xi \in T_{\mathfrak{p}}^* \mathcal{P} \simeq (\mathfrak{g}/\mathfrak{p})^*. \tag{2.4}$$

The moment map $\mu_{\mathcal{P}}$ from $T^*\mathcal{P}$ to $\mathfrak{g}^*$ maps a point $(\mathfrak{p}, \xi)$ in $T^*\mathcal{P}$ simply to ξ. Consider now the conormal variety for K orbits on $\mathcal{P}$,

$$T_K^* \mathcal{P} = \bigcup_{Q \in K \backslash \mathcal{P}} T_Q^* \mathcal{P},$$

where $T_Q^* \mathcal{P}$ denotes the conormal bundle to the K orbit Q. (In the special case $G = G_1 \times G_1$ and $\mathcal{P} = \mathcal{B}$ mentioned in the introduction, the conormal variety is the usual Steinberg variety of triples). In general, we may identify

$$T_Q^* \mathcal{P} = \{(\mathfrak{p}, \xi) \mid \mathfrak{p} \in Q, \xi \in (\mathfrak{g}/(\mathfrak{k} + \mathfrak{p}))^*\}, \tag{2.5}$$

and hence the image of $T_K^* \mathcal{P}$ under $\mu_{\mathcal{P}}$ is simply $\mathcal{N}_{\mathcal{P}}^\theta$. Moreover, the image of $T_Q^* \mathcal{P}$ under $\mu_{\mathcal{P}}$ is nothing but the space in (1.1). Hence, $\Phi_{\mathcal{P}}(Q)$ is simply the unique dense K orbit in the moment map image of $T_Q^* \mathcal{P}$.

Here, are some elementary properties of $\Phi_{\mathcal{P}}$.

Proposition 2.6. *1. Fix $Q \in K\backslash\mathcal{P}$ and suppose $Q' \in K\backslash\mathcal{B}$ is dense in $\pi_{\mathcal{P}}^{-1}(Q)$. Then*

$$\Phi_{\mathcal{B}}(Q') = \Phi_{\mathcal{P}}(Q).$$

2. The map $\Phi_{\mathcal{P}}$ is order reversing from the weak closure order in $K\backslash\mathcal{P}$ to the closure order on $K\backslash\mathcal{N}_{\mathcal{P}}^\theta$; that is, if $Q < Q'$ in the weak closure order on $K\backslash\mathcal{P}$, then

$$\overline{\Phi_{\mathcal{P}}(Q)} \supset \Phi_{\mathcal{P}}(Q').$$

Proof. Part (1) is clear from the definitions. Part (2) reduces to the assertion for $Q \xrightarrow{\alpha} Q'$. In that case, it amounts to a rank one calculation where it is obvious. □

Example 2.7. Proposition 2.6(2) fails for the full closure order on $K\backslash\mathcal{P}$. The first example which exhibits this failure is $G_{\mathbb{R}} = \mathrm{Sp}(4, \mathbb{R})$ and $\mathcal{P} = \mathcal{B}$. Let α denote the short simple root in Δ^+ and β the long one. The closure order for $K\backslash\mathcal{B}$ is as in the diagram in (2.8). Orbits on the same row of the diagram below all have the same dimension. (The bottom row consists of orbits of dimension one, the next row consists of orbits of dimension two, and so on.) Dashed lines represent relations in the full closure order, which are not in the weak order.

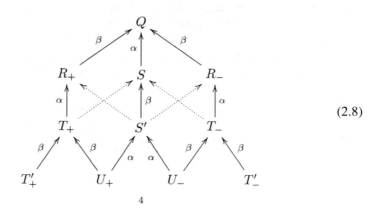

$$(2.8)$$

Adopt the parameterization of $K\backslash\mathcal{N}^\theta$ given in [CM, Theorem 9.3.5] in terms of signed tableau. Let $(i_1)_{\epsilon_1}^{j_1}(i_2)_{\epsilon_2}^{j_2}\cdots$ denote the tableau with j_k rows of length i_k beginning with sign ϵ_k for each k. Then the closure order on $K\backslash\mathcal{N}^\theta$ is given by

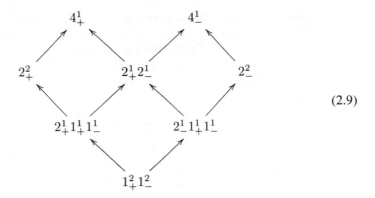

$$(2.9)$$

Then $\Phi_\mathcal{B}$ maps Q to $1^2_+ 1^2_-$; $R_\pm$ to $2^1_\pm 1^1_+ 1^1_-$; S and S' to $2^1_+ 2^1_-$; $T_\pm$ and $T'_\pm$ to $2^2_\pm$; and $U_\pm$ to $4^1_\pm$. Note that $\Phi_\mathcal{B}$ reverses all closure relations *except* the two dashed edges indicating $T_\pm \subset \overline{S}$.

We are now in a position to determine the size of the fiber $\Phi_\mathcal{P}^{-1}(\mathcal{O}_K)$ for $\mathcal{O}_K \in K\backslash\mathcal{N}_\mathcal{P}^\theta$. For $\xi \in \mathcal{O}_K$, let $A_K(\xi)$ (resp. $A_G(\xi)$) denote the component group of the centralizer in K (resp. G) of ξ. Obviously, there is a natural map

$$A_K(\xi) \to A_G(\xi),$$

which we often invoke implicitly. Write $\mathrm{Sp}(\xi)$ for the Springer representation of $W \times A_G(\xi)$ on the top homology of the Springer fiber over ξ (normalized so that $\xi = 0$ gives the sign representations of W). Let

$$\mathrm{Sp}(\xi)^{A_K} = \mathrm{Hom}_{A_K(\xi)}\left(\mathrm{Sp}(\xi), \mathbb{1}\right).$$

Proposition 2.10. *Fix $\xi \in \mathcal{O}_K$. Then*

$$\#\Phi_\mathcal{P}^{-1}(\mathcal{O}_K) = \dim \mathrm{Hom}_{W(\mathcal{P})}\left(\mathrm{sgn}, \mathrm{Sp}(\xi)^{A_K}\right)$$

$$= \dim \mathrm{Hom}_W(\mathrm{ind}_{W(\mathcal{P})}^W(\mathrm{sgn}), \mathrm{Sp}(\xi)^{A_K}).$$

Proof. The second equality follows by Frobenius reciprocity. For the first, set

$$S_\mathcal{P} = \{Q \in K\backslash\mathcal{B} \mid Q \text{ is dense in } \pi_\mathcal{P}^{-1}(\pi_\mathcal{P}(Q))\}.$$

According to the discussion around (2.3) and Proposition 2.6(1), $\pi_\mathcal{P}$ implements a bijection

$$S_\mathcal{P} \cap \Phi_\mathcal{B}^{-1}(\mathcal{O}_K) \to \Phi_\mathcal{P}^{-1}(\mathcal{O}_K).$$

We will count the left-hand side if K is connected. If K is disconnected, there are a few subtleties (none of which are very serious), which are best treated later.

Consider the top integral Borel–Moore homology of the conormal variety $T_K^*\mathcal{P}$. Since we have assumed K is connected, the closures of the individual conormal bundles exhaust the irreducible components of $T_K^*\mathcal{P}$, and their classes form a basis of the homology,

$$\mathrm{H}_{\text{top}}^{\infty}\left(T_K^*\mathcal{P}, \mathbb{Z}\right) = \bigoplus_{Q \in K\backslash\mathcal{P}} [\overline{T_Q^*\mathcal{P}}].$$

If $\mathcal{P} = \mathcal{B}$, Rossmann [R] (extending earlier work of Kazhdan–Lusztig [KL]) described a construction giving an action of the Weyl group W on this homology space. The action is graded in the following sense that if $Q \in \Phi_{\mathcal{B}}^{-1}(\mathcal{O}_K)$, then

$$w \cdot [\overline{T_Q^*\mathcal{B}}]$$

is a linear combination of conormal bundles to orbits in fibers $\Phi_{\mathcal{B}}^{-1}(\mathcal{O}_K')$ with $\mathcal{O}_K' \subset \overline{\mathcal{O}_K}$. Hence, if we set

$$\Phi_{\mathcal{B}}^{-1}(\mathcal{O}_K, \leq) = \bigcup_{\mathcal{O}_K' \subseteq \overline{\mathcal{O}_K}} \Phi_{\mathcal{B}}^{-1}(\mathcal{O}_K')$$

and

$$\Phi_{\mathcal{B}}^{-1}(\mathcal{O}_K, <) = \bigcup_{\mathcal{O}_K' \subsetneq \overline{\mathcal{O}_K}} \Phi_{\mathcal{B}}^{-1}(\mathcal{O}_K'),$$

then

$$\mathbf{M}(\mathcal{O}_K) := \bigoplus_{Q \in \Phi_{\mathcal{B}}^{-1}(\mathcal{O}_K, \leq)} [\overline{T_Q^*\mathcal{B}}] \Big/ \bigoplus_{Q \in \Phi_{\mathcal{B}}^{-1}(\mathcal{O}_K, <)} [\overline{T_Q^*\mathcal{B}}]$$

is a W module with basis indexed by $\Phi_{\mathcal{B}}^{-1}(\mathcal{O}_K)$. Rossmann's construction shows that

$$\mathbf{M}(\mathcal{O}_K) \simeq \mathrm{Sp}(\xi)^{A_K},$$

where $\xi \in \mathcal{O}_K$ as above. This proves the proposition for $\mathcal{P} = \mathcal{B}$. For the general case, we must identify $S_{\mathcal{P}}$ in terms of the Weyl group action. It follows from Rossmann's constructions that

$$s_\alpha \cdot [\overline{T_Q^*\mathcal{B}}] = -[\overline{T_Q^*\mathcal{B}}]$$

if and only if

$$\dim \pi_\alpha^{-1}\left(\pi_\alpha(Q)\right) = \dim(Q).$$

Thus, (2.3) implies that $S_\mathcal{P} \cap \Phi_\mathcal{B}^{-1}(\mathcal{O}_K)$ indexes exactly the basis elements of $\mathbf{M}(\mathcal{O}_K)$ which transform by the sign representation of the Weyl group of type $\mathcal{P}$. The proposition thus follows in the case of K connected. (A complete proof in the disconnected case is discussed after Proposition 2.15.) $\square$

The above proof is extrinsic in the sense that it is deduced from a statement about the $\mathcal{P} = \mathcal{B}$ case. We may argue more intrinsically (without reference to $\mathcal{B}$) using results of Borho–MacPherson [BM] as follows.

Fix $\xi \in \mathcal{N}_\mathcal{P}^\theta$ and consider $\mu_\mathcal{P}^{-1}(\xi)$. In terms of the identification around (2.4),

$$\mu_\mathcal{P}^{-1}(\xi) = \{(\mathfrak{p}, \xi) \mid \xi \in (\mathfrak{g}/\mathfrak{p})^*\}.$$

(Borho–MacPherson write $\mathcal{P}_\xi^0$ for $\mu_\mathcal{P}^{-1}(\xi)$ and call it a Spaltenstein variety.) Clearly $A_G(\xi)$, and hence $A_K(\xi)$, act on the set of irreducible components $\mathrm{Irr}(\mu_\mathcal{P}^{-1}(\xi))$. Fix $C \in \mathrm{Irr}(\mu_\mathcal{P}^{-1}(\xi))$, and consider $Z(C) := \overline{K \cdot C} \subset T^*\mathcal{P}$. Since $\xi \in \mathcal{N}_\mathcal{P}^\theta \subset \mathcal{N}(\mathfrak{g}/\mathfrak{k})^*$, it follows from (2.5) that $Z(C)$ is in fact contained in the conormal variety

$$Z(C) \subset T_K^*\mathcal{P},$$

which is of course pure-dimensional of dimension $\dim(\mathcal{P})$. Hence,

$$\dim(Z(C)) \le \dim(\mathcal{P}).$$

But clearly

$$\dim(Z(C)) = \dim(K \cdot \xi) + \dim(C),$$

and thus

$$\dim(C) \le \dim(\mathcal{P}) - \dim(K \cdot \xi). \tag{2.11}$$

Write $\mathrm{Irr}_{\max}(\mu_\mathcal{P}^{-1}(\xi))$ for those irreducible components whose dimensions actually achieve the upper bound. (This set could be empty, for instance, as we shall see in Example 3.3 below when $\mathcal{P} = \mathcal{P}_\beta$ and ξ is a representative of a minimal nilpotent orbit. Note, however, that it is a general theorem of Spaltenstein's that if $\mathcal{P} = \mathcal{B}$, the full flag variety, then $\mathrm{Irr}_{\max}(\mu_\mathcal{B}^{-1}(\xi)) = \mathrm{Irr}(\mu_\mathcal{B}^{-1}(\xi))$.)

Proposition 2.12. *Fix $\xi \in \mathcal{N}_\mathcal{P}^\theta$, set $\mathcal{O}_K = K \cdot \xi$, assume $\Phi_\mathcal{P}^{-1}(\mathcal{O}_K)$ is nonempty, and fix $Q \in \Phi_\mathcal{P}^{-1}(\mathcal{O}_K)$. Then*

$$C(Q) := \overline{T_Q^*\mathcal{P}} \cap \mu_\mathcal{P}^{-1}(\xi)$$

is the union of elements in an $A_K(\xi)$ orbit on $\mathrm{Irr}_{\max}(\mu_\mathcal{P}^{-1}(\xi))$. The assignment $Q \mapsto C(Q)$ gives a bijection

$$\Phi_\mathcal{P}^{-1}(\mathcal{O}_K) \longrightarrow A_K(\xi) \backslash \mathrm{Irr}_{\max}(\mu_\mathcal{P}^{-1}(\xi)). \tag{2.13}$$

Proof. Fix $C \in \mathrm{Irr}_{\max}(\mu_{\mathcal{P}}^{-1}(\xi))$. Then $\dim(Z(C)) = \dim(\mathcal{P})$ by definition. Notice that $Z(C)$ is nearly irreducible (and it is if K is connected). In general, the component group of K (which is finite by hypothesis) acts transitively on the irreducible components of $Z(C)$. But from the definition of $T_K^*\mathcal{P}$, the closure of each conormal bundle $T_Q^*\mathcal{P}$ consists of a subset of irreducible components of $T_K^*\mathcal{P}$ on which the component group of K acts transitively. Since $\dim(Z(C)) = \dim(T_K^*\mathcal{P})$, it follows that there is some Q such that

$$Z(C) = \overline{T_Q^*\mathcal{P}};$$

moreover, Q must be an element of $\Phi_{\mathcal{P}}^{-1}(\mathcal{O}_K)$. Clearly $Z(C) = Z(C')$ if and only if C and C' are in the same $A_K(\xi)$ orbit. The assignment $C \mapsto Q$ gives a bijection $A_K(\xi)\backslash \mathrm{Irr}_{\max}(\mu_{\mathcal{P}}^{-1}(\xi)) \to \Phi_{\mathcal{P}}^{-1}(\mathcal{O}_K)$ which, by construction, is the inverse of the map in (2.13). This completes the proof. $\square$

Corollary 2.14. *Let* $\xi_1, \ldots, \xi_k$ *be representatives of the* K *orbits on* $\mathcal{N}_{\mathcal{P}}^{\theta}$. *Then the map*

$$Q \longrightarrow \left(\Phi_{\mathcal{P}}(Q), \overline{T_Q^*\mathcal{P}} \cap \mu_{\mathcal{P}}^{-1}(\xi_i) \right)$$

for i *the unique index such that* $K \cdot \xi_i$ *dense in* $\Phi_{\mathcal{P}}(Q)$ *implements a bijection*

$$K\backslash\mathcal{P} \longrightarrow \coprod_i A_K(\xi_i)\backslash \mathrm{Irr}_{\max}(\mu_{\mathcal{P}}^{-1}(\xi_i)).$$

Thus everything reduces to understanding the irreducible components of $\mu_{\mathcal{P}}^{-1}(\xi)$ of maximal possible dimension. For this we need some nontrivial results of Borho–MacPherson. [BM, Theorem 3.3] shows that the fundamental classes of the elements of $\mathrm{Irr}_{\max}(\mu_{\mathcal{P}}^{-1}(\xi))$ index a basis of $\mathrm{Hom}_{W(\mathcal{P})}(\mathrm{sgn}, \mathrm{Sp}(\xi))$. Actually, to be precise, their condition for C to belong to $\mathrm{Irr}_{\max}(\mu_{\mathcal{P}}^{-1}(\xi))$ is that

$$\dim(C) = \dim(\mathcal{P}) - \frac{1}{2}\dim(G \cdot \xi).$$

To square with (2.11), we need to invoke the result of Kostant–Rallis [KR] that $K \cdot \xi$ is Lagrangian in $G \cdot \xi$. In any case, because $A_G(\xi)$ acts on $\mathrm{Sp}(\xi)$ and commutes with the W action, $A_G(\xi)$ also acts on $\mathrm{Hom}_{W(\mathcal{P})}(\mathrm{sgn}, \mathrm{Sp}(\xi))$, and [BM, Theorem 3.3] shows that this action is compatible with the action of $A_G(\xi)$ on $\mathrm{Irr}(\mu_{\mathcal{P}}^{-1}(\xi))$. In particular, this implies the following result.

Proposition 2.15. *Fix* $\xi \in \mathcal{N}_{\mathcal{P}}^{\theta}$. *Then the number of* $A_K(\xi)$ *orbits on* $\mathrm{Irr}_{\max}(\mu_{\mathcal{P}}^{-1}(\xi))$ *equals the dimension of*

$$\mathrm{Hom}_{W(\mathcal{P})}\left(\mathrm{sgn}, \mathrm{Sp}(\xi)^{A_K}\right).$$

Combining Propositions 2.12 and 2.15, we obtain an alternate proof of Proposition 2.10, which makes no assumption on the connectedness of K.

Remark 2.16. The $\mathcal{P} = \mathcal{B}$ case of Corollary 2.14 is due to Springer (unpublished). In this case, $W(\mathcal{B})$ is trivial, and thus $\Phi_{\mathcal{B}}^{-1}(\mathcal{O}_K)$ has order equal to the W-representation $\mathrm{Sp}(\xi)^{A_K}$.

It is of interest to compute the bijection of Corollary 2.14 as explicitly as possible. For instance, if $G_{\mathbb{R}} = \mathrm{GL}(n, \mathbb{C})$ and $\mathcal{P} = \mathcal{B}$ consists of pairs of flags, the left-hand side of the bijection in Corollary 2.14 consists of elements of the symmetric group S_n. On the right-hand side, all A-groups are trivial, and the irreducible components in question amount to pairs of irreducible components of the usual Springer fiber. Such pairs are parameterized by same-shape pairs of standard Young tableaux. Steinberg [St] showed that the bijection of the corollary amounts to the classical Robinson–Schensted correspondence.

A few other classical cases have been worked out explicitly [vL, Mc1, T1, T3]. But general statements are lacking. For instance, given Q and Q', there is no known effective algorithm to decide if $\Phi_{\mathcal{P}}(Q) = \Phi_{\mathcal{P}}(Q')$. The next section is devoted to special cases of the parameterization, which lead to nice general statements. It might appear that these special cases are too restrictive to be of much use. But it turns out that they encode exactly the geometry needed for the Adams–Barbasch–Vogan definition of Arthur packets. This is explained in Sect. 4.

3 $\mathcal{P}$-Regular K Orbits

The main results of this section are Proposition 3.7(b) and Remark 3.10, which together give an effective computation of a portion of the bijection of Proposition 2.12 under the assumption that $\mu_{\mathcal{P}}$ is birational.

Definition 3.1 (see [ABV, Definition 20.17]). A nilpotent orbit $\mathcal{O}_K$ of K on $\mathcal{N}_{\mathcal{P}}^{\theta}$ is called $\mathcal{P}$-regular (or simply regular, if $\mathcal{P}$ is clear from the context) if $G \cdot \mathcal{O}_K$ is dense in $\mu_{\mathcal{P}}(T^*\mathcal{P})$. Since $\mathcal{O}_K$ is Lagrangian in $G \cdot \mathcal{O}_K$ [KR], this condition is equivalent to

$$\dim(\mathcal{O}_K) = \frac{1}{2} \dim \mu(T^*\mathcal{P}) = \dim(\mathfrak{g}/\mathfrak{p}),$$

for any $\mathfrak{p} \in \mathcal{P}$. In other words, $\mathcal{P}$-regular nilpotent K-orbits meet the complex Richardson orbit induced from $\mathfrak{p}$. An orbit Q of K on $\mathcal{P}$ is called $\mathcal{P}$-regular (or simply regular) if $\Phi_{\mathcal{P}}(Q)$ is a $\mathcal{P}$-regular nilpotent orbit. Note that regular $\mathcal{P}$-orbits need not exist in general (for instance, if $G_{\mathbb{R}}$ is compact and $\mathcal{P}$ is not trivial).

Since regular nilpotent K orbits are automatically maximal in the closure order on $\mathcal{N}_{\mathcal{P}}^{\theta}$, Proposition 2.6(2) shows that regular K orbits on $\mathcal{P}$ are minimal in the weak closure order:

Proposition 3.2. *Suppose Q is a regular K orbit on $\mathcal{P}$. Then Q is minimal in the weak closure order on $K \backslash \mathcal{P}$.*

The next example shows that regular K orbits on $\mathcal{P}$ need not be minimal in the full closure order (i.e., they need not be closed).

Example 3.3. Retain the notation of Example 2.7. Let $\mathcal{P}_\alpha$ (resp. $\mathcal{P}_\beta$) consist of parabolic subalgebras of type α (resp. β) and write π_α and π_β in place of $\pi_{\mathcal{P}_\alpha}$ and $\pi_{\mathcal{P}_\beta}$, and similarly for μ_α and μ_β. Then the closure order on $K\backslash\mathcal{P}_\alpha$ is obtained by the appropriate restriction from (2.8). (Subscripts now indicate dimensions; dashed edges are those covering relations present in the full closure order but not the weak one.)

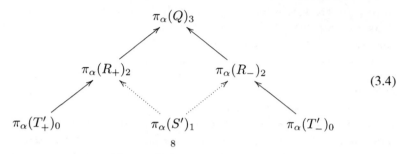

$$(3.4)$$

The closure order on $K\backslash\mathcal{P}_\beta$ is again obtained by restriction from (2.8). (Once again subscripts indicate dimensions.)

$$\pi_\beta(Q)_3$$
$$\uparrow$$
$$\pi_\beta(S)_2$$

$$\pi_\beta(T_+)_1 \qquad\qquad \pi_\beta(T_-)_1$$

$$(3.5)$$

In this case $\mathcal{N}_\alpha^\theta = \mathcal{N}_\beta^\theta$, and the closure order on $K\backslash\mathcal{N}_\mathcal{P}^\theta$ is just the bottom three rows of (2.9),

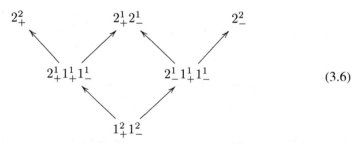

$$(3.6)$$

From Proposition 5.2 below (for instance), both $\Phi_\alpha = \Phi_{\mathcal{P}_\alpha}$ and $\Phi_\beta = \Phi_{\mathcal{P}_\beta}$ are injective. There are enough edges in the weak closure order on $K \backslash \mathcal{P}_\alpha$ so that Proposition 2.6(1) allows one to conclude that Φ_α reverses the full closure order. In fact, Φ_α is the obvious order reversing bijection of (3.4) onto (3.6). Hence, $\pi_\alpha(T'_\pm)$ and $\pi_\alpha(S')$ are $\mathcal{P}_\alpha$-regular.

By contrast, Φ_β does not invert the dashed edges in (3.5): Φ_β maps $\pi_\beta(Q)$ to the zero orbit, and the three remaining orbits to the three orbits of maximal dimension in $\mathcal{N}_{\mathcal{P}}^\theta$. Hence, $\pi_\beta(T'_\pm)$ and $\pi_\beta(S)$ are $\mathcal{P}_\beta$-regular. In particular, $\pi_\beta(S)$ is a $\mathcal{P}_\beta$-regular orbit which is not closed.

Finally note that the fiber of Φ_α over $2^1_\pm 1^1_+ 1^1_-$ consists of a single element, while the corresponding fiber for Φ_β is empty. This is consistent with Proposition 2.10 since $\mathrm{Sp}(\xi)$ (for ξ a representative of these orbits) is a one-dimensional representation on which the simple reflection s_α (resp. s_β) acts nontrivially (resp. trivially). $\qquad\qquad\square$

An essential difference in the two cases considered in Example 3.3 is that μ_α is birational, but μ_β has degree two.

Proposition 3.7 ([ABV, Theorem 20.18]). *Suppose $\mu_\mathcal{P}$ is birational onto its image. Then:*

(a) *Any regular K orbit on $\mathcal{P}$ consists of θ-stable parabolic subalgebras (and hence is closed).*

(b) *$\Phi_\mathcal{P}$ is a bijection from the set of regular K orbits on $\mathcal{P}$ to the set of regular nilpotent K orbits on $\mathcal{N}_{\mathcal{P}}^\theta$.*

Proof. Fix a $\mathcal{P}$-regular nilpotent K orbit $\mathcal{O}_K$ in $\mathcal{N}_{\mathcal{P}}^\theta$, $\xi \in \mathcal{O}_K$, and $Q \in \Phi_\mathcal{P}^{-1}(\mathcal{O}_K)$. Since $\mu_\mathcal{P}$ is birational, the set $\mathrm{Irr}_{\max}(\mu_\mathcal{P}^{-1}(\xi))$ is a single point, and so Proposition 2.12 shows that Q is the unique orbit in $\Phi_\mathcal{P}^{-1}(\mathcal{O}_K)$. This gives (b).

Again since $\mu_\mathcal{P}$ is birational, there is a unique parabolic $\mathfrak{p} \in Q$ such that $\xi \in (\mathfrak{g}/\mathfrak{p})^*$. Since $\theta(\xi) = -\xi$, $\theta(\mathfrak{p})$ is also such a parabolic. So $\theta(\mathfrak{p}) = \mathfrak{p}$. Thus, $Q = K \cdot \mathfrak{p}$ consists of θ-stable parabolic subalgebras. This gives the first part of (a). The same (well-known) proof of the fact that K orbits of θ-stable Borel subalgebras are closed (for example, [Mi, Lemma 5.8]), also applies to show that orbits of θ-stable parabolics are closed. (It is no longer true that a closed K orbit on $\mathcal{P}$ consists of θ-stable parabolic subalgebras. But if a θ-stable parabolic algebra in $\mathcal{P}$ exists, all closed orbits do indeed consist of θ-stable parabolic subalgebras.) $\qquad\square$

Because of the good properties in Proposition 3.7, we will mostly be interested in $\mathcal{P}$-regular orbits when $\mu_\mathcal{P}$ is birational. For orientation (and later use in Sect. 4), it is worth recalling a sufficient condition for birationality from [He]; see also [CM, Theorem 7.1.6] and [ABV, Lemma 27.8].

Proposition 3.8. *Suppose $\mathcal{O}$ is an even complex nilpotent orbit. Let $\mathcal{P}$ denote the variety of parabolic subalgebras in $\mathfrak{g}$ corresponding to the subset of the simple roots labeled 0 in the weighted Dynkin diagram for $\mathcal{O}$ (e.g. [CM, Sect. 3.5]). Then $\mathcal{O}$ is dense in $\mu_\mathcal{P}(T^*\mathcal{P})$ and $\mu_\mathcal{P}$ is birational.* $\qquad\square$

Return to Proposition 3.7(a). Example 5.12 below shows that if $\mu_{\mathcal{P}}$ is birational, then not every (necessarily closed) K orbit of θ-stable parabolic subalgebras on $\mathcal{P}$ need be regular. (A good example to keep in mind is the case when K and G have the same rank and $\mathcal{P} = \mathcal{B}$. Then the closed K orbits on $\mathcal{B}$ parameterize discrete series representations with a fixed infinitesimal character. But the regular orbits are the ones which parameterize large discrete series.) So the question becomes: can one give an effective procedure to select the regular K orbits on $\mathcal{P}$ from among all orbits of θ-stable parabolics (when $\mu_{\mathcal{P}}$ is birational)? This is only a small part of computing the parameterization of Corollary 2.14, so it is perhaps surprising that the answer we give after Proposition 3.9 depends on the power of the Kazhdan–Lusztig–Vogan algorithm for $G_{\mathbb{R}}$, the real form of G with complexified Cartan involution θ.

We need a few definitions. Recall that the associated variety of a two-sided ideal I in $U(\mathfrak{g})$ is the subvariety of $\mathfrak{g}^*$ cut out by the associated graded ideal $\mathrm{gr}I$ (with respect to the standard filtration on $U(\mathfrak{g})$) in $\mathrm{gr}U(\mathfrak{g}) = S(\mathfrak{g})$. (From [BB1], if I is primitive, then $AV(I)$ is the closure of a single nilpotent coadjoint orbit.) Finally if $\mathfrak{p}$ is a θ-stable parabolic subalgebra of $\mathfrak{g}$, recall the irreducible $(\mathfrak{g}, K)$-module $A_{\mathfrak{p}}$ constructed in [VZ]. (It would be more customary to denote these modules $A_{\mathfrak{q}}$, but we have already used the letter Q for another purpose.)

Proposition 3.9. *Suppose $\mu_{\mathcal{P}}$ is birational. Fix a closed K orbit Q on $\mathcal{P}$ consisting of θ-stable parabolic subalgebras. Fix $\mathfrak{p} \in Q$. Then Q is $\mathcal{P}$-regular in the sense of Definition 3.1 if and only if*

$$AV(\mathrm{Ann}(A_{\mathfrak{p}})) = \mu(T^*\mathcal{P}),$$

the closure of the complex Richardson orbit induced from $\mathfrak{p}$.

Remark 3.10. We remark that the condition of the proposition is effectively computable from a knowledge of the Kazhdan–Lusztig–Vogan polynomials for $G_{\mathbb{R}}$. More precisely, the results of Sect. 2 allow us to enumerate the closed orbits of K on $\mathcal{P}$ from the structure of K orbits on $\mathcal{B}$. In turn, the description of $K\backslash\mathcal{B}$ has been implemented in the command kgb in the software package atlas (available for download from www.liegroups.org). Moreover, it is not difficult to determine which closed orbits consist of θ-stable parabolic subalgebras; in fact, if one of closed orbit does, then they all do. (Alternatively, one may implement the algorithms of [BH, Sect. 3.3], at least if K is connected.) For a representative $\mathfrak{p}$ of each such orbit, one then uses the command wcells to enumerate the cell of Harish–Chandra modules containing the Vogan–Zuckerman module $A_{\mathfrak{p}}$. (The computation of cells relies on computing Kazhdan–Lusztig–Vogan polynomials.) Finally $AV(\mathrm{Ann}(A_{\mathfrak{p}})) = \mu(T^*\mathcal{P})$ if and only if the cell containing $A_{\mathfrak{p}}$ affords the Weyl group representation $\mathrm{Sp}(\xi)^{A_G}$ (with notation as in Sect. 2), where ξ is an element of the Richardson orbit induced from $\mathfrak{p}$. Again, this is an effectively computable condition and is easy to implement from the output of atlas. Hence *if $\mu_{\mathcal{P}}$ is birational, there is an effective algorithm to enumerate the $\mathcal{P}$-regular orbits of K on $\mathcal{P}$.*

Remark 3.11. Suppose $\mathcal{O}$ is an even complex nilpotent orbit, so that Proposition 3.8 applies. Then Proposition 3.7(b) shows that the algorithm of Remark 3.10 also enumerates the K orbits in $\mathcal{O} \cap (\mathfrak{g}/\mathfrak{k})^*$. Using the Kostant–Sekiguchi correspondence, this amounts to the enumeration of the real forms of $\mathcal{O}$, i.e. $G_\mathbb{R}$ orbits on $\mathcal{O} \cap \mathfrak{g}_\mathbb{R}^*$. By contrast, if $\mathcal{O}$ is not even, the only known way to enumerate the real forms of $\mathcal{O}$ involves case-by-case analysis.

Proposition 3.9 is known to experts, but we sketch a proof (of more refined results) below; see also [ABV, Chap. 20]. We begin with some representation-theoretic preliminaries. Let $\mathcal{D}_\mathcal{P}$ denote the sheaf of algebraic differential operators on $\mathcal{P}$, and let $D_\mathcal{P}$ denote its global section. Since the enveloping algebra $\mathrm{U}(\mathfrak{g})$ acts on $\mathcal{P}$ by differential operators, we obtain a map $\mathrm{U}(\mathfrak{g}) \to D_\mathcal{P}$. Let $I_\mathcal{P}$ denote its kernel, and $R_\mathcal{P}$ its image. By choosing a base-point $\mathfrak{p}_o \in \mathcal{P}$, it is easy to see that $I_\mathcal{P}$ is the annihilator of the irreducible generalized Verma module induced from $\mathfrak{p}_o \in \mathcal{P}$ with trivial infinitesimal character. We will be interested in studying Harish–Chandra modules whose annihilators contain $I_\mathcal{P}$, i.e. $(R_\mathcal{P}, K)$-modules. For orientation, note that if $\mathcal{P} = \mathcal{B}$, $I_\mathcal{B}$ is a minimal primitive ideal, and thus any Harish–Chandra module with trivial infinitesimal character contains it.

Unlike the case of $\mathcal{P} = \mathcal{B}$, $\mathrm{U}(\mathfrak{g})$ need not surject onto $D_\mathcal{P}$ in general, and so $R_\mathcal{P} \simeq \mathrm{U}(\mathfrak{g})/I_\mathcal{P}$ is generally a proper subring of $D_\mathcal{P}$. Thus, the localization functor

$$R_\mathcal{P}\text{-mod} \longrightarrow \mathcal{D}_\mathcal{P}\text{-mod}$$

$$X \longrightarrow \mathcal{X} := \mathcal{D}_\mathcal{P} \otimes_{R_\mathcal{P}} X.$$

need not be an equivalence of categories. But, nonetheless, we have that the appropriate irreducible objects match. (Much more conceptual statements of which the following proposition is a consequence have recently been established by S. Kitchen.)

Proposition 3.12. *Suppose X is an irreducible $(D_\mathcal{P}, K)$-module. Then its restriction to $R_\mathcal{P}$ is irreducible.*

Sketch. Irreducible $(D_\mathcal{P}, K)$-modules are parameterized by irreducible K equivariant flat connections on $\mathcal{P}$. We show that the irreducible $(R_\mathcal{P}, K)$-modules are also parameterized by the same set. The parameterizations have the property that support of the localization of either type of module parameterized by such a connection $\mathcal{L}$ is simply the closure of the support of $\mathcal{L}$. This implies there are the same number of such irreducible modules and hence implies the proposition.

Let X be an irreducible $(R_\mathcal{P}, K)$-module. Hence, we may consider X as an irreducible $(\mathfrak{g}, K)$-module, say X', whose annihilator contains $I_\mathcal{P}$. By localizing on $\mathcal{B}$, we may consider the corresponding irreducible K equivariant flat connection on $\mathcal{B}$, say $\mathcal{L}'$, parameterizing X'. The condition that $\mathrm{Ann}(X') \supset I_\mathcal{P}$ can be translated into a geometric condition on $\mathcal{L}'$ using [LV, Lemma 3.5], the conclusion of which is that $\mathcal{L}'$ fibers over an irreducible flat K-equivariant connection on $\mathcal{P}$ (with fiber equal to the trivial connection on $\mathcal{B}_l$). This implies that irreducible $(R_\mathcal{P}, K)$-modules are also parameterized by K equivariant flat connections on $\mathcal{P}$, as claimed, and the proposition follows.   □

Remark 3.13. Proposition 3.12 need not hold when considering twisted sheaves of differential operators corresponding to singular infinitesimal characters.

Next suppose X is an irreducible $R_\mathcal{P}$ module. Let $(\mathcal{X}^i)$ denote a good filtration on its localization $\mathcal{X}$ compatible with the degree filtration on $\mathcal{D}_\mathcal{P}$. Let $\mathrm{CV}(X)$ denote the support of $\mathrm{gr}(\mathcal{X})$. This is well-defined independent of the choice of filtration. Moreover, there is a subset $\mathrm{cv}(X) \subset K\backslash\mathcal{P}$ such that

$$\mathrm{CV}(X) = \bigcup_{Q \in \mathrm{cv}(X)} \overline{T_Q^*\mathcal{P}}.$$

The set $\mathrm{cv}(X)$ is difficult to understand, but there are two easy facts about it. First, if X is irreducible, there is a dense K orbit, say $\mathrm{supp}_\mathrm{o}(X)$ in the support of $\mathcal{X}$; then $\mathrm{supp}_\mathrm{o}(X) \in \mathrm{cv}(X)$. Moreover if $Q \in \mathrm{cv}(X)$, then $Q \in \overline{\mathrm{supp}_\mathrm{o}(X)}$. So, for example, if $\mathrm{supp}_\mathrm{o}(X)$ is closed, then $\mathrm{cv}(X) = \{\mathrm{supp}_\mathrm{o}(X)\}$.

Finally, we define

$$\mathrm{AV}(X) = \mu(\mathrm{CV}(X)).$$

(Alternatively one may define $\mathrm{AV}(X)$ as in [V3] without localizing. The fact that the two definitions agree follows from [BB3, Theorem 1.9(c)].) Clearly, $\mathrm{AV}(X)$ is the union of closures of K orbits on $\mathcal{N}_\mathcal{P}^\theta$. We let $\mathrm{av}(X)$ denote the set of these orbits.

Here is how these invariants are tied together.

Theorem 3.14. *Retain the setting above. Then*

1. $\mathrm{AV}(I_\mathcal{P}) = \mu(T^*\mathcal{P})$.
2. If X is an irreducible $(R_\mathcal{P}, K)$-module, then

$$G \cdot \mathrm{AV}(X) = \mathrm{AV}(\mathrm{Ann}(X)) \subset \mathrm{AV}(I_\mathcal{P}).$$

Proof. Part (1) is Theorem 4.6 in [BB1]. The equality in part (2) is proved in [V3, Sect. 6]; the inclusion follows because X is an $R_\mathcal{P} = \mathrm{U}(\mathfrak{g})/I_\mathcal{P}$ module.   □

Proposition 3.15. *Suppose X is an irreducible $(R_\mathcal{P}, K)$-module such that there exists a $\mathcal{P}$-regular K orbit $Q \in \mathrm{cv}(X)$. (For instance, suppose $\mathrm{supp}_\mathrm{o}(X)$ is $\mathcal{P}$-regular.) Then $\Phi_\mathcal{P}(Q)$ is a K orbit of maximal dimension in $\mathrm{AV}(X)$; that is, $\Phi_\mathcal{P}(Q) \in \mathrm{av}(X)$.*

Proof. Since $\mathrm{AV}(X) = \mu(\mathrm{CV}(X))$ and since $Q \in \mathrm{cv}(X)$,

$$\Phi_\mathcal{P}(Q) \subset \mathrm{AV}(X) \tag{3.16}$$

for any $(R_\mathcal{P}, K)$-module. If Q is $\mathcal{P}$-regular, then the G saturation of the left-hand side of (3.16) is dense in $\mu(T^*\mathcal{P})$. But by Theorem 3.14 the right-hand side of (3.16) is also contained in $\mu(T^*\mathcal{P})$. So the current proposition follows.   □

Corollary 3.17. *Suppose X is an irreducible $(R_\mathcal{P}, K)$-module. Then the following are equivalent.*

(a) There exists a $\mathcal{P}$-regular orbit $Q \in \mathrm{cv}(X)$
(b) There exists a $\mathcal{P}$-regular orbit $\mathcal{O}_K \in \mathrm{av}(X)$
(c) $\mathrm{Ann}(X) = I_\mathcal{P}$
(d) $\mathrm{AV}(\mathrm{Ann}(X)) = \mathrm{AV}(I_\mathcal{P})$, i.e. $\mathrm{AV}(\mathrm{Ann}(X)) = \mu(T^*\mathcal{P})$

Proof. The equivalence of (a) and (b) follows from the definitions above. Since the annihilator of any $R_\mathcal{P}$ module contains $I_\mathcal{P}$, the equivalence of (c) and (d) follows from [BKr, 3.6]. Theorem 3.14 and the definitions gives the equivalence of (b) and (d).   □

Proof of Proposition 3.9. If $\mathfrak{p} \in \mathcal{P}$ is a θ-stable parabolic, then the Vogan–Zuckerman module $A_\mathfrak{p}$ is the unique irreducible $(R_\mathcal{P}, K)$-module whose localization is supported on the closed orbit $K \cdot \mathfrak{p}$ and thus, as remarked above, $\mathrm{cv}(A_\mathfrak{p}) = \{K \cdot \mathfrak{p}\}$. So Proposition 3.9 is a special case of Corollary 3.17.   □

4 Applications to Special Unipotent Representations

The purpose of this section is to explain how the algorithm of Remark 3.10 produces special unipotent representations. Much of this section is implicit in [ABV, Chap. 27].

Fix a nilpotent adjoint orbit $\mathcal{O}^\vee$ for $\mathfrak{g}^\vee$, the Langlands dual of $\mathfrak{g}$. Fix a Jacobson–Morozov triple $\{e^\vee, h^\vee, f^\vee\}$ for $\mathcal{O}^\vee$, and set

$$\chi(\mathcal{O}^\vee) = (1/2)h^\vee.$$

Then $\chi(\mathcal{O}^\vee)$ is an element of some Cartan subalgebra $\mathfrak{h}^\vee$ of $\mathfrak{g}^\vee$. There is a Cartan subalgebra $\mathfrak{h}$ of $\mathfrak{g}$ such that $\mathfrak{h}^\vee$ canonically identifies with $\mathfrak{h}^*$. Hence we may view

$$\chi(\mathcal{O}^\vee) \in \mathfrak{h}^*.$$

There were many choices made in the definition of $\chi(\mathcal{O}^\vee)$. But, nonetheless, the infinitesimal character corresponding to $\chi(\mathcal{O}^\vee)$ is well defined; i.e. $\chi(\mathcal{O}^\vee)$ is well-defined up to $G^\vee$ conjugacy and thus (via Harish–handra's theorem) specifies a well-defined maximal ideal $Z(\mathcal{O}^\vee)$ in the center of $U(\mathfrak{g})$. We call $\chi(\mathcal{O}^\vee)$ the unipotent infinitesimal character attached to $\mathcal{O}^\vee$.

By a result of Dixmier [Di], there exists a unique maximal primitive ideal in $U(\mathfrak{g})$ containing $Z(\mathcal{O}^\vee)$. Denote it by $I(\mathcal{O}^\vee)$, and let $d(\mathcal{O}^\vee)$ denote the dense nilpotent coadjoint orbit in $AV(I(\mathcal{O}^\vee))$. The orbit $d(\mathcal{O}^\vee)$ is called the Spaltenstein dual of $\mathcal{O}^\vee$ (after Spaltenstein who first defined it in a different way); see [BV, Appendix A].

Fix $G_\mathbb{R}$ as above, and define

$$\mathrm{Unip}(\mathcal{O}^\vee) = \{X \text{ an irreducible } (\mathfrak{g}, K) \text{ module} \mid \mathrm{Ann}(X) = I(\mathcal{O}^\vee)\}.$$

This is the set of special unipotent representations for $G_\mathbb{R}$ attached to $\mathcal{O}^\vee$. Since the annihilator of such a representation X is the maximal primitive ideal containing $Z(\mathcal{O}^\vee)$, X is as small as the (generally singular) infinitesimal character $\chi(\mathcal{O}^\vee)$ allows. These algebraic conditions are conjectured to have implications about unitarity.

Conjecture 4.1 (Arthur, Barbasch–Vogan [BV]). *The set* $\mathrm{Unip}(\mathcal{O}^\vee)$ *consists of unitary representations.*

We are going to produce certain special unipotent representations from the regular orbits of Definition 3.1. In order to do so, we need to shift our perspective and work on side of the Langlands dual $\mathfrak{g}^\vee$. So let $G'_\mathbb{R}$ be a real form of a connected reductive algebraic group with Lie algebra $\mathfrak{g}^\vee$ and let K' denote the complexification of a maximal compact subgroup in $G'_\mathbb{R}$. Fix an *even* nilpotent coadjoint orbit $\mathcal{O}^\vee$. (This is equivalent to requiring that $\chi(\mathcal{O}^\vee)$ is integral.) Define $\mathcal{P}^\vee$ as in Proposition 3.8. Thus, the main results of Sect. 3 are available in this setting.

Let X' denote an irreducible $(R_{\mathcal{P}^\vee}, K')$-module, and let X denote the Vogan dual of X' in the sense of [V2]. Thus, X is an irreducible Harish–Chandra module for a group $G_\mathbb{R}$ arising as the real points of a connected reductive algebraic group with Lie algebra $\mathfrak{g}$. Moreover, X has trivial infinitesimal character.

Recall that we are interested in representations with infinitesimal character $\chi(\mathcal{O}^\vee)$. In order to pass to this infinitesimal character, we need to introduce certain translation functors. There are technical complications which arise in this setting for two reasons. First, $G_\mathbb{R}$ need not be connected (although it is in Harish–Chandra's class by our hypothesis). Second, $G_\mathbb{R}$ may not have enough finite-dimensional representations to define all of the translations one would like. Both of these complications disappear if we assume G is simply connected, and we shall do so here in the interest of streamlining the exposition. (It is of course possible to relax this assumption, as in [ABV, Chap. 27].)

Fix a representative $\rho \in \mathfrak{h}^*$ representing the trivial infinitesimal character. Choose a representative $\chi \in \mathfrak{h}^*$ representing the (integral) infinitesimal character $\chi(\mathcal{O}^\vee)$ so that χ and ρ lie in the same closed Weyl chamber. Let $\nu = \rho - \chi$. Let F^ν denote the finite-dimensional representation of $G_\mathbb{R}$ with extremal weight ν; this exists since we have assumed G is simple connected. Using it, define the translation functor $\psi = \psi_\rho^\chi$ (as in [KnV, Sect. VII.13]) from the category of Harish–Chandra modules with trivial infinitesimal character to the category of Harish–Chandra modules with infinitesimal character $\chi(\mathcal{O}^\vee)$.

Theorem 4.2 (cf. [ABV, Chap. 27]). *Retain the notation introduced after Conjecture 4.1. In particular, fix an even nilpotent orbit $\mathcal{O}^\vee$, and let $\mathcal{P}^\vee$ denote the variety of parabolic subalgebras corresponding to the nodes labeled 0 in the weighted*

Dynkin diagram for $\mathcal{O}^\vee$. *Let* X' *be an irreducible* $(R_{\mathcal{P}^\vee}, K')$-*module, assume* G *is simply connected, and let* $Z = \psi(X)$ *denote the translation functor to infinitesimal character* $\chi(\mathcal{O}^\vee)$ *applied to the Vogan dual* X *of* X'. *Then the following are equivalent:*

(a) Z *is a (nonzero) special unipotent representation attached to* $\mathcal{O}^\vee$.
(b) *There exists a* $\mathcal{P}^\vee$-*regular orbit* $Q^\vee \in \mathrm{cv}(X')$.

Proof. From the properties of the duality explained in [V2, Sect. 14] (and the translation principle), Z is nonzero with infinitesimal character $\chi(\mathcal{O}^\vee)$ if and only if X' is annihilated by $I_{\mathcal{P}^\vee}$, i.e. if and only if X' descends to a $(R_{\mathcal{P}^\vee}, K)$-module. Moreover, Z is annihilated by a maximal primitive ideal if and only if the $R_{\mathcal{P}^\vee}$-module X' has minimal possible annihilator, namely $I_{\mathcal{P}^\vee}$. The conclusion is that Z is special unipotent attached to $\mathcal{O}^\vee$ if and only if X' is a $(R_{\mathcal{P}^\vee}, K)$-module annihilated by $I_{\mathcal{P}^\vee}$. So the theorem follows from the equivalence of (a) and (c) in Corollary 3.17. $\square$

Since the duality of [V2] is effectively computable, and since the same is true of the translation functors ψ, the theorem shows Remark 3.10 translates into an effective construction of special unipotent representations. More precisely, one uses Remark 3.10 to enumerate the relevant $\mathcal{P}^\vee$-regular orbits, and for each one constructs the representation $X' = A_{\mathfrak{p}}$ of Proposition 3.9. As remarked in the proof of Proposition 3.9, X' satisfies condition (b) of Theorem 4.2. Applying the construction of the theorem gives special unipotent representations.

In fact, this construction may be understood further in light of the following refinement. In the setting of Theorem 4.2, fix a $\mathcal{P}^\vee$-regular orbit $Q^\vee$, and define $\mathbb{A}(Q^\vee)$ be the set of special unipotent representations attached to $\mathcal{O}^\vee$ produced by applying Theorem 4.2 to all modules X' with $Q^\vee \in \mathrm{cv}(X')$. Then the theorem implies

$$\mathrm{Unip}(\mathcal{O}^\vee) = \bigcup \mathbb{A}(Q^\vee),$$

where the (not necessarily disjoint) union is over all $\mathcal{P}^\vee$-regular orbits.

The sets $\mathbb{A}(Q^\vee)$ are the Arthur packets defined in [ABV, Chap. 27]. While there are effective algorithms to enumerate $\mathrm{Unip}(\mathcal{O}^\vee)$, there are no such algorithms for individual packets $\mathbb{A}(Q^\vee)$ (except in favorable cases). In any event, the discussion of the previous paragraph shows that *Remark 3.10 leads to an effective algorithm to enumerate one element of each Arthur packet of integral special unipotent representations.* These representatives are necessarily distinct.

5   Examples

Example 5.1 (Maximal parabolic subalgebras for classical groups). Suppose G is classical and $\mathcal{P}$ consists of maximal parabolic subalgebra. Then it is well known that

$$\mathrm{ind}^W_{W(\mathcal{P})}(\mathrm{sgn})$$

decomposes multiplicity freely as a W-module. Thus if $\mathrm{Sp}(\xi)^{A_K}$ is irreducible as a W-module, then Proposition 2.10 implies $\Phi_{\mathcal{P}}^{-1}(\mathcal{O}_K)$ is a single orbit. In particular, if the orbits of $A_K(\xi)$ and $A_G(\xi)$ on irreducible components of the Springer fiber $\mu_B^{-1}(\xi)$ coincide (for instance, if $A_K(\xi)$ surjects onto $A_G(\xi)$ for each ξ), then $\mathrm{Sp}(\xi)^{A_K} = \mathrm{Sp}(\xi)^{A_G}$ is irreducible and $\Phi_{\mathcal{P}}$ is injective.

Proposition 5.2. *Suppose the real form $G_{\mathbb{R}}$ of G corresponding to θ is a classical semisimple Lie group with no complex factors whose Lie algebra has no simple factor isomorphic to $\mathfrak{so}^*(2n)$ or $\mathfrak{sp}(p,q)$. If $\mathcal{P}$ consists of maximal parabolic subalgebras, then $\Phi_{\mathcal{P}}$ is injective.*

Proof. Unfortunately, this follows from a case-by-case analysis of the classical groups. First note that the orbits of $A_K(\xi)$ and $A_G(\xi)$ on $\mu_B^{-1}(\xi)$ are insensitive to the isogeny class of $G_{\mathbb{R}}$. So, by the remarks preceding the proposition, it is enough to examine when the two kinds of orbits coincide for a simply connected group $G_{\mathbb{R}}$ with simple Lie algebra. In type A, all A-groups are trivial (up to isogeny) so there is nothing to check. It follows from direct computation that $A_K(\xi)$ surjects on $A_G(\xi)$ for $G_{\mathbb{R}} = \mathrm{Sp}(2n,\mathbb{R})$ and $\mathrm{SO}(p,q)$, but that the image of $A_K(\xi)$ in $A_G(\xi)$ is always trivial for $\mathrm{Sp}(p,q)$ and $\mathrm{SO}^*(2n)$. This completes the case-by-case analysis and hence the proof.

Remark 5.3. For the groups in Proposition 5.2, the map $\Phi_{\mathcal{B}}$ is computed explicitly in [T1] and [T3]. Using Proposition 2.6(1), this gives one (rather roundabout) way to compute $\Phi_{\mathcal{P}}$ in these cases. For exceptional groups, the injectivity of the proposition fails. See Example 5.12 below.

Example 5.4. Suppose now $G_{\mathbb{R}} = \mathrm{Sp}(2n,\mathbb{R})$ and $\mathcal{P}$ consists of maximal parabolic of type corresponding to the subset of simple roots obtained by deleting the long one. (So if $n = 2$, $\mathcal{P} = \mathcal{P}_\alpha$ in Example 3.3.) Then the analysis of the preceding example extends to show that $\Phi_{\mathcal{P}}$ is an order-reversing bijection. The closure order on $K \backslash \mathcal{N}_{\mathcal{P}}^\theta$ (and hence $K \backslash \mathcal{P}$) is as follows.

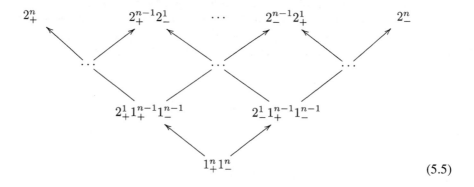

$$\tag{5.5}$$

Here, as before, we are using the parameterization of $K \backslash \mathcal{N}_{\mathcal{P}}^{\theta}$ given in [CM, Theorem 9.3.5]. There are thus $n + 1$ orbits which are $\mathcal{P}$-regular, all of which are closed according to Proposition 3.7(a) (which applies since $\mathcal{P}$ is attached via Proposition 3.8 to the even complex orbit with partition 2^n).

In this setting, we may now apply Theorem 4.2. (Notationally, the roles of the group and dual group must unfortunately be inverted: for the application, we should take $G^{\vee} = \mathrm{Sp}(2n, \mathbb{C})$ in the statement of the theorem.) Even though $\mathrm{SO}(n, n+1)$ is not simply connected, the complications involving the relevant translation functors are absent, and the construction of the theorem, nonetheless, applies and produces $n + 1$ special unipotent representations for $\mathrm{SO}(n, n+1)$.

Example 5.6. Suppose $G_{\mathbb{R}} = \mathrm{U}(n, n)$ and $\mathcal{P}$ corresponds to the subset of simple roots obtained by deleting the middle simple root in the Dynkin diagram of type A_{2n-1}. Then $\Phi_{\mathcal{P}}$ is an order reversing bijection, and the partially ordered sets in question again look like that (5.5) using the parameterization of $K \backslash \mathcal{N}_{\mathcal{P}}^{\theta}$ given in [CM, Theorem 9.3.3]. Again, there are $n + 1$ orbits which are $\mathcal{P}$-regular. The construction of Theorem 4.2 produces $n + 1$ special unipotent representation for $\mathrm{GL}(2n, \mathbb{R})$, each of which turns out to be a constituent of maximal Gelfand–Kirillov dimension in the degenerate principal series for $\mathrm{GL}(2n, \mathbb{R})$ induced from a one-dimensional representation of a Levi factor isomorphic to a product of n copies of $\mathrm{GL}(2, \mathbb{R})$.

In terms of representation theory of $G_{\mathbb{R}} = \mathrm{U}(n, n)$, it is well known that the enveloping algebra in this case does surject on the ring of global differential operators on $\mathcal{P}$ (e.g., the discussion of [T2, Remark 3.3]) and localization is an equivalence of categories. Because all Cartan subgroups in $\mathrm{U}(n, n)$ are connected, the only irreducible flat K-equivariant connections on $\mathcal{P}$ are the trivial ones supported on single K orbits. The map $Q \mapsto \Phi_{\mathcal{P}}(Q)$ coincides with the map which sends the unique irreducible $(R_{\mathcal{P}}, K)$-module supported on the closure of Q to the dense orbit in its (irreducible) associated variety, and is a bijection between such irreducible modules and the K orbits on $\mathcal{N}_{\mathcal{P}}^{\theta}$. It would be interesting to see if this observation could be used to give a geometric explanation of the computation of composition series of certain degenerate principal series for $\mathrm{U}(n, n)$ first given in [Sa] and later reproved in [Le]. (See, for instance, Sahi's module diagrams reproduced in [Le, Fig. 7], for example.)

Example 5.7. Suppose $G_{\mathbb{R}} = \mathrm{Sp}(1, 1)$, a real form of $G = \mathrm{Sp}(4, \mathbb{C})$. If $\mathcal{O}$ is the subregular nilpotent orbit for $\mathfrak{g}$ and $\xi \in \mathcal{O} \cap (\mathfrak{g}/\mathfrak{k})^*$, then $A_K(\xi)$ is trivial, but $A_G(\xi) \simeq \mathbb{Z}/2$. So the proof of Proposition 5.2 does not apply. Let α denote the short simple root and β the long one. The closure order on $K \backslash \mathcal{B}$ is given by

$$
\begin{array}{c}
Q \\
\beta \uparrow \\
R \\
{}_{\alpha}\nearrow \quad \nwarrow_{\alpha} \\
S_+ \qquad\qquad S_-
\end{array}
\qquad (5.8)
$$

The picture for $K\backslash\mathcal{P}_\alpha$ is

$$\begin{array}{c} \pi_\alpha(Q)_3 \\ \uparrow \\ \pi_\alpha(R)_2 \end{array}$$ (5.9)

and for $K\backslash\mathcal{P}_\beta$

$$\pi_\beta(Q)_3$$
$$\nearrow \qquad \nwarrow$$ (5.10)
$$\pi_\beta(S_+)_2 \qquad\qquad \pi_\beta(S_-)_2$$

Here, $\mathcal{N}_\alpha^\theta = \mathcal{N}_\beta^\theta = \mathcal{N}_\mathsf{B}^\theta$, and the closure order of K orbits is simply

$$\begin{array}{c} 2_+^1 2_-^1 \\ \uparrow \\ 1_+^2 1_-^2 \end{array}$$ (5.11)

in the notation of [CM, Theorem 9.3.5]. Then Φ_α is an order reversing bijection, but Φ_β is two-to-one over $2_+^1 2_-^1$. The reason is that

$$\mathrm{Sp}(\xi) = \mathrm{std} \oplus \chi,$$

where std is the two-dimensional standard representation of W and χ is a character on which the simple reflection s_α acts trivially and on which s_β acts nontrivially. The orbit $\pi_\alpha(R)$ is $\mathcal{P}_\alpha$-regular, and the orbits $\pi_\beta(S_\pm)$ are $\mathcal{P}_\beta$-regular.

Example 5.12. As an example of what can happen in the exceptional cases, let G be the (simply connected) connected complex group of type F_4 and θ correspond to the split real form $G_\mathbb{R}$ of G. (So K is a quotient of $\mathrm{Sp}(3, \mathbb{C}) \times \mathrm{SL}(2, \mathbb{C})$ by $\mathbb{Z}/2$.) Then the corresponding real form $G_\mathbb{R}$ is split. Let $\mathcal{P}$ denote the variety of maximal parabolic obtained by deleting the middle long root from the Dynkin diagram, and let $\mathcal{O}$ denote the corresponding Richardson orbit. Then $\mathcal{O}$ is 40 dimensional and is labeled $F_4(A_3)$ in the Bala–Carter classification. Moreover, $\mathcal{O}$ is the unique orbit which is fixed under Spaltenstein duality. (Here, we are of course identifying $\mathfrak{g}$ and $\mathfrak{g}^\vee$.) For $\xi \in \mathcal{O}$, $A_G(\xi) = S_4$, the symmetric group on four letters. The weighted Dynkin diagram of $\mathcal{O}$ has the middle long root labeled 2 and all others nodes labeled 0. So $\mathcal{P}$ corresponds to $\mathcal{O}$ as in Proposition 3.8.

From results of Djoković (recalled in [CM, Sect. 9.6]), there are 19 orbits of K on $\mathcal{N}_P^\theta$. They are labeled 0–18; the orbit corresponding to label i will be denoted $\mathcal{O}_K^i$, and ξ^i will denote an element of $\mathcal{O}_K^i$. Orbits $\mathcal{O}_K^{16}$, $\mathcal{O}_K^{17}$, and $\mathcal{O}_K^{18}$ are the three K orbits on $\mathcal{O} \cap (\mathfrak{g}/\mathfrak{k})^*$. From the discussion leading to [Ki, Table 2], it follows that $A_K(\xi^i)$ surjects onto $A_G(\xi^i)$ for $i = 0, \ldots, 15$. In each of these cases, $A_G(\xi)$ is either trivial or $\mathbb{Z}/2$. We also have $A_K(\xi^{16}) = A_G(\xi^{16}) = S_4$. But $A_K(\xi^{17}) = D_4$, the dihedral group with eight elements, and $A_K(\xi^{17}) \to A_G(\xi^{17})$ is the natural inclusion into S_4. Finally, $A_K(\xi^{18}) = \mathbb{Z}/2 \times \mathbb{Z}/2$ which injects into $A_G(\xi^{18})$.

For $i = 17$ and 18, it is not immediately obvious how to read off $\mathrm{Sp}(\xi^i)^{A_K(\xi^i)}$ from, say, the tables of [Ca]. But for $i = 0, \ldots, 16$, the component group calculations of the previous paragraph imply that $\mathrm{Sp}(\xi^i)^{A_K(\xi^i)} = \mathrm{Sp}(\xi^i)^{A_G(\xi^i)}$, and such representations are indeed tabulated in [Ca]. Applying Proposition 2.10, it is then not difficult to show that

$$\#\Phi^{-1}(\mathcal{O}_K^i) = 1 \text{ if } i \in \{0, 1, 2, 3\} \cup \{9, 10, \ldots, 16\}$$

and

$$\#\Phi^{-1}(\mathcal{O}_K^i) = 2 \text{ if } i \in \{4, 5, 6, 7, 8\}.$$

In more detail, the G-saturation of $\mathcal{O}_K^4$ and $\mathcal{O}_K^5$ is the complex orbit $A_1 \times \widetilde{A_1}$ in the Bala–Carter labeling, while $\mathcal{O}_K^6$, $\mathcal{O}_K^7$, and $\mathcal{O}_K^8$ have G saturation labeled by A_2. The corresponding irreducible Weyl group representations in these two cases both appear with multiplicity two in $\mathrm{ind}_{W(P)}^W(\mathrm{sgn})$. All other relevant multiplicities are one.

We thus conclude that there are 22 orbits of K on $\mathcal{P}$ which map via $\Phi_{\mathcal{P}}$ to some $\mathcal{O}_K^i$ for $i = 0, \ldots, 15$. Meanwhile, using the software program atlas, one can compute the closure order of K on $\mathcal{B}$, and thus (as explained in Sect. 2), the closure order on $K \backslash \mathcal{P}$. Fig. 4.1 gives the full closure order for $K \backslash \mathcal{P}$. Vertices are labeled according to their dimensions. (The edges in Fig. 4.1 do *not* distinguish between the weak and full closure order. Doing so would make the picture significantly more complicated and difficult to draw.) There are thus 24 orbits of K on $\mathcal{P}$. Since 22 have been shown to map to $\mathcal{O}_K^i$ for $i = 0, \ldots, 15$, one concludes that the the fiber of $\Phi_{\mathcal{P}}$ over $\mathcal{O}^i$ for $i = 16$ and 17 must consist of just one element in each case.

In particular, there are three $\mathcal{P}$-regular K orbits on $\mathcal{P}$ which are bijectively matched via Proposition 3.7(b) to $\mathcal{O}_K^{16}$, $\mathcal{O}_K^{17}$, and $\mathcal{O}_K^{18}$. But from the atlas computation of the closure order on $K \backslash \mathcal{P}$, there are *four* closed orbits of K on $\mathcal{P}$. (These are in fact exactly the four orbits, which are minimal in the weak closure order.) See Fig. 5.12. The atlas labels of the closed orbits are 3, 22, 31, and 47. Their respective dimensions are 0, 1, 2, and 3. Applying the algorithm of Remark 3.10, one deduces that the three $\mathcal{P}$-regular orbits are 3, 31, and 47. Theorem 4.2 thus produces three distinct special unipotent representations, one in each of the three Arthur packets for $\mathcal{O} = d(\mathcal{O})$.

84

D. Ciubotaru et al.

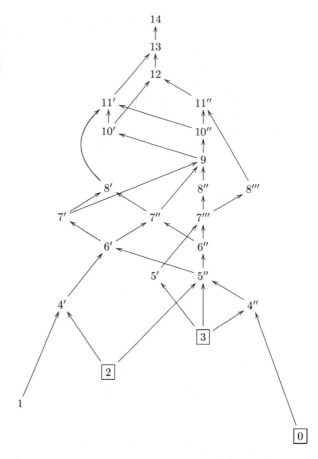

Fig. 1 The full closure ordering of K-orbits on $\mathcal{P}$ for $G_{\mathbb{R}} = F_4$ and $\mathcal{O} = F_4(A_3)$. Vertices are labeled according to their dimensions and boxed vertices are $\mathcal{P}$-regular. Note, in particular, that not every closed orbit is $\mathcal{P}$-regular

Acknowledgements KN and PT would like to thank the Hausdorff Research Institute for Mathematics for its hospitality during their stay in 2007. They would also like to thank the organizers of the joint MPI-HIM program devoted to representation theory, complex analysis and integral geometry. KN is partially supported by JSPS Grant-in-Aid for Scientific Research (B) #21340006. PT was supported by NSA grant MSPF-06Y-096 and NSF grant DMS-0532393.

References

[ABV] Adams, J., D. Barbasch, and D. A. Vogan, Jr., *The Langlands Classification and Irreducible Characters for Real Reductive Groups*, Progress in Math, Birkhäuser (Boston), **104** (1992).
[BT] D. Barbasch, P. Trapa, Stable combinations of special unipotent representations, *Contemp. Math*, **557** (2011).
[BV] D. Barbasch, D. A. Vogan, Unipotent representations of complex semisimple groups, *Ann. of Math. (2)*, **121** (1985), no. 1, 41–110.

[BB1] W. Borho, J.-L. Brylinski, Differential operators on homogeneous spaces I. Irreducibility of the associated variety for annihilators of induced modules, *Invent. Math,* **69** (1982), no. 3, 437–476.

[BB3] W. Borho, J.-L. Brylinski, Differential operators on homogeneous spaces III. Characteristic varieties of Harish-Chandra modules and of primitive ideals, *Invent. Math,* **80** (1985), no. 1, 1–68.

[BKr] W. Borho and H. Kraft, Über die Gelfand-Kirillov Dimension, *Math. Ann.* **220** (1976), 1-24.

[BM] W. Borho and R. MacPherson, Partial resolutions of nilpotent varieties, in *Analysis and topology on singular spaces, II, III (Luminy, 1981),* Astérisque **101** (1983), 2374.

[BH] M. Brion and A. Helminck, On Orbit Closures of Symmetric Subgroups in Flag Varieties, *Canad. J. Math.,* **52** (2000), no. 2, 265–292.

[Ca] R. Carter, *Finite Groups of Lie Type,* Wiley Interscience (New York), 1985.

[CM] D. Collingwood and W. M. McGovern, *Nilpotent orbits in semisimple Lie algebras,* Van Nostrand Reinhold (New York), 1993.

[Di] J. Dixmier, Idéaux primitifs dans l'algébre enveloppante d'une de Lie semisimple complexe, *C. R. Acad. Sci. Paris (A),* **271** (1970), 134–136.

[Dj] D. Djoković, The closure order of nilpotent orbits for real forms of F_4 and G_2, *J. Lie Theory,* **10** (2000), 471–490

[He] W. Hesselink, Polarizations in classical groups, *Math. Z.,* **160** (1978), 217–234.

[KL] D. Kazhdan and G. Lusztig, A topological approach to Springer's representations, *Adv. in Math.,* **38** (1980), 222–228.

[Ki] D. R. King, The component groups of nilpotents in exceptional simple real Lie algebras, *Comm. Algebra,* **20** (1992), no. 1, 219–284.

[KnV] A. W. Knapp and D. A. Vogan, *Cohomological Induction and Unitary Representations,* Princeton Mathematical Series, number 45, Princeton University Press, 1995.

[KR] B. Kostant and S. Rallis, Orbits and representations associated with symmetric spaces, *Amer. J. Math.,* **93** (1971), 753–809.

[Le] S. T. Lee, On some degenerate principal series representations of $U(n, n)$. *J. Funct. Anal.,* **126** (1994), no. 2, 305–366.

[LV] G. Lusztig and D. A. Vogan, Singularities of closures of K-orbits on flag manifolds, *Invent. Math.,* **71** (1983), no. 2, 365–379.

[M] T. Matsuki, Closure relations for orbits on affine symmetric spaces under the action of minimal parabolic subgroups, *Representations of Lie groups, Kyoto, Hiroshima, 1986,* Adv. Stud. Pure Math., **14** (1988), 541–559.

[Mc1] W. M. McGovern, On the Spaltenstein-Steinberg map for classical Lie algebras, *Comm. Algebra,* **27** (1999), no. 6, 2979–2993.

[Mi] D. Miličić, Algebraic D-modules and representation theory of semisimple Lie groups, in *Analytic Cohomology and Penrose Transform,* Contemporary Mathematics, **154** (1993), 133–168.

[R] W. Rossmann, Invariant eigendistributions on a semisimple Lie algebra and homology classes on the conormal variety II: Representations of Weyl groups, *J. Funct. Anal.,* **96** (1991), no. 1, 155–193.

[RS] R. W. Richardson and T. A. Springer, The Bruhat order on symmetric varieties, *Geom. Dedicata,* **35** (1990), no. 1-3, 389–436.

[Sa] S. Sahi, Unitary representations on the Shilov boundary of a symmetric tube domain, in *Representation theory of groups and algebras,* Contemp. Math., **145** (1993), 275–286.

[St] R. Steinberg, An occurrence of the Robinson-Schensted correspondence, *J. Algebra,* **113** (1988), no. 2, 523–528.

[T1] P. E. Trapa, Generalized Robinson-Schensted algorithms for real Lie groups, *Internat. Math. Res. Notices,* **1999,** no. 15, 803–834.

[T2] P. E. Trapa, Annihilators and associated varieties of $A_q(\lambda)$ modules for $U(p, q)$, *Compositio Math.,* **129** (2001), no. 1, 1–45.

[T3]   P. E. Trapa, Symplectic and Orthogonal Robinson-Schensted algorithms, *J. Algebra*, **286** (2005), 386–404.

[vL]   M. van Leeuwen, A Robinson-Schensted algorithm in the geometry of flags for classical groups, Rijksuniversiteit Thesis, Utrecht, 1989.

[V1]   D. A. Vogan, Irreducible characters of semisimple Lie groups III. Proof of Kazhdan-Lusztig conjecture in the integral case, *Invent. Math.*, **71** (1983), no 2, 381–417.

[V2]   D. A. Vogan, Irreducible characters of semisimple Lie groups IV. Character-multiplicity duality, *Duke Math. J.*, **49** (1982), no. 4, 943–1073.

[V3]   D. A. Vogan, Associated varieties and unipotent representations, in *Harmonic analysis on reductive groups (Brunswick, ME, 1989)*, 315–388, Progress in Mathematics, **101**, Birkhäuser(Boston), 1991.

[VZ]   D. A. Vogan and G. Zuckerman, Unitary representations with nonzero cohomology, *Compositio Math.*, **53** (1984), no. 1, 51–90.

Helgason's Conjecture in Complex Analytical Interior

Simon Gindikin

Abstract We discuss Helgason's conjecture in the language of complex analysis and integral geometry on symmetric Stein manifolds.

Keywords Symmetric manifold • Complex horospherical transform • Integral Cauchy formula • Poisson integral formula • Dolbeault cohomology • Penrose transform

Mathematics Subject Classification (1991): 22E30, 22E46, 32A45, 33T15, 44A12

The Helgason conjecture [He74] is one of the most fundamental facts of harmonic analysis on Riemannian symmetric manifolds of noncompact type. Let $X = G/K$ be such a manifold; G be a connected real semisimple Lie group with a finite center and K be its maximal compact subgroup. The conjecture states that joint eigenfunctions of invariant differential operators on X can be reconstructed through their hyperfunction's boundary values and that for all eigenvalues, except an explicitly described set, the operator of boundary values is surjective on the space of hyperfunctions on the boundary F (the real flag manifold). Helgason proved this for manifolds of rank 1. In the general case, the conjecture was proven in [KKMOOT] using a very deep technology of differential operators with singularities, and this proof continues to be one of the most analytically challenging in the theory of representations.

There were several approaches to understand this result in a broader context of differential operators. For example, let us give mention to nontrivial results of Penney on similar facts for nonsymmetric homogeneous manifolds. Nevertheless,

S. Gindikin
Department of Mathematics, Hill Center, Rutgers University, 110 Frelinghysen Road,
Piscataway, NJ 08854-8019, USA
e-mail: gindikin@math.rutgers.edu

B. Krötz et al. (eds.), *Representation Theory, Complex Analysis, and Integral Geometry*,
DOI 10.1007/978-0-8176-4817-6_5, © Springer Science+Business Media, LLC 2012

from my point of view, it continues to be a rather isolated fact which to a large extent depends on the structure of symmetric manifolds and semisimple groups. I want to discuss here an idea that the intrinsic nature of this fact lies in integral geometry and complex analysis rather than in differential operators.

In the beginning, I want to explain the ideology of integral geometry in the context of theory of representations, as I understand it. Of course, this point of view is subjective. The crucial moments in the theory of representations are different kinds of equivalencies of representations of different nature. In Helgason's conjecture, it is the equivalency of representations in eigenspaces of invariant differential operators on the symmetric manifold X and hyperfunction's sections of some line bundles on the boundary F. The idea of integral geometry is to consider an intertwining operator not for individual representations (eigenvalues in this example) but a generating operator whose restrictions give individual equivalences of representations. It turns out that often this universal intertwining operator has a geometrical nature and is similar to Radon's transform. The transition to individual representations often is an elementary step since it is reduced to commutative harmonic analysis (for a Cartanian subgroup). Of course, this situation is not universal, but it is observed in many important cases, hopefully, in an appropriate interpretation, for all semisimple symmetric manifolds. Integral geometry on Riemannian symmetric manifolds of noncompact type is one of the primary examples where this structure was realized, but all considered functional spaces on X did not include eigenfunctions of invariant differential operators. We will discuss below how to make this, quite a substantial, adjustment, but we will explain in the beginning how to construct the integral geometrical picture corresponding to finite dimensional representations (this is a relatively new result [Gi06]).

Following the unitary trick of H.Weyl, finitely dimensional representations for complex semisimple Lie groups $G_{\mathbb{C}}$ and the maximal compact subgroups U coincide. Of course, it does not mean that harmonic analyses on $G_{\mathbb{C}}$ and U are identical. Let us start from the complex picture. Let $Z = G_{\mathbb{C}}/H$ be a complex symmetric manifold, H be an involutive subgroup corresponding to a holomorphic involution. Let us remind that Z is a Stein manifold. The group $G_{\mathbb{C}}$ itself is symmetric relative to the action of $G_{\mathbb{C}} \times G_{\mathbb{C}}$. Let A and N be Cartanian and maximal unipotent subgroups transversal to H so that HAN is Zariski open in $G_{\mathbb{C}}$. Let M be the centralizer of A at H. We call $\Xi = G_{\mathbb{C}}/MN$ the horospherical manifold. There is a geometrical duality between Z and Ξ through the double fibering:

$$Z \leftarrow G_{\mathbb{C}}/M \rightarrow \Xi$$

($z \in Z$ and $\zeta \in \Xi$ are incidental if they have a joint preimage at $G_{\mathbb{C}}/M$). The horosphere $E(\zeta), \zeta \in \Xi$, is the set of points $z \in Z$ incidental to ζ. Horospheres are orbits of all maximal unipotent subgroups. Correspondingly, there are defined dual submanifolds – pseudospheres $S(z), z \in Z$, – of points on Ξ incidental to z. Let us remark that Z and Ξ have the same dimension (let it be denoted as n) and horospheres and pseudospheres have the dimension $n - l$, where l is the rank of the symmetric space Z (the dimension of A).

On Ξ there is a natural action of the Abelian group A, which commutates with the action of $G_{\mathbb{C}}$. It corresponds to the fibering

$$\Xi \to F,$$

where the base is the flag manifold $F = G_{\mathbb{C}}/AMN$ and the fibers are A-orbits. Let us consider the space of holomorphic functions $\mathcal{O}(\Xi)$ on Ξ (in the algebraic setting we could consider regular functions – "polynomials"). Let us decompose it relative to the action of A ("Taylor series"):

$$\mathcal{O}(\Xi) = \bigoplus \mathcal{O}_m(\Xi).$$

The subspaces $\mathcal{O}_m$ are the same in analytic and algebraic pictures. The Borel–Weil theorem means that in $\mathcal{O}_m$ there are realized irreducible representations of the group $G_{\mathbb{C}}$ and the (simple) spectrum is described. Indeed, the spaces of sections of different line bundles on F, which participate in the Borel–Weil theorem can be identified with the spaces $\mathcal{O}_m$. On the other side, Helgason [He94] described the (simple) spectrum of the $G_{\mathbb{C}}$-representation in Z. It turns out that these spectrums coincide. It is connected with the fact that the action of $G_{\mathbb{C}}$ on Ξ is a contraction of the action on Z and Popov [Po87] gave a general conceptual proof of the coincidence of spectrums in such situations. It is possible, using standard technology, to construct isomorphisms of irreducible components, but the position of integral geometry is that there must be a generating intertwining operator. Indeed, it turns out that the spaces $\mathcal{O}(Z)$ and $\mathcal{O}(\Xi)$ are isomorphic as $G_{\mathbb{C}}$-modules [Gi06]. In the algebraic case – for regular functions – it is equivalent to the spectral result, but for holomorphic functions it is a stronger statement. What is important is not so much the fact of the isomorphism but the explicit structure of the intertwining operator.

Let us define the objects which participate in the formula for this operator. First, we define the basic special functions on $Z \times \Xi$ which participate in the formula. The points of Ξ parameterize maximal unipotent subgroups. For $\zeta \in \Xi$, we consider the corresponding maximal unipotent subgroup $N(\zeta)$ and its Iwasawa decomposition HAN on the open part of $G_{\mathbb{C}}$. We lift characters of A on the open part of $G_{\mathbb{C}}$ and take only characters for which these functions holomorphically extend on the whole $G_{\mathbb{C}}$ and push them down on Z. Let $\delta_1(a), \ldots, \delta_l(a)$ be generating characters of A and $\Delta_j(z|\zeta), j \leq l$, be corresponding functions on Z depending on $\zeta \in \Xi$. They are dominant highest weight functions. We call these functions *Sylvester's functions* since they are principal minors in the case of the manifold of nondegenerated symmetric matrices. The horosphere $E(\zeta)$ is an orbit of $N(\zeta)$. In an appropriate normalization, the horosphere $E(\zeta)$ is defined by the equations

$$\Delta_j(z|\zeta) = 1, \quad j \leq l. \tag{1}$$

Let $\mu(z, dz)$ be the holomorphic invariant differential form on Z of the maximal degree (it is defined up to a constant factor). The pseudospheres $S(z)$ are homogeneous manifolds relative to the isotropy subgroup $H(z) = H$. Let $v(z|\zeta, d\zeta)$ for each $z \in Z$ be the form μ on $S(z)$, which is holomorphic on z.

Let us define *the horospherical Cauchy transform*

$$\hat{f}(\zeta) = \int_\Gamma \frac{f(z)}{\prod_{1 \le j \le l}(\Delta_j(z|\zeta) - 1)} \mu(z, dz), \quad f \in \mathcal{O}(Z). \tag{2}$$

Here, Γ is a n-dimensional cycle which does not intersect the singularities of the kernel. It is possible to choose as such a cycle a compact real form of the manifold Z. This is an intertwining operator between $\mathcal{O}(Z)$ and $\mathcal{O}(\Xi)$.

Let us construct its inversion. We consider on Abelian group A the differential operators $P(D)$ with the polynomial symbols $P(m)$ in logarithmic coordinates. Since A acts on the fibers of the fibering $\Xi \to F$, these operators will act also on functions on Ξ. Let

$$W(m) = \prod_{\alpha \in \Sigma_+} \frac{\langle m + \rho, \alpha \rangle}{\langle \rho, \alpha \rangle}$$

be Weyl's polynomial for the dimensions of representations; ρ be the sum of positive roots. Then

$$f(z) = c \int_{\gamma(z)} W(D)\hat{f}(\zeta) v(z|\zeta, d\zeta). \tag{3}$$

Here, $\gamma(z)$ is a cycle (of the dimension $n-l$) in the pseudosphere $S(z)$. Let us remind that $S(z)$ is a homogeneous Stein manifold (with the group H) and we can take as the cycle $\gamma(z)$ a compact real form of $S(z)$ (a flag manifold). Again it reminds very much of Radon's inversion formula, but in a holomorphic environment. The composition of these two integral operators gives an integral formula representing f through the integration of a differential form on $Z \times \Xi$ along a cycle, which is fibered on cycles $\gamma(z)$ over the cycle Γ. This form-integrand is a closed meromorphic form and we can integrate it along any cycle avoiding the singularities. It is natural to interpret it as Cauchy integral formula on Z. It can be deduced for the special cycles from the Plancherel formula for compact symmetric manifolds, but the conceptual proof goes through a generalization of the Cauchy–Fantappie formula for cycles of higher codimension [Gi06']. Let us pay attention to the fact that the application of the operator $W(D)$ to the kernel in (2) gives a complicated combination of the monomials of the factors in the denominator of the degree $-(n - l)$. The remarkable property of Weyl's operator is that this combination gives the closed form – a quite nontrivial combinatorial fact.

If the function $\hat{f}$ lies in $\mathcal{O}_m(\Xi)$, it is homogeneous relative to A-action: it is multiplied on the character $\delta^m(a) = \delta_1^{m_1}(a) \cdots \delta^{m_l}(a)$. Then the application of Weyl's differential operator $W(D)$ just multiplies $\hat{f}$ on $W(m)$:

$$f(z) = cW(m) \int_{\gamma(z)} \hat{f}(\zeta)v(z|\zeta, d\zeta), \quad f \in \mathcal{O}(m). \tag{4}$$

In this case, $f \in \mathcal{O}_m(Z)$ are eigenfunctions of invariant differential operators; restrictions of the sections $\hat{f}$ of the line bundle on F on $S(z^0)$ for a fixed z^0 can be interpreted as boundary values of f. We can rewrite (4) as

$$f(z) = cW(m) \int_{\gamma(z^0)} P(z, z^0, \zeta) \hat{f}(\zeta)v(z|\zeta, d\zeta), \quad f \in \mathcal{O}(m), \tag{4'}$$

where the kernel $P(z, z^0, \zeta)$ for $\zeta \in \gamma(z^0)$ is the value of the character δ on such $a \in A$ that $a\zeta \in S(z)$. Of course, we need to choose the cycle $\gamma(z^0)$ so that such a would exist (fibers of $\Xi \to F$ through points of the cycle must intersect $S(z)$). The formula (4') can be interpreted as a holomorphic analog of the Poisson formula for finite dimensional representations. It has a very interesting structure compared to the real version. When we work with complex groups, we replace the averaging along of orbits of compact groups by an integration of closed holomorphic forms along cycles on orbits of complex groups.

The aim of the harmonic analysis on pseudo Riemannian symmetric manifolds (real forms X of Z), from the points of view of integral geometry, is the search for real forms of the horospherical Cauchy transform, in other words, to extend this transform from $\mathcal{O}(Z)$ to appropriate functional spaces on X. A choice of such a space is not unique and it is a substantial, informal part of the problem. It must be big enough to include considering functions, but not so big as to obscure the specifics of the problem. The first interesting example is the compact form $X = U/K$ – the compact Riemannian symmetric space. The natural maximal functional space is the space of hyperfunctions on X (functionals on the space of functions holomorphic in a neighborhood of X). Then a natural extension of the horospherical Cauchy transform can be defined [Gi06] such that the image is the space of holomorphic functions in the domain $D \subset \Xi$ parameterizing horospheres which do not intersect X. In such a way, we see that completely real objects – compact symmetric manifolds (including the real sphere) have canonical dual complex objects – the domains D (a real horospherical transform there can not be defined). We will not go into details here.

Now let $X = G/K$ be a Riemannian symmetric manifold of noncompact type; G be a real form of $G_{\mathbb{C}}$, K be its maximal compact subgroup (the compact form of H). Here, the real horospherical transform is well known: it is defined on C_0^∞ or another space of decreasing functions through the integration along of (real) horospheres on X. As we mentioned, this transform does not satisfy us since we want to work with eigenfunctions of invariant differential operators which are analytic.

Let us remind that there is a canonical Stein neighborhood of X in $Z - \text{Crown}(X)$ [AG90]. All eigenfunctions of invariant differential operators on X admit holomorphic extension in $\text{Crown}(X)$, and it is the maximal domain with this property. The crown was defined in [AG90] explicitly, but for our aims an equivalent description from [GK02] is more convenient. Let $A_{\mathbb{R}}$ be the real form of the Cartanian subgroup $A \subset G_{\mathbb{C}}$, corresponding to X, and $\mathfrak{a}$ be its Lie algebra. We consider the convex polyhedral

$$\Omega = \{a \in \mathfrak{a}; |\alpha(a)| < \pi/2, \forall \alpha \in \Sigma\},$$

where Σ is the restricted system of roots and let $t(\Omega) = A_{\mathbb{R}} \exp(i\Omega))$ be the tube domain in A. We saw that the parallel horospheres on Z are parameterized by elements of A (by a fiber in Ξ). Let $T(\Omega)$ be the union of parallel horospheres $E(\zeta)$ corresponding to elements of the tube $t(\Omega)$. Then

$$\text{Crown}(X) = \left(\bigcap_{g \in G} gT(\Omega)\right)_0. \tag{5}$$

We take here the connected component of the intersection, which contains X. This explicit description is not so important for us as a geometrical corollary: the domain $\text{Crown}(X)$ *is horospherically convex* [Gi08]. It means that the compliment to $\text{Crown}(X)$ in Z is horospherically concave: it is the union of horospheres $E(\zeta), \zeta \in Y$, which do not intersect $\text{Crown}(X))$. This set $Y \subset \Xi$ admits an explicit description. Let

$$F_{\mathbb{R}} = G/A_{\mathbb{R}} M_{\mathbb{R}} N_{\mathbb{R}} \subset F$$

be the real flag manifold. Then the projection of Y on F is $F_{\mathbb{R}}$ and preimages of points in $F_{\mathbb{R}}$ are compliments at A to the tube $t(\Omega)$. To be more exact, the manifold $\Xi_{\mathbb{R}} = G/M_{\mathbb{R}} N_{\mathbb{R}}$ of real horospheres on X admits the canonical imbedding in Ξ and there is the fibering $\Xi_{\mathbb{R}} \to F_{\mathbb{R}}$ with fibers $A_{\mathbb{R}}$: real horospheres have complex forms. On $F_{\mathbb{R}}$ the group K acts transitively – $F_{\mathbb{R}} = K/M_{\mathbb{R}}$ – and we can identify $F_{\mathbb{R}}$ with the manifold $S_{\mathbb{R}}(x)$ of real horospheres passing through a point $x \in X$ (the real form of the pseudosphere $S(x)$). Then the orbit Y of the action of the tube $A_{\mathbb{R}} \subset t(\Omega) \subset A$ on $F_{\mathbb{R}}$ in F is welldefined.

Let us remark that the domain $\Xi \setminus Y$ is concave relative to pseudospheres: it is the union of $S(z), z \in \text{Crown}(X)$. We can interpret this domain as a dual object for X. We will need one modification. The horospherical manifold Ξ is not a Stein one. It admits the extension up to a Stein space (not a manifold) but for us it is convenient to consider the holomorphically complete (but not holomorphically separable) extension of Ξ which is smooth. An essential point is that the fibering $\Xi \to F$ has no global sections (as different from the real case): pseudospheres $S(z)$ are sections over the Zariski open orbits of the isotropy subgroups $H(z) = H$ on F.

We define homogeneous coordinates on the fibers $(\zeta, u), \zeta \in \Xi, u \in \mathbb{C}^l$, such that

$$(\zeta, u) \sim (a\zeta, \delta(a)^{-1}u), \quad a \in A,$$

where $\delta_j(a), 1 \leq j \leq$ are characters of A corresponding to Δ_j. Let $\tilde{\Xi}$ be the factorization of $\Xi \times \mathbb{C}^l$ relative to this equivalency relation. Let us extend the equivalency relation on the manifold of triplets $(z, \zeta, u) \in Z \times \Xi \times \mathbb{C}^l$ such that

$$\Delta_j(z|\zeta) = u_j, \quad 1 \leq j \leq l$$

and let L be the result of the factorization. It is a Stein manifold and its points are incidental pairs $(z, \tilde{\zeta}), z \in Z, \tilde{\zeta} \in \tilde{\Xi}$: we have the double fibering

$$Z \leftarrow L \rightarrow \tilde{\Xi}.$$

Left fibers $E(\zeta, u)$ are defined by the equations

$$\Delta_j(z|\zeta) = u_j, \quad 1 \leq j \leq l$$

and they coincide with the horospheres if $u \in (\mathbb{C}^*)^l (u_j \neq 0)$. If some $u_j = 0$, then we have some unions of degenerated orbits (of nonmaximal dimension) of the maximal unipotent subgroups, which sometimes are called degenerated horospheres. These fibers can be singular, but it is possible to prove that they have the dimension $n - l$ as the horospheres (unpublished result with Vinberg). On $\tilde{\Xi}$ we obtain the compactifications $\tilde{S}(z)$ of hyperspheres $S(z)$ isomorphic to F. Let $\hat{X}$ be the extension in $\tilde{\Xi}$ of the domain $\Xi \setminus Y \subset \Xi$; it is the union of compacts $\tilde{S}(z), z \in \text{Crown}(X)$.

Let $Q = \text{Crown}(X) \times \hat{X} \subset L$. It is a Stein manifold. The manifold $\hat{X}$ is $(n - l)$-pseudoconcave and we consider cohomology $H^{(n-l)}(\hat{X}, \mathcal{O}(\delta^m))$ with the coefficients in line bundles corresponding to characters $\delta^m(a)$ of A using the homogeneous coordinates (ζ, u). The dimension of this space of cohomology is infinite and we will use for the description of this cohomology the holomorphic language developed in [Gi93, EGW95]. We consider the complex of holomorphic differential forms $\omega(z, \zeta, u, dz)$ on Q with differentials only on z; on (ζ, u) they are sections of the bundle $\mathcal{O}(\delta^m)$ (homogeneous relative to the action of A with the character δ^m). The differential in the complex also acts only along z. The corresponding cohomology is isomorphic to Dolbeault cohomology. Let us describe this isomorphism. We consider the fibering $Q \rightarrow \hat{X}$ with contractible fibers $E(\zeta, u)$, restrict a d_z-closed form ω on a section Γ and take $(0, n - l)$-part of this form, considered as a form on $\hat{X}$. The result is $\bar{\partial}$-closed form and we have a map from holomorphic cohomology on Dolbeault cohomology $H^{(0,n-l)}(\hat{X}, \mathcal{O}(\delta^m))$.

We need two special holomorphic forms. The invariant holomorphic form of maximal degree $\lambda(z, \zeta, u, dz)$ has coefficients from $\mathcal{O}(\delta^{-1})$ on $(\zeta, u))$, where

$\delta^{-1}(a) = \prod(\delta_j(a))^{-1}$. It is the residue of the form $\mu(z, dz)/\prod(\Delta_j(\zeta) - u_j)$ on the fibers. Also on $\tilde{\Xi}$ there is an invariant holomorphic $(n - l)$-form $\kappa(\zeta, u, d\zeta, du)$ with coefficients in $\mathcal{O}(\delta^{-2\rho})$. Now we are ready to define the horospherical transform as an operator from holomorphic functions $f \in \mathcal{O}(\text{Crown}(X))$ to cohomology from $H^{(n-l)}(\hat{X}, \mathcal{O}(\delta^{-1}))$:

$$\hat{f} = f(z)\lambda(z, \zeta, u, dz). \tag{6}$$

We obtain here the $(n - l)$-form on Q with differentials on Z (along fibers), which is closed since it has the maximal degree on fibers. So it defines a cohomology class (in the holomorphic language). It may look strange that in this form of the horospherical transform there is no integration at all but it is typical for representations of inverse Penrose transforms in the holomorphic language for analytic cohomology (cf. [Gi93, Gi07]). The principal fact is the following theorem.

Theorem. *The horospherical transform*

$$f \in \mathcal{O}(\text{Crown}(X)) \mapsto \hat{f} \in H^{(n-l)}(\hat{X}, \mathcal{O}(\delta^{-1}))$$

is injective.

This fact follows from the next inversion formula

$$f(z) = c \int_{\tilde{S}(z)} W(D)\hat{f} \wedge \kappa(\zeta, u, d\zeta, du).$$

In this integral after the application of the Weyl's differential operator to $\hat{f}$, we obtain the coefficients in $\mathcal{O}(\delta^{-2\rho})$ instead of $\mathcal{O}(\delta^{-1})$. Then we push down this form on $\hat{X}$ as $\bar{\partial}$-closed $(n - l)$-form (using a section Γ) and after the multiplication on κ we obtain an $(n - l, n - l)$-form with coefficients in $\mathcal{O}$. We integrate this form along cycles $\tilde{S}(z)$. Thus, the inverse operator is a Penrose-type transform. The fact that this operator reproduces f is another version of the integral Cauchy formula at Z. We need in the kernel of the integral Cauchy formula, which we discussed above, to take the residue on the edge of the singular set of the denominator – the horosphere $E(\zeta, u)$. Such formulas in the case of codimension 1 Leray [Le00] called 2nd Cauchy–Fantappie formulas. They were considered in [GH90, Gi07]. We will consider details in another paper dedicated to the integral Cauchy formulas on symmetric manifolds. Let us again emphasize that in this consideration the explicit form of the initial domain $\mathcal{O}(\text{Crown}(X))$ was not important; it was essential only that it was horospherically convex. Also let us remark that the space $\mathcal{O}(\Xi)$ admits a canonical imbedding in $H^{(n-l)}(\hat{X}, \mathcal{O}(\delta^{-1}))$ such that both horospherical transforms are compatible. This fact is again a consequence of Cauchy formulas.

To return to the usual form of Helgason's conjecture, we need to consider the action of A on $H^{(n-l)}(\hat{X}, \mathcal{O}(\delta^{-1}))$ and investigate homogeneous cohomology classes relative to this action since just these classes correspond to eigenfunctions of invariant differential operators. We will not discuss this in this paper. It is interesting to investigate the image of the horospherical transform.

Conjecture. The kernel of Weyl's differential operator $W(D)$ on $H^{(n-l)}(\hat{X}, \mathcal{O}(\delta^{-1}))$ is complimentary to the image of the horospherical transform.

We can see that in this approach to Helgason's conjecture, compared with the usual considerations, there are no difficult analytic constructions. Instead, we use integral Cauchy formulas on symmetric manifolds and holomorphic language for analytic cohomology. We also avoid a reduction to functions on the Cartanian subgroup, which decreases the number of variables, but brings singularities which did not exist in the initial problem.

References

[AG90]   D. N. Akhiezer and S. G. Gindikin., *On Stein extensions of real symmetric spaces*, Math. Ann. **286** (1990), 1–12.

[EGW95]   M. G. Eastwood, S. G. Gindikin, and H.-W. Wong, *Holomorphic realization of $\bar{\partial}$-cohomology and constructions of representations*, Jour. Geom. Phys. **17** (1995), 231–244.

[Gi93]   S. Gindikin, *Holomorphic language for $\bar{\partial}$-cohomology and representations of real semisimple Lie groups*, The Penrose Transform and Analytic Cohomology in Representation Theory (M. Eastwood, J. Wolf, R. Zierau, eds.), vol. 154 Cont. Math., Amer. Math. Soc., 1993, pp. 103–115.

[Gi06]   S. Gindikin, *Harmonic analysis on symmetric manifolds from the point of view of complex analysis*, Japanese J. Math. **1** (2006), no. 1, 87–105.

[Gi06']   S. Gindikin, *The integral Cauchy formula on symmetric Stein manifolds*, Colloquium de Giorgi (2006), Edizione della Normale, 19–28.

[Gi07]   S. Gindikin, *Second Cauchy-Fantappie formula and the inversion of the Martineau-Penrose transform*, J. Algebra **313** (2007), 199–207.

[Gi08]   S. Gindikin, *The horospherical duality*, Science in China Series A: Mathematics **51** (2008), no. 4, 562-567.

[GH90]   S. Gindikin and G.Henkin, *The Cauchy-Fantappie on projective space*, Amer. Math. Soc. Transl. (2) **146** (1990), 23–32.

[GK02]   S. Gindikin and B.Krötz, *Invariant Stein domains in Stein symmetric spaces and a non-linear complex convexity theorem*, Intern. Math. Res. Not. **18** (2002), 959–971.

[He74]   S. Helgason, *Eigenspaces of Laplacian: integral representation and irreducibility*, J. Funct. Anal. **17** (1974), 328–353.

[He94]   S. Helgasson, *Geometric analysis on symmetric spaces*, AMS, 1994.

[KKMOOT]   M. Kashiwara, A. Kowata, K. Minemura, K. Okamoto, T. Oshima, M. Tanaka, *Eigenfunctions of invariant differential operators on a symmetric space*, Ann. Math. **107** (1978), 1–39.

[Po87]   V. Popov, *Contraction of the actions of reductive algebraic groups*, Math. USSR Sbornik **58** (1987), 311–335.

Lectures on Lie Algebras

Joseph Bernstein

Abstract This is a lecture course for beginners on representation theory of semisimple finite dimensional Lie algebras. It is shown how to use infinite dimensional representations (Verma modules) to derive the Weyl character formula. We also provide a proof for Harish–Chandra's theorem on the center of the universal enveloping algebra and for Kostant's multiplicity formula.

Keywords Lie algebra • Verma module • Weyl character formula • Kostant multiplicity formula • Harish–Chandra center

1991 Mathematics Subject Classification. 17A70 (Primary) 17B35 (Secondary)

Introduction

These notes originally were a draft of the transcript of my lectures in the Summer school in Budapest in 1971. For the lectures addressed to the advanced part of the audience, see [Ge]. The beginners' part was released a bit later see [Ki]. It contains a review by Feigin and Zelevinsky, which *expands* my lectures. Therefore, the demand in a short and informal guide for the beginners still remains, I was repeatedly told. So here it is.

We will consider finite dimensional representations of semisimple finite dimensional complex Lie algebras. The facts presented here are well known ([Bu], [Di], [Se]) and in a more rigorous setting. But our presentation of these facts is comparatively new (at least, it was so in 1971) and is based on the systematic usage of the Verma modules M_χ.

J. Bernstein (✉)
Department of Mathematics, Tel-Aviv University, Tel-Aviv, Israel
e-mail: bernstei@math.tau.ac.il

B. Krötz et al. (eds.), *Representation Theory, Complex Analysis, and Integral Geometry*,
DOI 10.1007/978-0-8176-4817-6_6, © Springer Science+Business Media, LLC 2012

The reader is supposed to be acquainted with the main notions of Linear Algebra ([Pr] will be just fine). The knowledge of the first facts and notions from the theory of Lie algebra will not hurt but is not required.

The presentation is arranged as follows:

In Sect. 1, we discuss general facts regarding Lie algebras, their universal enveloping algebras, and their representations.

In Sect. 2, we discuss in detail the case of the simplest simple Lie algebra $\mathfrak{g} = \mathfrak{sl}(2)$. The results of this section provide essential tools for treating the general case.

In Sect. 3, we provide without proofs a list of results on the structure of semisimple Lie algebras and their root systems.

In Sect. 4, we introduce some special category of $\mathfrak{g}$-modules, so-called category $\mathcal{O}$. We construct basic objects of this category – Verma modules M_χ – and describe some of their properties.

In Sect. 5, we construct, for every semisimple Lie algebra $\mathfrak{g}$, a family of irreducible finite dimensional representations A_λ.

In Sect. 6, we formulate one of the central results – Harish–Chandra's description of the algebra $\mathfrak{Z}(\mathfrak{g})$ – center of the enveloping algebra of $\mathfrak{g}$. For the proof see Sect. 9.

In Sect. 7, we describe various properties of the category $\mathcal{O}$ that follow from the Harish–Chandra theorem.

In Sect. 8, we prove Weyl's character formula for irreducible $\mathfrak{g}$-modules A_λ and derive Kostant's formula for the multiplicities of weights for these representations. We also prove that every finite dimensional $\mathfrak{g}$-module is decomposable into a direct sum of irreducible modules isomorphic to A_λ.

In Sect. 9, we present a proof of the Harish–Chandra theorem.

1   General Facts About Lie Algebras

All vector spaces considered in what follows are defined over a ground field $\mathbb{K}$. We assume that $\mathbb{K}$ is algebraically closed of characteristic 0. The reader can assume $\mathbb{K} = \mathbb{C}$.

1.1   Lie Algebras

Definition. A *Lie algebra* is a $\mathbb{K}$-vector space $\mathfrak{g}$ equipped with a bilinear multiplication $[\,,\,] : \mathfrak{g} \otimes \mathfrak{g} \to \mathfrak{g}$ (it is called *bracket*) that satisfies the following identities:

$$[X, Y] + [Y, X] = 0 \quad \text{for any } X, Y \in \mathfrak{g} \qquad \text{(S - S)}$$

$$[X, [Y, Z]] + [Y, [Z, X]] + [Z, [X, Y]] = 0 \text{ for any} \quad X, Y, Z \in \mathfrak{g}. \qquad \text{(J.I.)}$$

The identity (S-S) signifies skew-symmetry of the bracket, (J.I.) is called the *Jacobi identity*.

Example. Let A be an associative algebra. By means of the subscript L we will denote the Lie algebra $\mathfrak{g} = A_L$ whose underlying vector space is a copy of A and the bracket is given by the formula $[X, Y] = XY - YX$. Clearly, A_L is a Lie algebra: (S-S) and (J.I.) are subject to a direct verification.

If V is a vector space, we denote $\mathfrak{gl}(V)$ its *general linear Lie algebra* that is defined as $\mathfrak{gl}(V) = (\mathrm{End}_{\mathbb{K}}(V))_L$.

We abbreviate $\mathfrak{gl}(\mathbb{K}^n)$ to $\mathfrak{gl}(n)$. Note that this is just the algebra $Mat(n)$ of $n \times n$-matrices with the operation $[X, Y] = XY - YX$.

1.2 Representations of Lie Algebra

A *representation* γ of a Lie algebra $\mathfrak{g}$ in a vector space V is a morphism of Lie algebras $\gamma : \mathfrak{g} \to \mathfrak{gl}(V)$. We will denote by the same symbol γ the corresponding morphism of vector spaces $\mathfrak{g} \otimes V \to V$.

We will also use the following equivalent terms for representations: "γ is an action of Lie algebra $\mathfrak{g}$ on V"; "V is $\mathfrak{g}$-module".

Morphisms of $\mathfrak{g}$-modules are defined as usual. The category of $\mathfrak{g}$-modules will be denoted by $\mathcal{M}(\mathfrak{g})$.

An important example of a representation is the adjoint representation ad of a Lie algebra $\mathfrak{g}$ on the vector space $V = \mathfrak{g}$. It is defined by formula $ad(X)(Y) := [X, Y]$. The fact that this is a representation follows from Jacobi identity.

1.3 Tensor Product Representation

Given representations γ, δ of a Lie algebra $\mathfrak{g}$ in spaces V and E we construct the tensor product representation $\eta = \gamma \otimes \delta$ in the space $V \otimes E$ via Leibnitz rule $\eta(X) = \gamma(X) \otimes Id + Id \otimes \delta(X)$.

Lemma. *Let $\gamma : \mathfrak{g} \otimes V \to V$ be any representation of a Lie algebra $\mathfrak{g}$. Consider on the space $\mathfrak{g} \otimes V$ the structure of $\mathfrak{g}$-module given by representation $Ad \otimes \gamma$. Then $\gamma : \mathfrak{g} \otimes V \to V$ is a morphism of $\mathfrak{g}$-modules.*

The verification is left to the reader.

1.4 Some Examples of Lie Algebras

Example 1. Let $\mathfrak{n}^-$, $\mathfrak{n}_-$, and $\mathfrak{h}$ be the subspaces of $\mathfrak{g} = \mathfrak{gl}(n)$ consisting of all strictly upper triangular, strictly lower triangular and diagonal matrices, respectively.

Clearly, $\mathfrak{n}_+, \mathfrak{n}_-,$ and $\mathfrak{h}$ are Lie subalgebras of $\mathfrak{gl}(n)$. Important role in representation theory plays a triangular decomposition $\mathfrak{gl}(n) = \mathfrak{n}_- \oplus \mathfrak{h} \oplus \mathfrak{n}_+$ (this is a direct sum decomposition of vector spaces, but not of Lie algebras).

Example 2. The space of $n \times n$ matrices with trace zero is a Lie algebra; it is called the *special linear algebra* and denoted by $\mathfrak{sl}(n)$.

Example 3. Let B be a bilinear form on a vector space V. Consider the space $Der(B)$ of all operators $X \in \mathfrak{gl}(V)$ that preserve B, i.e., $B(Xu, v) + B(u, Xv) = 0$ for any $u, v \in V$.

It is easy to see that this subspace is closed under the bracket and so is a Lie subalgebra of $\mathfrak{gl}(V)$.

If B is nondegenerate, we distinguish two important subcases:

- B is symmetric, then $Der(B)$ is called the *orthogonal Lie algebra* and denoted by $\mathfrak{o}(V, B)$.
- B is skew-symmetric, then $Der(B)$ is called the *symplectic Lie algebra* and denoted by $\mathfrak{sp}(V, B)$.

It is well known that over $\mathbb{C}$ all nondegenerate symmetric forms on V are equivalent to each other and the same applies to skew-symmetric forms. So Lie algebras $\mathfrak{o}(V, B)$ and $\mathfrak{sp}(V, B)$ actually depend only on the dimension of V, and we will sometimes denote them by $\mathfrak{o}(n)$ and $\mathfrak{sp}(2m)$.

The Lie algebras $\mathfrak{gl}(n), \mathfrak{o}(n),$ and $\mathfrak{sp}(2m)$ are called *classical Lie algebras*. For the proof of the statements of this section, see ([Bu], [Di], [OV], [Se]).

1.5 Universal Enveloping Algebra

Let $\mathfrak{g}$ be a Lie algebra over $\mathbb{K}$. To $\mathfrak{g}$ we assign an *associative* $\mathbb{K}$-algebra with unit, $U(\mathfrak{g})$, called the *universal enveloping* algebra of the Lie algebra $\mathfrak{g}$. Namely, consider the tensor algebra $T(\mathfrak{g})$ of the *space* $\mathfrak{g}$, i.e.,

$$T^{\boldsymbol{\cdot}}(\mathfrak{g}) = \underset{n \geq 0}{\oplus} T^n(\mathfrak{g}),$$

where $T^0(\mathfrak{g}) = \mathbb{K}, T^n(\mathfrak{g}) = \mathfrak{g} \otimes \cdots \otimes \mathfrak{g}$ (n factors). Consider also the two-sided ideal $I \subset T(\mathfrak{g})$ generated by the elements $X \otimes Y - Y \otimes X - [X, Y]$ for any $X, Y \in \mathfrak{g}$. Set $U(\mathfrak{g}) = T(\mathfrak{g})/I$.

We will identify the elements of $\mathfrak{g}$ with their images in $U(\mathfrak{g})$. Under this identification, any $\mathfrak{g}$-module may be considered as a (left, unital) $U(\mathfrak{g})$-module and, conversely, any $U(\mathfrak{g})$-module may be considered as a $\mathfrak{g}$-module. We will not distinguish the $\mathfrak{g}$-modules from the corresponding $U(\mathfrak{g})$-modules.

The algebra $U(\mathfrak{g})$ has a natural increasing filtration $U(\mathfrak{g})_n = \sum_{i \leq n} T^i(\mathfrak{g})$. We denote by $grU(\mathfrak{g})$ the associated graded algebra $grU(\mathfrak{g}) = \oplus_{n \geq 0} gr_n U(\mathfrak{g})$, where $gr_n U(\mathfrak{g}) := U(\mathfrak{g})_n / U(\mathfrak{g})_{n-1}$. This algebra is clearly commutative and hence the natural morphism $i : \mathfrak{g} \to \mathrm{gr}_1 U(\mathfrak{g})$ extends to a morphism of graded commutative algebras $i : S^{\cdot}(\mathfrak{g}) \to \mathrm{gr} U(\mathfrak{g})$, where $S^{\cdot}(\mathfrak{g})$ is the symmetric algebra of the linear space $\mathfrak{g}$. The following result will be used repeatedly in the lectures.

Theorem (Poincaré–Birkhoff–Witt). The morphism $i : S^{\cdot}(\mathfrak{g}) \longrightarrow \mathrm{gr} U(\mathfrak{g})$ is an isomorphism of graded commutative algebras.

Corollary. *(1) $U(\mathfrak{g})$ is a Noetherian ring without zero divisors.*
(2) Let symm' : $S^{\cdot}(\mathfrak{g}) \longrightarrow T^{\cdot}(\mathfrak{g})$ be the map determined by the formula

$$X_1 \otimes \cdots \otimes X_k \mapsto \frac{1}{k!} \sum_{\sigma \in \mathfrak{S}_k} X_{\sigma(1)} \otimes \cdots \otimes X_{\sigma(k)}.$$

Denote by symm: $S^{\cdot}(\mathfrak{g}) \longrightarrow T^{\cdot}(\mathfrak{g}) \longrightarrow U(\mathfrak{g})$ the composition of symm' and the projection onto $U(\mathfrak{g})$. The map symm is an isomorphism of linear spaces (not algebras).
(3) If $X_1, \ldots, X_k$ is a basis of $\mathfrak{g}$, then the set of monomials $X_1^{n_1} X_2^{n_2} \ldots X_k^{n_k}$, where the n_i run over the set $\mathbb{Z}_{\geq 0}$ of nonnegative integers, is a basis of $U(\mathfrak{g})$.

Remark. The version of PBW as stated above is in [Di] 2.3.6. A direct proof can be found in [BG]. For point one of the corollary, see 2.3.8 and 2.3.9 of loc. cit. For the second point, see 2.4 in [Di] and for the last point see 2.1.8 in [Di].

1.6 Some Finiteness Results

In order to analyze finite dimensional representations of a Lie algebra, we will often use infinite dimensional representations that satisfy some finiteness assumptions.

1.6.1 Locally Finite Representations

Definition. Let A be an associative algebra. An A-module V is called *locally finite* if it is a union of finite dimensional A-submodules.

Notice that the subcategory $\mathcal{M}(A)^{lf} \subset \mathcal{M}(A)$ of locally finite A-modules is closed with respect to subquotients. It is easy to check that if algebra A is finitely generated then $\mathcal{M}(A)^{lf}$ is also closed under extensions.

If V is an arbitrary A-module, then the sum of all locally finite submodules is the maximal locally finite submodule of V. We denote it $V^{A\text{-finite}}$.

We use the same definitions for a module V over a Lie algebra $\mathfrak{a}$. In particular, we denote by $V^{\mathfrak{a}\text{-finite}}$ the maximal locally finite $\mathfrak{a}$-submodule of V.

Lemma. *(i) Let $\mathfrak{a}$ be a Lie algebra. Then the tensor product of locally finite representations is locally finite.*

*(ii) Let $\mathfrak{g}$ be a finite-dimensional Lie algebra and $\mathfrak{a} \subset \mathfrak{g}$ its Lie subalgebra. Given a
$\mathfrak{g}$-module V consider its maximal $\mathfrak{a}$-locally finite submodule $L = V^{\mathfrak{a}\text{-finite}}$. Then
L is a $\mathfrak{g}$-submodule of V.*

Proof. The proof of (i) is straightforward. Then (i) implies that the morphism of
$\gamma : \mathfrak{g} \otimes V \to V$ maps $\mathfrak{g} \otimes L$ into L, i.e., L is a $\mathfrak{g}$-submodule.   □

Exercise. Show that the same result is true under weaker assumptions. Namely, it
is enough to assume that the adjoint action of the Lie algebra $\mathfrak{a}$ on the space $\mathfrak{g}/\mathfrak{a}$ is
locally finite.

1.6.2 Locally Nilpotent Representations

Definition. Let $\mathfrak{a}$ be a Lie algebra. An $\mathfrak{a}$-module V is called *nilpotent* if for some
natural number k we have $\mathfrak{a}^k(V) = 0$. An $\mathfrak{a}$-module V is called *locally nilpotent* if
it is a sum of nilpotent submodules.

As before we denote by $V^{\mathfrak{a}\text{-nilp}}$ the maximal locally nilpotent submodule of V.

Lemma. *(i) Tensor product of locally nilpotent representations is locally
    nilpotent.*
*(ii) Let $\mathfrak{g}$ be a Lie algebra and $\mathfrak{a} \subset \mathfrak{g}$ its Lie subalgebra such that the adjoint
    action of $\mathfrak{a}$ on $\mathfrak{g}$ is locally nilpotent. Given a $\mathfrak{g}$-module V consider its maximal
    $\mathfrak{a}$-locally nilpotent submodule $L = V^{\mathfrak{a}-nilp}$. Then L is a $\mathfrak{g}$-submodule of V.*

The proof is the same as in Lemma 1.6.1.

1.7 Representations of Abelian Lie Algebras

Let $\mathfrak{a}$ be an abelian Lie algebra (i.e., the bracket on $\mathfrak{a}$ is identically 0). Let V be a
locally finite $\mathfrak{a}$-module.

For every character $\chi \in \mathfrak{a}^*$, we denote by $V(\chi)$ the space of generalized
eigenvectors of $\mathfrak{a}$ with eigencharacter χ.

Proposition. *V is a direct sum of the subspaces $V(\chi)$.*

This is a standard result of linear algebra, see Proposition A.1 in the appendix.

Definition. A module V over an abelian Lie algebra $\mathfrak{a}$ is called *semisimple* if it is
spanned by eigenvectors of $\mathfrak{a}$.

For any $\mathfrak{a}$-module V, we denote by $V^{\mathfrak{a}\text{-ss}}$ the maximal semisimple $\mathfrak{a}$-submodule
of V.

Lemma. *(i) Tensor product of semisimple representations is semisimple.*
*(ii) Let $\mathfrak{g}$ be a Lie algebra and $\mathfrak{a} \subset \mathfrak{g}$ its abelian Lie subalgebra such that the
    adjoint action of $\mathfrak{a}$ on $\mathfrak{g}$ is semisimple. Given a $\mathfrak{g}$-module V consider its maximal
    $\mathfrak{a}$-semisimple submodule $L = V^{\mathfrak{a}\text{-ss}}$. Then L is a $\mathfrak{g}$-submodule of V.*

Again, the proof is the same as in Lemma 1.6.1.

2 The Representations of $\mathfrak{sl}(2)$

In this section, we will describe representations of the simplest simple Lie algebra $\mathfrak{g} = \mathfrak{sl}(2)$.

2.1 The Lie Algebra $\mathfrak{sl}(2)$

The Lie algebra $\mathfrak{sl}(2)$ consists of matrices $\mathbf{x} = \begin{pmatrix} a & b \\ c & d \end{pmatrix}$ over field $\mathbb{K}$ such that $\operatorname{tr} \mathbf{x} = a + d = 0$. In $\mathfrak{sl}(2)$, select the following basis

$$E_+ = \begin{pmatrix} 0 & 1 \\ 0 & 0 \end{pmatrix}, \quad H = \begin{pmatrix} 1 & 0 \\ 0 & -1 \end{pmatrix}, \quad E_- = \begin{pmatrix} 0 & 0 \\ 1 & 0 \end{pmatrix}.$$

The commutation relations between the elements of the basis are:

$$[H, E_+] = 2E_+ \; ; [H, E_-] = -2E_- \; ; [E_+, E_-] = H.$$

Remark. We will see that in any semisimple Lie algebra $\mathfrak{g}$ we can find many triples of elements (E_+, H, E_-) of $\mathfrak{g}$ that satisfy above relation. We call such a triple an $\mathfrak{sl}(2)$-*triple*. In this way, the study of the representations of the Lie algebra $\mathfrak{sl}(2)$ provides us with lots of information on the representations of any semisimple Lie algebra $\mathfrak{g}$.

The above relations between E_-, H, and E_+ and a simple inductive argument yield the following relations in $U(\mathfrak{sl}(2))$:

$$[H, E_+^k] = 2kE_+^k, \; [H, E_-^k] = -2kE_-^k, \; [E_+, E_-^k] = kE_-^{k-1}(H - (k-1)).$$

Besides, it is easy to verify that the element

$$C = 4E_- E_+ + H^2 + 2H$$

belongs to the center of $U(\mathfrak{sl}(2))$. The element C is called the *Casimir operator*.

Let V be an $\mathfrak{sl}(2)$-module. A vector $v \in V$ is called a *weight vector* if it is an eigenvector of the operator H, i.e. $Hv = \chi v$; the number $\chi \in \mathbb{K}$ is called the *weight* of v.

We denote by $V^{ss}(\chi)$ the subspace of all such vectors. Similarly, we define $V(\chi)$ to be the space of generalized weight vectors for H (see appendix for definitions).

Lemma.

$$E_+(V^{ss}(\chi)) \subset V^{ss}(\chi + 2), \; E_+(V(\chi)) \subset V(\chi + 2)$$
$$E_-(V^{ss}(\chi)) \subset V^{ss}(\chi - 2), \; E_-(V(\chi)) \subset V(\chi - 2).$$

Proof. Let $v \in V^{ss}(\chi)$. Then $(H - \chi - 2)E_+v = E_+(H - \chi)v = 0$, i.e. $E_+v \in V^{ss}(\chi + 2)$. Similarly if $v \in V(\chi)$ then $(H - \chi - 2)^n E_+v = E_+(H - \chi)^n v = 0$ for large n, i.e. $E_+v \in V(\chi + 2)$.

The proof for E_- is similar. $\qquad\square$

A nonzero vector v is called *a highest weight vector* if it is a weight vector with some weight χ and $E_+v = 0$.

2.2 A Key Lemma

Lemma 1. *Let V be a representation of $\mathfrak{sl}(2)$ and $v \in V$ a highest weight vector of weight χ. Consider the sequence of vectors $v_k = E_-^k v$, $k = 0, 1 \ldots$. Then*

1) $Hv_k = (\chi - 2k)v_k$, $E_+v_{k+1} = (k + 1)(\chi - k)v_k$
2) *The subspace $L \subset V$ spanned by vectors v_k is an $\mathfrak{sl}(2)$-submodule and all non-zero vectors v_k are linearly independent.*
3) *Suppose that $v_k = 0$ for large k. Then $\chi = l \in \mathbb{Z}_{\geq 0}$, $v_k \neq 0$ for $0 \leq k \leq l$ and $v_k = 0$ for $k > l$.*

Proof. (1) is proved by induction in k.
(2) follows from 1) since v_k are eigenvectors of H with distinct eigenvalues.
(3) Let l be the first index such that $v_{l+1} = 0$. Then $0 = E_+v_{l+1} = (l+1)(\chi-l)v_l$ and hence $\chi = l$. $\qquad\square$

2.3 Construction of Representations A_l

Let us now describe a family of irreducible finite dimensional representations of $\mathfrak{sl}(2)$. For every $l \in \mathbb{Z}_{\geq 0}$, we construct a representation A_l of dimension $l + 1$. This representation is generated by a highest weight vector v_l of weight l.

First we describe this representations geometrically. Consider the natural action of the group $G = SL(2, \mathbb{K})$ on the plane $\mathbb{K}^2$ with coordinates (x, y). It induces the action of G on the space V of polynomial functions on $\mathbb{K}^2$.

The action of the group G on V induces a representation of its Lie algebra $\mathfrak{g} = \mathfrak{sl}(2)$. It can be described via explicit formulas using differential operators

$$E_+ = x\partial_y, H = x\partial_x - y\partial_y, E_- = y\partial_x.$$

The representation V is a direct sum of invariant subspaces $A_l, l \in \mathbb{Z}_{\geq 0}$, where A_l is the space of homogeneous polynomials of degree l.

In particular, the representations A_l extend to representations of the group $G = SL(2, \mathbb{K})$.

Let us describe these representations explicitly. The space A_l has a basis consisting of monomials $\{a_{-l}, a_{-l+2}, \ldots, a_{l-2}, a_l\}$, where $a_i = x^{(l+i)/2} y^{(l-i)/2}$. The action of the algebra $\mathfrak{sl}(2)$ is as follows:

$$Ha_i = ia_i, \; E_-a_i = \frac{l+i}{2} a_{i-2}, \; E_+a_i = \frac{l-i}{2} a_{i+2}.$$

Exercise. (i) Show that the module A_ℓ is irreducible.
(ii) Consider the $\mathfrak{sl}(2)$-module M generated by a vector m subject to the relations $H(m) = \ell m$ (i.e. m has weight ℓ), $E_+(m) = 0$ and $E_-^{\ell+1}(m) = 0$. Prove that M is isomorphic to the module A_ℓ described above.

2.4 Classification of Irreducible Finite Dimensional Modules of the Lie Algebra $\mathfrak{sl}(2)$

Proposition. *(1) In any finite dimensional nonzero $\mathfrak{sl}(2)$-module V, there is a submodule isomorphic to one of A_l.*
(2) The Casimir operator C acts on A_l as the scalar $l(l + 2)$.
(3) The modules A_l are irreducible, distinct, and exhaust all (isomorphism classes of) finite dimensional irreducible $\mathfrak{sl}(2)$-modules.

Proof. (1) Consider all eigenvalues of H in V and choose an eigenvalue χ such that $\chi + 2$ is not an eigenvalue. Let v_0 be a corresponding eigenvector. Then $Hv_0 = \chi v_0$, $E_+v_0 = 0$. Since V is finite dimensional, Key Lemma implies that $\chi = \ell \in \mathbb{Z}_{\geq 0}$, $E_-^{\ell+1}v_0 = 0$ and the space spanned by $E_-^r v_0$, where $r = 0, 1, \ldots, \ell$, forms an $\mathfrak{sl}(2)$-submodule $L \subset V$. The Exercise above implies that L is isomorphic to A_ℓ.
(2) It is quite straightforward that $Ca_l = l(l + 2)a_l$. If $a \in A_l$, then $a = Xa_l$ for a certain $X \in U(\mathfrak{sl}(2))$. Hence, $Ca = CXa_l = XCa_l = l(l + 2)a$.
(3) If A_l contains a nontrivial submodule V, then it contains A_k, where $k < l$, contradicting the fact that $C = l(l + 2)$ on A_l and $C = k(k + 2)$ on A_k.

Heading (1) implies that A_l, where $l \in \mathbb{Z}_{\geq 0}$, exhaust all irreducible $\mathfrak{sl}(2)$-modules. $\qquad\square$

2.5 Complete Reducibility of $\mathfrak{sl}(2)$-Modules

Proposition. *Any finite dimensional $\mathfrak{sl}(2)$-module V is isomorphic to a direct sum of modules of type A_l. In other words, finite dimensional representations of $\mathfrak{sl}(2)$ are completely reducible.*

Proof. We will use the following general lemma that we prove below.

Lemma. *Let C be an abelian category. Suppose that any object $V \in C$ of length 2 is completely reducible. Then any object $V \in C$ of finite length is completely reducible.*

This implies that it is enough to prove the proposition for a module V of length two. Let $S \simeq A_k$ be an irreducible submodule of V and $Q = V/S \simeq A_l$ a quotient module.

If $k \neq l$, then the Casimir operator has two distinct eigenvalues on V and hence V splits as a direct sum of generalized eigenvectors of C and this decomposition is $\mathfrak{sl}(2)$-invariant. Thus, we can assume that $k = l$.

Consider now the decomposition of $V = \oplus V(i)$ with respect to generalized eigenspaces of the operator H. Since V is glued from two copies of representation A_l, it is clear that $\dim V(i) = 2$ if $i = -l, -l + 2, ..., l$ and there are no other summands. Also, it is clear that $E_-^l : V(l) \to V(-l)$ is an isomorphism.

Let us show that the action of H on the space $V(l)$ is given by a scalar operator. Indeed consider the identity $E_+ E_-^{l+1} - E_-^{l+1} E_+ = E_-^l (H - l)$. The left-hand side is 0 on the space $V(l)$ so the right-hand side is 0. Since the operator E_-^l does not have kernel on $V(l)$, we conclude that $H = l$ on $V(l)$.

Now let us choose a vector $v \in V(l)$ that does not lie in the submodule S. Then it is a highest weight vector and by the Key Lemma it generates a submodule $Q' \subset V$ isomorphic to A_l. It is clear that this submodule isomorphically maps to the quotient module $Q = V/S$, i.e. $V \simeq S \oplus Q'$.   □

Proof of lemma. We proceed by induction on the length of the object V. Find a simple submodule $S \subset V$ and consider the quotient module $Q = V/S$. By the induction assumption, we can write the quotient module $Q = V/S$ as a direct sum of simple objects $Q = \oplus W_i$. It is enough to show that the natural projection $p : V \to Q$ has a section $\nu : Q \to V$. We construct this section ν separately on every summand W_i. Namely, consider the module $V_i = p^{-1}(W_i)$. This module has length two and by assumption is completely reducible. Hence, the projection $p_i : V_i \to W_i$ has a section $\nu_i : W_i \to V_i \subset V$.

Corollary. *Let V be a finite dimensional $\mathfrak{sl}(2)$-module. Then*

(1) H is diagonalizable and each of the operators E_-^i and E_+^i gives an isomorphism between $V(i)$ and $V(-i)$.

(2) The action of $\mathfrak{sl}(2)$ uniquely extends to the action ρ of the group $SL(2, \mathbb{K})$ on V that satisfies the following condition: Let X equal E_+ or E_-, $t \in \mathbb{K}$ and $g = exp(tX) \in SL(2, \mathbb{K})$. Then the operator $\rho(g)$ in V equals $exp(tX)$.

Remark. The same conclusion holds under the weaker assumption that the module V is $\mathfrak{sl}(2)$-finite. This is left as an exercise to the reader.

3 A Crash Course on Semi-Simple Lie Algebras

3.1 Killing Form

Any Lie algebra $\mathfrak{g}$ admits a unique maximal solvable ideal called the radical $Rad(\mathfrak{g})$. The Lie algebra $\mathfrak{g}$ is called *semisimple* iff its *radical* is zero.

For finite dimensional Lie algebras over a field $\mathbb{K}$ of characteristic 0, there is an equivalent definition, often more convenient. It is given in terms of the *Killing* form, which is the symmetric bilinear form on $\mathfrak{g}$ defined by

$$(X, Y) = \mathrm{tr}(\mathrm{ad}\ X \cdot \mathrm{ad}\ Y).$$

Theorem (Cartan–Killing). $\mathfrak{g}$ is semisimple iff its Killing form is nondegenerate.

3.2 Cartan Subalgebra

There exists a maximal commutative subalgebra $\mathfrak{h} \subset \mathfrak{g}$ such that the adjoint action of $\mathfrak{h}$ on $\mathfrak{g}$ is semisimple.

Such subalgebra is called a *Cartan* subalgebra of $\mathfrak{g}$. In what follows we will fix a Cartan subalgebra $\mathfrak{h}$. One can show that any two Cartan subalgebras are conjugate, so we do not lose information fixing one of them. The number $r = \dim \mathfrak{h}$ is called the *rank* of $\mathfrak{g}$.

3.3 Root System

Consider the adjoint action of the Cartan subalgebra $\mathfrak{h}$ on $\mathfrak{g}$. We obtain a decomposition $\mathfrak{g} = \oplus \mathfrak{g}_\chi$, where for $\chi \in \mathfrak{h}^*$ we have

$$\mathfrak{g}_\chi = \{X \in \mathfrak{g} : [H, X] = \chi(H)X\}.$$

This is called the weight decomposition of $\mathfrak{g}$. Since Killing form is $\mathfrak{h}$-invariant, we see that $(\mathfrak{g}_\chi, \mathfrak{g}_\nu) = 0$ unless $\chi + \nu = 0$. Since this form is nondegenerate, it gives a nondegenerate pairing between $\mathfrak{g}_\chi$ and $\mathfrak{g}_{-\chi}$. In particular, the restriction of the Killing form to $\mathfrak{g}_0$ is nondegenerate.

Proposition. *(1)* $\mathfrak{g}_0 = \mathfrak{h}$
(2) For $\chi \neq 0$, we have $\dim_\mathbb{K}(\mathfrak{g}_\chi) \leq 1$.

Let

$$R = \{\chi \in \mathfrak{h}^* - \{0\} : \mathfrak{g}_\chi \neq \{0\}\}$$

Then $R \subset \mathfrak{h}^*$ is a finite subset of nonzero elements of the dual space $\mathfrak{h}^*$.
Elements of R are called *roots*.

For every $\gamma \in R$, we fix a nonzero element $E_\gamma \in \mathfrak{g}$. It is called a *root vector*.
We will see later that if $\gamma \in R$, then $-\gamma \in R$ and $\lambda\gamma \notin R$ for $\lambda \neq \pm 1$.

3.4 $\mathfrak{sl}(2)$-*Triples*

Proposition. *We can choose root vectors E_γ for all roots $\gamma \in R$ in such a way that for every root γ the triple of elements $E_\gamma \in \mathfrak{g}_\gamma$, $H_\gamma := [E_\gamma, E_{-\gamma}] \in \mathfrak{h}$ and $E_{-\gamma} \in \mathfrak{g}_{-\gamma}$ form an $\mathfrak{sl}(2)$-triple.*

Essentially, this means that we can find an element $H_\gamma \in [\mathfrak{g}_\gamma, \mathfrak{g}_{-\gamma}] \subset \mathfrak{h}$ such that $\gamma(H_\gamma) = 2$.

The vector $H_\gamma \in \mathfrak{h}$ is called a *coroot* corresponding to the root $\gamma \in \mathfrak{h}^*$.

Corollary. *Let γ, δ be roots. If $\delta + \gamma \notin R$, then $[E_\gamma, E_\delta] = 0$. If $\delta + \gamma \in R$, then $[E_\gamma, E_\delta] = CE_{\gamma+\delta}$, where $C \neq 0$.*

3.5 *Integral Structure: Weight Lattice and Root Lattice*

From properties of $\mathfrak{sl}(2)$-representations, we see that all eigenvalues of the operator H_γ are integers. In particular for any root δ we have $\delta(H_\gamma) \in \mathbb{Z}$.

Let $\check{Q}$ denote the subgroup of $\mathfrak{h}$ generated by all coroots H_γ (it is called a *coroot lattice*).

For elements $H \in \check{Q}$, we have $(H, H) = \sum \delta(H)^2 \geq 0$, i.e. the Killing form is positive on $\check{Q}$. In fact, it is strictly positive since for any vector H in its kernel we have $\delta(H) = 0$ for all $\delta \in R$ and hence H acts trivially in the adjoint representation. The same reason shows that $\check{Q}$ is a lattice in $\mathbb{K}$-vector space $\mathfrak{h}$.

Let us denote by P the lattice in $\mathfrak{h}^*$ dual to the lattice $\check{Q}$ (it is usually called the *weight lattice*; the elements of P are called *integral weights*). It contains a sublattice Q generated by all roots (it is called *root lattice*).

Since the restriction of the Killing form to $\mathfrak{h}$ is nondegenerate, it induces a bilinear form $\langle \cdot, \cdot \rangle$ on $\mathfrak{h}^*$.

One can describe the coroot $H_\gamma \in \mathfrak{h}$, with $\gamma \in R$ by the property

$$\chi(H_\gamma) = \frac{2\langle \chi, \gamma \rangle}{\langle \gamma, \gamma \rangle} \text{ for any } \chi \in \mathfrak{h}^*.$$

3.6   The Weyl Group of the Lie Algebra $\mathfrak{g}$

We will consider the $\mathbb{R}$ vector space $\mathfrak{a} = \mathbb{R} \otimes \check{Q}$ equipped with Euclidean structure defined by positive definite Killing form on it. It is convenient to use convex geometry of this space to state and prove many results about roots and weights.

For any root $\gamma \in R$, consider the linear transformation in the space $\mathfrak{h}^*$ defined by the formula

$$\sigma_\gamma(\chi) = \chi - \chi(H_\gamma)\gamma.$$

The transformation σ_γ is the reflection in the hyperplane defined by the equation $\langle \chi, \gamma \rangle = 0$. In particular $\sigma_\gamma^2 = \mathrm{Id}$ and $\det(\sigma_\gamma) = -1$. The corresponding reflection on the space $\mathfrak{h}$ is given by the formula $\sigma_\gamma(H) = H - \gamma(H)H_\gamma$.

The group of linear transformations of $\mathfrak{h}^*$ generated by operators σ_γ, where $\gamma \in R$, is called the *Weyl group* of $\mathfrak{g}$ and will be denoted by W.

The group W is a group of orthogonal transformations of the space $\mathfrak{h}^*$. It naturally acts on the space $\mathfrak{h}$. The action of W preserves the Killing form, the set of roots R, the set of coroots, lattices P, Q, and $\check{Q}$. Since the Killing form on the lattice P is positive definite, the Weyl group W is finite.

If $\chi_1, \chi_2 \in \mathfrak{h}^*$, then we write $\chi_1 \sim \chi_2$ whenever χ_1 and χ_2 belong to the same orbit of the Weyl group, i.e., when $\chi_1 = w\chi_2$ for a certain $w \in W$.

We also consider the induced actions of W on the Euclidean space $\mathfrak{a}$ and on its dual. In this realization, the Weyl group is a finite group generated by reflections and we can use many geometric facts about actions of such groups.

3.7   Weyl Chamber

For every root $\gamma \in R$ consider the hyperplane Π_γ in the space $\mathfrak{a}^*$ orthogonal to γ, i.e. the set of weights that vanish on H_γ. Consider in $\mathfrak{a}^*$ an open subset $\mathfrak{a}^* \backslash \bigcup_{\gamma \in R} \Pi_\gamma$ obtained by removing all root hyperplanes and fix a connected component C of this set. We denote by $\overline{C}$ the closure of C in $\mathfrak{a}$. The set $\overline{C}$ is called the *Weyl chamber*.

The choice of this set plays central role in the theory. We will see that all Weyl chambers are conjugate under the action of W.

We have the following

Proposition. *$\overline{C}$ is a fundamental domain for the W-action on $\mathfrak{a}$. More precisely:*

(1) If $\chi \in \mathfrak{a}$, then $w\chi \in \overline{C}$ for a certain $w \in W$.
(2) If $\chi, w\chi \in \overline{C}$, then $\chi = w\chi$. If, moreover, $\chi \in C$, then $w = e$.

3.8 Positive Roots and Simple Roots

In what follows we fix a Weyl chamber C. A root γ is called *positive* if the coroot H_γ is positive on C, i.e. if $(\chi, \gamma) > 0$ for all $\chi \in C$.

We denote by R^+ the subset of positive roots. It is clear that R is a disjoint union of sets R^+ and $R^- = -R^+$. Also R^+ is closed under addition, i.e. if γ, δ are positive roots and their sum is a root then this root is positive.

A positive root α is called a *simple root* if it cannot be written as a sum of two positive roots. We denote by $B \subset R^+$ the subset of simple roots .

Proposition. *(1) B is a base of the root lattice Q. Every positive root γ is a sum of simple roots with nonnegative integer coefficients.*

(2) Simple roots correspond to hyperplanes in $\mathfrak{a}^*$ that are walls of the Weyl chamber $\bar{C}$.

(3) The Weyl group W is generated by reflections σ_α corresponding to simple roots (they are called simple reflections).

(4) Let α be a simple root. Then for any positive root γ different from α the root $\sigma_\alpha(\gamma)$ is positive. In particular, if β is a simple root different from α, then $(\alpha, \beta) \le 0$.

(5) Let $\rho \in \mathfrak{h}^*$ be half of the sum of all positive roots. Then for any simple root α we have $\rho(H_\alpha) = 1$ and $\sigma_\alpha(\rho) = \rho - \alpha$. In particular, ρ lies in the lattice P.

Let us denote by Q^+ the subsemigroup of the root lattice Q generated by positive roots. In other words, Q^+ is a free semigroup generated by the set B.

Using this semigroup, we introduce a partial order $\prec$ on the space $\mathfrak{h}^*$ by $\chi \prec \psi$ if $\psi = \chi + q$ with $q \in Q^+$.

Note that a weight χ lies in P iff $\chi(H_\alpha) \in \mathbb{Z}$ for every simple root α. A weight χ is called *dominant* if $\chi(H_\alpha) \in \mathbb{Z}_{\ge 0}$ for every simple root α. Equivalent condition: $\sigma_\alpha(\chi) \prec \chi$.

We denote the semigroup of dominant weights by P^+. Note that the cone generated by P^+ in $\mathfrak{a}^*$ is usually much smaller than the cone generated by Q^+.

3.9 The Triangular Decomposition of a Lie Algebra $\mathfrak{g}$

From this description of the root system R, we derive the following decomposition :

$$\mathfrak{g} = \mathfrak{n}_- \oplus \mathfrak{h} \oplus \mathfrak{n}_+,$$

where $\mathfrak{n}_-$ and $\mathfrak{n}_+$ are subspaces generated by E_γ for $\gamma \in R^-$ and $\gamma \in R^+$, respectively. This is a decomposition of linear spaces (not of Lie algebras). We have

Lemma. *(i) $\mathfrak{n}_+$ (resp. $\mathfrak{n}_-$) is the Lie subalgebra of $\mathfrak{g}$ generated by E_α (resp. by $E_{-\alpha}$), where $\alpha \in B$.*

(ii) $[\mathfrak{h}, \mathfrak{n}_+] = \mathfrak{n}_+$ and $[\mathfrak{h}, \mathfrak{n}_-] = \mathfrak{n}_-$.

(iii) *The Lie algebras* $\mathfrak{n}_+$ *and* $\mathfrak{n}_-$ *are nilpotent. Moreover, if* $X \in \mathfrak{n}_+$ *or* $\mathfrak{n}_-$, *then* $\operatorname{ad} X$ *is a nilpotent operator on* $\mathfrak{g}$.

(iv) $U(\mathfrak{g}) \simeq U(\mathfrak{n}_-) \otimes U(\mathfrak{h}) \otimes U(\mathfrak{n}_+) \simeq U(\mathfrak{n}_-) \otimes U(\mathfrak{n}_+) \otimes U(\mathfrak{h})$.

4 Category $\mathcal{O}$ and Verma Modules M_χ

The aim of these lectures is the description of finite dimensional $\mathfrak{g}$-modules. In the sixties, it was noted that it is more natural to describe the finite dimensional modules in the framework of a wider class of $\mathfrak{g}$-modules. First, let us give several preparatory definitions.

4.1 Weight Spaces

Let V be a $\mathfrak{g}$-module. For any $\chi \in \mathfrak{h}^*$ denote by $V^{ss}(\chi)$ the space of vectors $v \in V$ such that $Hv = \chi(H)v$ for any $H \in \mathfrak{h}$ and call it the *weight space* of weight χ. If $V^{ss}(\chi) \neq 0$, then χ is called a *weight* of the $\mathfrak{g}$-module V and any $v \in V^{ss}(\chi)$ is called a *weight vector*. A module V is called $\mathfrak{h}$-*diagonalizable* if $V = \sum_{\chi \in \mathfrak{h}^*} V^{ss}(\chi)$.

Similarly, we introduce a generalized weight space $V(\chi)$ as the space of vectors $v \in V$ such that for any $H \in \mathfrak{h}$ one has $(H - \chi(H))^n = 0$ for large n. If V is $\mathfrak{h}$-finite, it has decomposition $V = \oplus V(\chi)$ (see appendix A). We denote by $P(V)$ the set of weights $\chi \in \mathfrak{h}^*$ such that $V(\chi) \neq 0$ (the weight support of V).

Lemma. *Let* V *be a* $\mathfrak{g}$-*module. For any* $\gamma \in R$, $\chi \in \mathfrak{h}^*$ *we have* $E_\gamma V^{ss}(\chi) \subset V^{ss}(\chi + \gamma)$ *and* $E_\gamma V(\chi) \subset V(\chi + \gamma)$

The proof is the same as in $\mathfrak{sl}(2)$ case.

4.2 The Category $\mathcal{O}$

Let us now introduce a class of $\mathfrak{g}$-modules that we will consider. The objects of *category* $\mathcal{O}$ are $\mathfrak{g}$-modules M satisfying the following conditions.

(1) M is a finitely generated $U(\mathfrak{g})$-module.
(2) M is $\mathfrak{h}$-diagonalizable.
(3) M is $\mathfrak{n}_+$-finite.

Clearly, if a $\mathfrak{g}$-module M belongs to $\mathcal{O}$, then so does any submodule of M and any quotient module of M, and if $M_1, M_2 \in \mathcal{O}$, then $M_1 \oplus M_2 \in \mathcal{O}$.

Lemma. *Let* $\mathfrak{g}$ *be a semisimple Lie algebra. Then any finite dimensional* $\mathfrak{g}$-*module* V *lies in* $\mathcal{O}$.

Proof. It suffices to verify that V is $\mathfrak{h}$-diagonalizable. Since the operators H_γ, where $\gamma \in R$, generate $\mathfrak{h}$ and commute, it suffices to verify that V is H_γ-diagonalizable.

Now V is a finite dimensional $\mathfrak{s}_\gamma$-module, with $\mathfrak{s}_\gamma \subset \mathfrak{g}$ generated by E_γ, H_γ and $E_{-\gamma}$. Since $\mathfrak{s}_\gamma$ is isomorphic to $\mathfrak{sl}(2)$, the result follows from Corollary 2.5.   □

4.3 Highest Weight

A nonzero weight vector $m \in M$ is called *a highest weight vector* if $\mathfrak{n}_+ m = 0$.

Since $\mathfrak{n}_+$ is generated by E_α for $\alpha \in B$, we have

Lemma. *A weight vector m is a highest weight vector if and only if $E_\alpha m = 0$ for every $\alpha \in B$.*

Proposition. *Let $M \in \mathcal{O}$ be nonzero. Then M contains a nonzero highest weight vector.*

Proof. The proof is the same as in the case of $\mathfrak{sl}(2)$. We choose an $\mathfrak{h}$-invariant finite dimensional vector subspace $V \subset M$ that generates M. Replacing it by $U(\mathfrak{n}_+)V$ we can assume that it is also $\mathfrak{n}_+$-invariant. Consider all weights χ of $\mathfrak{h}$ in V. Since this is a finite set, there exists a weight χ in V such that for every positive root γ the weight $\chi + \gamma$ is not a weight in V. Any nonzero vector $v \in V(\chi)$ is a highest weight vector.   □

4.4 Verma Modules

We now introduce a family of central objects in the category $\mathcal{O}$. These are the *Verma modules M_χ*.

Lemma. *Let $\chi \in \mathfrak{h}^*$. There exists a pair (M_χ, m_χ) of a $\mathfrak{g}$-module and a highest weight vector $m_\chi \in M_\chi(\chi - \rho)$ that satisfies the following universality condition.*

For any $\mathfrak{g}$-module M and highest weight vector $v \in M$ of weight $\chi - \rho$, there exists a unique morphism of $\mathfrak{g}$-modules $i_v : M_\chi \to M$ with $i(m_\chi) = v$.

Remark. By abstract nonsense, if such a module exists it is unique up to a canonical isomorphism.

Proof. Let $\chi \in \mathfrak{h}^*$. In $U(\mathfrak{g})$, consider the left ideal I_χ generated by E_γ, where $\gamma \in R^+$, and by $H + \rho(H) - \chi(H)$, where $H \in \mathfrak{h}$. Define the $\mathfrak{g}$-module M_χ setting $M_\chi = U(\mathfrak{g})/I_\chi$. Let m_χ stand for the natural generator of M_χ (over $\mathfrak{g}$), i.e., the image of $1 \in U(\mathfrak{g})$ under the mapping $U(\mathfrak{g}) \longrightarrow M_\chi$. The module M_χ and the vector m_χ clearly satisfy the universal property.   □

Since Verma module is generated by a highest weight vector, the results of Sect. 1.6 imply that it lies in category $\mathcal{O}$.

Lemma. *Let $\chi \in \mathfrak{h}^*$. Then M_χ is a free $U(\mathfrak{n}_-)$ – module with one generator m_χ.*

Proof. The statement follows from the decomposition $U(\mathfrak{g}) = U(\mathfrak{n}_-) \otimes U(\mathfrak{h}) \otimes U(\mathfrak{n}_+)$. □

Corollary. *(1) The set of weights $P(M_\chi)$ of the module M_χ equals to $(\chi-\rho)-Q^+$, i.e. weights of M_χ are of the form $\chi - \rho - q$ for $q \in Q^+$.*
(2) Let M be an arbitrary $\mathfrak{g}$-module, $m \in M$ a highest weight vector of weight $\chi - \rho$ and $i_m : M_\chi \longrightarrow M$ be the corresponding unique morphism. Then i_m is an embedding if and only if $Xm \neq 0$ for any nonzero $X \in U(\mathfrak{n}_-)$.

4.5   The Irreducible Objects L_χ

The next lemma provides a precise parametrization of isomorphism classes of irreducible objects in the category $\mathcal{O}$ in terms of characters of $\mathfrak{h}$.

Lemma. *(1) Let $\chi \in \mathfrak{h}^*$. Then Verma module M_χ has a unique irreducible quotient L_χ and $\mathrm{Hom}(M_\chi, L_\chi) \approx \mathbb{K}$.*
(2) Any irreducible module $L \in \mathcal{O}$ is isomorphic to a module L_χ for a unique weight $\chi \in \mathfrak{h}^*$.

In other words, up to isomorphism L_χ is the unique irreducible $\mathfrak{g}$-module that has highest weight vector of weight $\chi - \rho$. Modules L_χ for different χ are not isomorphic and every irreducible object L in category $\mathcal{O}$ is isomorphic to one of the modules L_χ.

Proof. (1) Consider the weight decomposition $M = M^{\mathrm{top}} \oplus M'$ where M^{top} is the one-dimensional space $M(\chi - \rho)$ and $M' = \oplus M(\mu)$ with sum over $\mu \lneq \chi - \rho$. Any $\mathfrak{g}$-submodule $N \subset M_\chi$ splits with respect to this decomposition, i.e $N = N \bigcap M^{\mathrm{top}} \oplus N \bigcap M'$. Since any non-zero vector of the space M^{top} generates the module M_χ, we see that any proper $\mathfrak{g}$-submodule of M_χ is contained in M'. Thus, the sum of all proper submodules is contained in M'. This shows that M_χ has a unique maximal proper submodule and hence it has unique simple quotient.
(2) Lemma 4.3. implies that every simple module L in $\mathcal{O}$ has a highest weight vector. Using 4.4 we construct a non-zero morphism $M_\chi \to L$ and this implies that L is isomorphic to the module L_χ.
  Note that the set of weights $P(L_\chi)+\rho$ has χ as the unique maximal element. This shows how to reconstruct the weight χ from the simple module L. □

Remark. An alternative argument that yields the uniqueness of an irreducible module with highest weight $\chi - \rho$ is as follows. Let M_1, M_2 be two irreducible modules of highest weight $\chi - \rho$ and m_1, m_2 be their highest weight vectors. Then $N = U(\mathfrak{n}_-)(m_1 \oplus m_2) \subset M_1 \oplus M_2$ is a $U(\mathfrak{g})$-submodule. Since both projections $N \to M_1$ and $N \to M_2$ are non zero we see that $M_1 \approx M_2$.

4.6   Characters

In the study of modules from the category $\mathcal{O}$ we will use the notion of the character of a $\mathfrak{g}$-module M.

More generally, let M be a $\mathfrak{g}$-module such that it is $\mathfrak{h}$-finite and in the weight decomposition all the weight spaces $M(\chi)$ are finite-dimensional. In this case, we define the *character* π_M to be the function on $\mathfrak{h}^*$ defined by the equation

$$\pi_M(\chi) = \dim M(\chi).$$

On $\mathfrak{h}^*$, define the *Kostant function* K by the equality

$$K(\chi) = \text{ the number of presentations of the weight } \chi \text{ in the form}$$

$$\chi = -\sum_{\gamma \in R^+} n_\gamma \gamma, \text{ where } n_\gamma \in \mathbb{Z}_{\geq 0}.$$

For any function u on $\mathfrak{h}^*$ set supp $u = \{\chi \in \mathfrak{h}^* \mid u(\chi) \neq 0\}$. Denote by $\mathcal{E}$ the space of $\mathbb{Z}$-valued functions u on $\mathfrak{h}^*$ such that supp u is contained in the union of a finite number of sets of the form $\psi - Q^+$, where $\psi \in \mathfrak{h}^*$. For example, supp $K = -Q^+$, hence, $K \in \mathcal{E}$.

Lemma.   (i) $\pi_{M_\chi}(\psi) = K(\psi - \chi + \rho)$.
(ii) If $M \in \mathcal{O}$, then π_M is defined and $\pi_M \in \mathcal{E}$.

Proof. (1) Let us enumerate the elements of R^+, e.g., $\gamma_1, \ldots, \gamma_s$. Then the elements $E_{-\gamma_1}^{n_1} \ldots E_{-\gamma_s}^{n_s} m_\chi$, where $n_1, \ldots, n_s \in \mathbb{Z}_{\geq 0}$, form a basis in M_χ. Hence, $\pi_{M_\chi}(\psi) = K(\psi - \chi + \rho)$.
(2) Choose a finite-dimensional $\mathfrak{h}$-invariant subspace $V \subset M$ that generates M. Replacing V by $U(\mathfrak{n}_+)V$ we can assume that V is also $\mathfrak{n}_+$-invariant. This implies that $M = U(\mathfrak{n}_-)V$. Thus we can write $M = \sum U(\mathfrak{n}_-)(v_i)$, where v_i is a basis of V consisting of weight vectors.

As in heading (i) we have $\dim M(\psi) \leq \sum_{1 \leq i \leq k} K(\psi - \chi_i + \rho)$ implying lemma. $\square$

Exercise. Prove the converse statement: Let M is a finitely generated $U(\mathfrak{g})$-module such that it is $\mathfrak{h}$-diagonalizable, its character π_M is defined and lies in $\mathcal{E}$. Then $M \in \mathcal{O}$.

5   The Weyl Modules A_λ, $\lambda \in P^+$

In this section we construct for every $\lambda \in P^+$ a finite dimensional $\mathfrak{g}$-module A_λ of highest weight λ. Later we will show that $A_\lambda \cong L_{\lambda+\rho}$ and that these modules exhaust all irreducible finite dimensional $\mathfrak{g}$-modules.

5.1 Injections Between Verma Modules

We begin with the following key Proposition:

Proposition. *Let M be a $\mathfrak{g}$-module and $m \in M$ a highest weight vector of weight $\chi - \rho$. Suppose that $k = \chi(H_\alpha) \in \mathbb{Z}_{\geq 0}$. Then the vector $m' = E_{-\alpha}^k m$ is either zero or a highest weight vector of weight $\sigma_\alpha(\chi) - \rho$.*

Proof. Clearly, the weight of the vector $m' = E_{-\alpha}^k m_\chi$ is equal to $\chi - \rho - k\alpha = \sigma_\alpha(\chi) - \rho$.

By Lemma 4.3. it suffices to show that $E_\beta m' = 0$ for $\beta \in B$. If $\beta \neq \alpha$, then

$$E_\beta m' = E_\beta E_{-\alpha}^k m_\chi = E_{-\alpha}^k E_\beta m_\chi = 0,$$

because $[E_\beta, E_{-\alpha}] = 0$. Further,

$$E_\alpha m' = E_\alpha E_{-\alpha}^k m_\chi = E_{-\alpha}^k E_\alpha m_\chi + k E_{-\alpha}^{k-1}(H_\alpha - (k-1))m_\chi = 0,$$

since $H_\alpha m_\chi = (\chi - \rho)(H_\alpha)m_\chi = (k-1)m_\chi$.   □

Remark. This last point is just a repetition of $\mathfrak{sl}(2)$ computation in 2.2.

Corollary. *Suppose $\chi \in \mathfrak{h}^*$ and $\alpha \in B$ are such that $\sigma_\alpha(\chi) < \chi$.*
Then there is a canonical embedding $M_{\sigma_\alpha(\chi)} \longrightarrow M_\chi$ that maps $m_{\sigma_\alpha(\chi)}$ to $E_{-\alpha}^k m_\chi$, where $k = \chi(H_\alpha)$

5.2 π_M is σ_α-Invariant

Lemma. *Let $\alpha \in B$ be a simple root and let $\mathfrak{s}_\alpha \subset \mathfrak{g}$ be the corresponding $\mathfrak{sl}(2)$-subalgebra. Let $M \in \mathcal{O}$ be a $\mathfrak{s}_\alpha$-finite module. Then the character π_M is σ_α invariant.*

Proof. Consider the decomposition $M = \oplus M(k)$ with respect to the action of $H_\alpha \in \mathfrak{g}_\alpha$. By $\mathfrak{sl}(2)$ theory, we have $E_{-\alpha}^k : M(k) \longrightarrow M(-k)$ is an isomorphism for any $k \geq 0$. Decomposing $M(k) = \oplus M(\chi)$, where $\chi \in \mathfrak{h}^*$ with $\chi(H_\alpha) = k$ it is clear that $E_{-\alpha}^k$ induces an isomorphism between $M(\chi)$ and $M(\sigma_\alpha(\chi))$.   □

5.3   Construction of the Weyl Modules

For any $\lambda \in P^+ = P \cap \overline{C}$ we have $\sigma_\alpha(\lambda + \rho) \lneq \lambda + \rho$ and hence by Corollary 5.1. we have the containment $M_{\sigma_\alpha(\lambda+\rho)} \subsetneq M_{\lambda+\rho}$

We now set

$$A_\lambda = M_{\lambda+\rho} / \sum_{\alpha \in B} M_{\sigma_\alpha(\lambda+\rho)}.$$

Theorem. *(1)* $\pi_{A_\lambda}(\lambda) = 1$.
(2) $P(A_\lambda) \subset \lambda - Q^+ \subset P$ and $\pi_{A_\lambda}(wv) = \pi_{A_\lambda}(v)$ for any $w \in W$ and $v \in P$.
(3) If v is a weight of A_λ, then either $v \sim \lambda$ or $|v| < |\lambda|$, where $|v|$ is the length of the weight v.
(4) $\dim A_\lambda < \infty$.

Proof. (1) The modules $M_{\sigma_\alpha(\lambda+\rho)}$ do not contain vectors of weight λ; hence, these modules are contained in $\sum_{\psi \in \mathfrak{h}^* \setminus \{\lambda\}} M_{\lambda+\rho}(\psi)$. Therefore, $\dim A_\lambda(\lambda) = \dim M_{\lambda+\rho}(\lambda) = 1$.
(2) Since W is generated by σ_α, where $\alpha \in B$, it is enough to verify heading (2) for these elements. Fix $\alpha \in B$. Since A_λ is generated by $\mathfrak{s}_\alpha$-finite vector, Lemma 1.6.1. implies that A_λ is $\mathfrak{s}_\alpha$-finite. The result now follows from Sect. 5.2.
(3) It is clear that

$$\operatorname{supp} \pi_{A_\lambda} \subset \operatorname{supp} \pi_{M_{\lambda+\rho}} = \lambda - Q^+.$$

Let $\pi_{A_\lambda}(v) \neq 0$. By replacing v with a W-equivalent element we can assume that $v \in \bar{C}$. Hence, $\lambda = v + q$, where $q \in Q^+$. Further on

$$|\lambda|^2 = |v|^2 + |q|^2 + 2(v,q) \geq |v|^2 + |q|^2.$$

Hence, either $|\lambda| > |v|$ or $q = 0$ and then $\lambda = v$.
(4) $\operatorname{supp} \pi_{A_\lambda}$ is contained in the intersection of the lattice P with the ball $|v| \leq |\lambda|$, and, therefore, is finite. Hence, $\dim A_\lambda < \infty$. $\square$

We can now deduce a few results concerning the modules L_χ that are finite dimensional.

Corollary. *An irreducible module L_χ is finite dimensional if and only if $\chi - \rho \in P^+$.*

What is missing is the irreducibility of the modules A_λ as this identifies them with $L_{\lambda+\rho}$. This will be proven in Sect. 8.

6 Statement of Harish–Chandra's Theorem on $\mathfrak{Z}(\mathfrak{g})$

The center of the associative algebra $U(\mathfrak{g})$ plays an important role in the study of representations of $\mathfrak{g}$. It is common to denote this commutative algebra by $\mathfrak{Z}(\mathfrak{g})$. In this section I formulate the Harish–Chandra theorem that describes the algebra $\mathfrak{Z}(\mathfrak{g})$. The description of the Harish–Chandra homomorphism is very simple when we consider the action of $\mathfrak{Z}(\mathfrak{g})$ on Verma modules. Indeed, it is easy to see that any element $z \in \mathfrak{Z}(\mathfrak{g})$ acts by a scalar on each of the modules M_χ. Thus we obtain, for each $z \in \mathfrak{z}(\mathfrak{g})$ a complex valued function on $\mathfrak{h}^*$. We show below that this is a polynomial function on $\mathfrak{h}^*$ that is invariant with respect to the Weyl group. The complete proof of Harish–Chandra's theorem is carried out in Sect. 9.

6.1 The Harish–Chandra Projection

In what follows we will identify the algebra $U(\mathfrak{h}) = S(\mathfrak{h})$ with the algebra $Pol(\mathfrak{h}^*)$ of polynomial functions on the space $\mathfrak{h}^*$.

For any element $X \in U(\mathfrak{g})$ we will construct a function $j(X)$ on the space $\mathfrak{h}^*$ as follows. Given a weight $\chi \in \mathfrak{h}^*$ consider the Verma module M_χ, its highest weight vector $m = m_\chi$ of weight $\chi - \rho$ and a functional $f = f_\chi$ on M_χ such that $f(m) = 1$ and f vanishes on the complementary subspace $M' = \oplus_{\psi \neq \chi - \rho} M_\chi(\psi)$.

It is clear that such functional f exists and is uniquely defined. Now we define
$$j(X)(\chi) := f_\chi(Xm_\chi).$$

Lemma. *For any $X \in U(\mathfrak{g})$ the function $j(X)$ is a polynomial function in χ.*

Proof. Using triangular decomposition we can write $X = X_0 + X_+ + X_-$, where $X_0 \in U(\mathfrak{h})$, $X_+ \in U(\mathfrak{g})\mathfrak{n}_+$ and $X_- \in \mathfrak{n}_- U(\mathfrak{g})$. This implies that $j(X)(\chi) = j(X_0)(\chi) = X_0(\chi - \rho)$ and this is a polynomial function in χ. □

This proof shows that up to a ρ shift the function $j(X)$ coincides with the "central" part X_0 of the element $X \in U(\mathfrak{g})$; this part is often called *the Harish–Chandra projection.*

6.2 The Harish–Chandra's Homomorphism

Lemma. *(1) For any $z \in \mathfrak{z}(\mathfrak{g})$ the operator z on the Verma module M_χ is a scalar operator $j(z)(\chi) \cdot Id_{M_\chi}$.*
(2) $j : \mathfrak{z}(\mathfrak{g}) \to Pol(\mathfrak{h}^*)$ is a morphism of algebras (it is called Harish–Chandra homomorphism).
(3) For any $z \in \mathfrak{z}(\mathfrak{g})$ the function $j(z) \in Pol(\mathfrak{h}^*)$ is W-invariant.

Proof. (1) Since z commutes with action of $\mathfrak{h}$ we see that $zm_\chi \in M_\chi(\chi - \rho)$ and hence $zm_\chi = cm_\chi$. Since vector m_χ generates M_χ we see that $z = c \cdot \mathrm{Id}$. It is clear that $c = j(z)(\chi)$.
(2) immediately follows from 1.
(3) We would like to show that for any $w \in W$ we have $j(z)(w\chi) = j(z)(\chi)$. It suffices to consider the case when $w = \sigma_\alpha$ for $\alpha \in B$.

Since $j(z)(\chi)$ and $j(z)(\sigma_\alpha(\chi))$ are polynomial functions in χ, it suffices to prove the equality for $\chi \in P^+$. But in this case $M_{\sigma_\alpha(\chi)} \subset M_\chi$, and that implies that the action of z on the Verma modules M_χ and $M_{\sigma_\alpha(\chi)}$ is given by the same scalar. □

6.3 The Harish–Chandra Theorem

By the previous lemmas, the correspondence $z \mapsto j_z$ defines a ring homomorphism $j : \mathfrak{Z}(\mathfrak{g}) \longrightarrow Pol(\mathfrak{h}^*)^W$. We can now state the following important result of Harish–Chandra.

Theorem. *The Harish–Chandra morphism* $j : \mathfrak{Z}(\mathfrak{g}) \longrightarrow Pol(\mathfrak{h}^*)^W$ *is an isomorphism of algebras.*

Remark. In [Di], the map j is described as a composition of the so-called Harish–Chandra projection with a shift. It is easy to trace both in our construction.

Remark. Our construction of the Harish–Chandra map appears to depend on a choice of ordering on the root system.

A different choice of ordering yields the same map, although this statement requires a proof.

7 Corollaries of the Harish–Chandra Theorem

7.1 Description of Infinitesimal Characters

Denote by $\Theta = Spec(\mathfrak{Z}(\mathfrak{g}))$ the set of all homomorphisms $\theta : \mathfrak{Z}(\mathfrak{g}) \to \mathbb{K}$ – such morphisms are usually called *infinitesimal characters.* The Harish–Chandra morphism $j : \mathfrak{Z}(\mathfrak{g}) \to Pol(\mathfrak{h}^*)$ defines a map of sets $\sigma : \mathfrak{h}^* \longrightarrow \Theta$. We usually denote the image $\sigma(\chi)$ by θ_χ.

One of the important corollaries of the Harish–Chandra theorem is the following.

Proposition. *The map* σ *gives a bijection* $\sigma : \mathfrak{h}^*/W \simeq \Theta$.

We have seen that $\sigma(w\chi) = \sigma(\chi)$ so σ defines a map of sets $\sigma : \mathfrak{h}^*/W \to \Theta$.

First let us show that this map is an imbedding.

Lemma. $\theta_{\chi_1} = \theta_{\chi_2}$ *only if* $\chi_1 \sim \chi_2$.

Proof. Let $\chi_1 \not\sim \chi_2$. Let us construct a polynomial $T \in Pol(\mathfrak{h}^*)^W$ such that $T(\chi_1) = 0$, while $T(\chi_2) \neq 0$. For this, take a polynomial $T' \in Pol(\mathfrak{h}^*)$ such that $T'(w\chi_1) = 0$ and $T'(w\chi_2) = 1$ for any $w \in W$ and set $T(\chi) = \sum_{w \in W} T'(w\chi)$.

As follows from the Harish–Chandra theorem, there is an element $z \in \mathfrak{Z}(\mathfrak{g})$ such that $j_z = T$. But then

$$j_z(\chi_1) = \theta_{\chi_1}(z) \neq \theta_{\chi_2}(z) = j_z(\chi_2).$$ $\square$

The proof of the surjectivity of the map $\sigma : \mathfrak{h}^*/W \to \Theta$ requires some knowledge of commutative algebra. In fact we will not need this statement so we leave it as an exercise for the reader.

Exercise. Show that any homomorphism of algebras $\theta : Pol(\mathfrak{h}^*)^W \longrightarrow \mathbb{K}$ is of the form θ_χ for a certain $\chi \in \mathfrak{h}^*$.

Hint. First show that $Pol(\mathfrak{h}^*)$ is finitely generated $Pol(\mathfrak{h}^*)^W$-module. Then using Nakayama lemma prove the following general fact from commutative algebra:

Let A be a commutative $\mathbb{K}$-algebra and $B \subset A$ is a $\mathbb{K}$-subalgebra such that A is finitely generated as B-module. Then any morphism of algebras $\theta : B \to \mathbb{K}$ can be extended to a morphism of algebras $A \to \mathbb{K}$ (see e.g. lemma 1.4.2 in [Ke]).

7.2   Decomposition of the Category $\mathcal{O}$

Lemma. *Let $M \in \mathcal{O}$. Then there exist an ideal $J \subset \mathfrak{Z}(\mathfrak{g})$ of finite codimension such that $JM = 0$.*

Proof. We can find finite family of weights $\chi_1, ..., \chi_r$ such that $V = \oplus M(\chi_i)$ generates M. The space V is $\mathfrak{Z}(\mathfrak{g})$-invariant. The ideal $J = ker(\mathfrak{Z}(\mathfrak{g}) \to End(V))$ has the desired property. $\square$

Corollary. *Any $M \in \mathcal{O}$ is $\mathfrak{Z}(\mathfrak{g})$-finite and hence has a direct sum decomposition $M = \oplus_\theta M(\theta)$. Moreover, the set of characters $\theta \in \Theta$ such that $M(\theta) \neq 0$ is finite.*

This follows from the Lemma and Proposition A.2 of the Appendix.

Remark. In our case, the submodule $M(\theta) \subset M$ can be described explicitly as

$$M(\theta) = Ker(I_\theta^n)$$

for sufficiently large n, where $I_\theta = Ker(\theta : \mathfrak{Z}(\mathfrak{g}) \to \mathbb{K})$.

Exercise. Show that the category $\mathcal{O}$ admits the following decomposition $\mathcal{O} = \oplus \mathcal{O}_\theta$, where the sum runs over $\theta \in \Theta = Spec(\mathfrak{Z}(\mathfrak{g}))$.
  Deduce that if N is a subquotient of M then $\Theta(N) \subset \Theta(M)$.

7.3   Finite Length

Proposition. *Any module $M \in \mathcal{O}$ has a finite length.*

Proof. We will prove a more precise statement. Fix $S \subset \Theta$ and consider the full subcategory $\mathcal{O}_S$ of $\mathcal{O}$ consisting of all objects M such that $\Theta(M) \subset S$. Consider the set $\Xi := \Xi(S) \subset \mathfrak{h}^*$ consisting of weights $\chi \in \mathfrak{h}^*$ such that $\theta_{\chi+\rho} \in S$. Consider the exact functor $Res_\Xi : \mathcal{O} \to Vect$, defined by

$$Res_\Xi(M) = \oplus_{\chi \in \Xi} M(\chi).$$

Lemma. *Res_Ξ is faithful on the subcategory $\mathcal{O}_S$*

The lemma follows from the fact that for any irreducible object L in $\mathcal{O}_S$ we have $Res_\Xi(L) \neq \{0\}$.

The lemma implies that for any $M \in \mathcal{O}_S$ we have that the length of M is bounded by $\dim Res_\Xi(M)$.   □

Exercise. Show that if $L_{\chi'}$ is a subquotient of M_χ, then $\chi' \sim \chi$. Furthermore, if $L_{\chi'}$ lies in the kernel of $M_\chi \to L_\chi$, then $\chi' \not\geqq \chi$

7.4   The Grothedieck Group of the Category $\mathcal{O}$

We will use the standard construction that assigns to every (small) abelain category $\mathcal{C}$ an abelian group $K(\mathcal{C})$ that is called **Grothendieck group** of $\mathcal{C}$.

Namely, denote by A the free abelian group generated by symbols $[M]$, where M runs through the isomorphism classes of objects of $\mathcal{C}$. Let B be the subgroup of A generated by expressions $[M_1] + [M_2] - [M]$ for all exact sequences

$$0 \longrightarrow M_1 \longrightarrow M \longrightarrow M_2 \longrightarrow 0.$$

By definition, the Grothendieck group $K(\mathcal{C})$ of the category $\mathcal{C}$ is the quotient A/B.

Exercise. Suppose we know that every object of an abelian category $\mathcal{C}$ is of finite length. Show that

(i)   The map $\mathbb{Z}[IrrC] \longrightarrow K(\mathcal{C})$ is an epimorphism. In other words, the classes of simple objects of $\mathcal{C}$ generate $K(\mathcal{C})$.
(ii) Prove that the map above is an isomorphism. In particular, $K(\mathcal{C})$ is a free abelian group. Hint: Jordan-Hoelder.

In what follows we will use the fact that the collection $\{[L_\chi]\}_{\chi \in \mathfrak{h}^*}$ forms a basis for $K(\mathcal{O})$.

Proposition. *The collection $\{[M_\chi]\}_{\chi \in \mathfrak{h}^*}$ forms a basis of $K(\mathcal{O})$.*

Proof. We can write $K(\mathcal{O}) = \oplus K(\mathcal{O}_\theta)$. We will show that for a given infinitesimal character θ the collection $\{[M_\chi] : \chi \in \mathfrak{h}^*$ is such that $\theta_\chi = \theta\}$ forms a basis for $K(\mathcal{O}_\theta)$. We note that the collection $\{[L_\chi] : \chi \in \mathfrak{h}^*$ is such that $\theta_\chi = \theta\}$ forms a basis $K(\mathcal{O}_\theta)$. Recall that for any $\psi \in \mathfrak{h}^*$

$$[M_\psi] = [L_\psi] + \sum_{\varphi \lneqq \psi, \, \varphi \sim \psi} n_\varphi [L_\varphi],$$

where $n_\varphi \in \mathbb{Z}$. Inverting this unipotent matrix yields the result. $\square$

7.5 Realization of the Grothendieck group $K(\mathcal{O})$

It will be convenient to have a realization of the group $K(\mathcal{O})$ by embedding it into the group $\mathcal{E}$, the group of $\mathbb{Z}$-valued functions on $\mathfrak{h}^*$ (see Sect. 4.6). Namely, we introduce the convolution product on $\mathcal{E}$ by setting

$$(u * v)(\chi) = \sum_{\varphi \subset \mathfrak{h}^*} u(\varphi) v(\chi - \varphi) \quad \text{for } u, v \in \mathcal{E}.$$

Note that only a finite number of the summands are non-zero. Since $u * v \in \mathcal{E}$, the convolution endows $\mathcal{E}$ with a commutative algebra structure.

For any $\chi \in \mathfrak{h}^*$ define $\delta_\chi \in \mathcal{E}$ by setting $\delta_\chi(\varphi) = 0$ for $\varphi \neq \chi$ and $\delta_\chi(\chi) = 1$. Clearly, δ_0 is the unit of $\mathcal{E}$.

Set

$$L = \prod_{\gamma \in R^+} (\delta_{\gamma/2} - \delta_{-\gamma/2}) = \delta_\rho \prod_{\gamma \in R^+} (\delta_0 - \delta_{-\gamma})$$

Here $\prod$ is the convolution product in $\mathcal{E}$.

We can now define a homomorphism $\tau : K(\mathcal{O}) \longrightarrow \mathcal{E}$ by the formula

$$\tau([M]) = L * \pi_M,$$

where $M \in \mathcal{O}$,

Theorem. *(1)* $\tau(M_\chi) = \delta_\chi$.
(2) The mapping $\tau : K(\mathcal{O}) \longrightarrow \mathcal{E}$ gives an isomorphism of $K(\mathcal{O})$ with the subgroup $\mathcal{E}_c \subset \mathcal{E}$ consisting of functions with compact support.

Proof. The second point is an immidiate consequence of the first in lieu of the fact that the family $\{[M_\chi]\}$ generates $K(\mathcal{O})$. The proof of the first point is based on Lemma 4.6 and the following Lemma.

Lemma. *Let K be the Kostant function, see Sect. 4.6. Then*

$$K * \delta_{-\rho} * L = \delta_0.$$

Proof. For any $\gamma \in R^+$ set $a_\gamma = \delta_0 + \delta_{-\gamma} + \ldots + \delta_{-n\gamma} + \ldots$. The definition of K implies that

$$K = \prod_{\gamma \in R^+} a_\gamma.$$

Further, $(\delta_0 - \delta_{-\gamma})a_\gamma = \delta_0$. Since L can be represented as $\prod_{\gamma \in R^+} (\delta_0 - \delta_{-\gamma})\delta_\rho$, we are done.   □

Remark. The theorem implies that finding the exact transition matrix between the basis $\{[M_\chi]\}$ and the basis $\{[L_\chi]\}$ is equivalent to the determination of $\tau(L_\chi)$. This is the subject of the Kazhdan–Lusztig conjecture.

8   Description of Finite Dimensional Representations

8.1   *Complete Reducibility of Finite Dimensional Modules*

In this section, we will describe all finite dimensional representations of a semisimple Lie algebra $\mathfrak{g}$. As was shown in Sect. 4.2, all such representations belong to $\mathcal{O}$. Recall that in Sect. 5 we constructed a collection of finite dimensional $\mathfrak{g}$-modules A_λ parameterized by weights $\lambda \in P^+$. We will now show that any finite dimensional module is isomorphic to a direct sum of such modules, and that these are irreducible. This yields complete reducibility.

Theorem. *(1) Let M be a finite dimensional $\mathfrak{g}$-module. Then M is isomorphic to a direct sum of modules of the form A_λ for $\lambda \in P^+$.*
(2) All the modules A_λ, where $\lambda \in P^+$, are irreducible.

Proof. (1) We may assume that $M = M(\theta)$, where $\theta \in \Theta$. Let m be any highest weight vector of M and λ its weight. Then $\theta = \theta_{\lambda+\rho}$. Besides, for any simple root α we have $E^k_{-\alpha}m = 0$ for large k, and hence by Lemma 2.2 $\lambda(H_\alpha) \in \mathbb{Z}_{\geq 0}$. Therefore, $\lambda \in P^+$.

Since $\lambda \in P^+$ the element $\lambda + \rho$ lies inside the interior of the Weyl chamber and thus is uniquely recovered from the infinitesimal character of the module A_λ.

Let $m_1, \ldots, m_l$ be a basis of $M(\lambda)$. Let us construct the morphism $p : \bigoplus_{1 \leq i \leq l} M(\lambda + \rho) \longrightarrow M$ so that each generator $(m_{\lambda+\rho})_i$ for $i = 1, 2, \ldots, l$ goes to m_i. As follows from Lemma 2.2, for any simple root α we have $E^{k_\alpha}_{-\alpha} m_i = 0$, where $k_\alpha = (\lambda + \rho)(H_\alpha)$

Hence p may be considered as the morphism $p : \bigoplus_{1 \leq i \leq l}(A_\lambda)_i \longrightarrow M$.

Let L_1 and L_2 be the kernel and cokernel of the morphism p. Then $\Theta(L_i) = \{\theta\}$ and $L_i(\lambda) = 0$, where $i = 1, 2$. As was shown above, $L_1 = L_2 = 0$, i.e., $M \cong \bigoplus_{1 \leq i \leq l}(A_\lambda)_i$.

(2) Let M be a nontrivial submodule of A_λ. Then $\Theta(M) = \theta_{\lambda+\rho}$, hence, $M(\lambda) \neq 0$, i.e., M contains a vector of weight λ. But then $M = A_\lambda$. Thus, the module A_λ is irreducible and the proof of the Theorem is complete. □

Corollary. $A_\lambda \cong L_{\lambda+\rho}$, where $\lambda \in P^+$.

Remark. The module A_λ is an irreducible module of highest weight λ. The strange shift in its numbering as an irreducible module corresponds to the Harish–Chandra shift.

8.2   Characters of Highest Weight Modules A_λ

Consider the natural action of the group W on the space of functions on $\mathfrak{h}^*$ defined by

$$(wu)(\chi) = u(w^{-1}\chi) \text{ for } w \in W, \chi \in \mathfrak{h}^*.$$

Lemma. $wL = \det w \cdot L$ for any $w \in W$.

Proof. It suffices to verify that $\sigma_\alpha L = -L$ for $\alpha \in B$. Since σ_α permutes the elements of the set $R^+ \setminus \{\alpha\}$ and transforms α into $-\alpha$, then

$$\sigma_\alpha L = (\delta_{-\alpha/2} - \delta_{\alpha/2}) \prod_{\gamma \in R^+ \setminus \{\alpha\}} (\delta_{\gamma/2} - \delta_{-\gamma/2}) = -L. \qquad \square$$

The next theorem provides a formula for the formal character of the finite dimensional irreducible module L_λ. This will give us Kostant multiplicity formulas, Weyl character formula and Weyl dimension formula.

Theorem. *Suppose L_λ is finite dimensional. Then*

$$L * \pi_{L_\lambda} = \sum_{w \in W} \det w \cdot \delta_{w(\lambda)}.$$

Proof. We have

$$[L_\lambda] = \sum_{\mu \sim \lambda} a_\mu [M_\mu]$$

with $a_\lambda = 1$.

Applying τ to this equation, we obtain

$$\tau([L_\lambda]) = \sum_{\mu \sim \lambda} a_\mu \delta_\mu.$$

Since π_{L_λ} is W-invariant and L is W-skew invariant, we see that $\tau([L_\lambda]) = L * \pi_{L_\lambda}$ is W-skew-invariant as well.

Thus,

$$L * \pi_{L_\lambda} = \sum_{w \in W} \det w \cdot \delta_{w(\lambda)} \tag{*}$$

Theorem 8.2 is proved.   □

Corollary. *(1) for any $\lambda \in P^+$ we have $[A_\lambda] = \sum_{w \in W} \det w \cdot [M_{w(\lambda+\rho)}]$*

(2) the Kostant formula for the multiplicity of the weight $\pi_{A_\lambda}(\psi) = \sum_{w \in W} \det w \cdot$

$K(\psi + \rho - w(\lambda + \rho))$ for any $\psi \in \mathfrak{h}^*$.

Proof. Since τ is an isomorphism, to verify the first item, we may apply τ to both sides. The second item is a reformulation of the first in view of the Lemma 4.6.   □

8.3   Weyl Character Formula

Denote by $F(\mathfrak{h})$ the ring of formal power series in $\mathfrak{h}$, i.e. the completion of the algebra of polynomial functions $Pol(\mathfrak{h})$ at the point zero. For any $\chi \in \mathfrak{h}^*$ set $e^\chi = \sum_{i \geq 0} \frac{\chi^i}{i!}$.

Clearly, $e^\chi \in F(\mathfrak{h})$ and $e^{\chi+\psi} = e^\chi e^\psi$ for $\chi, \psi \in \mathfrak{h}^*$. Let M be a finite dimensional $\mathfrak{g}$-module. Define the *character* $\mathrm{ch}_M \in F(\mathfrak{h})$ of M by the formula

$$\mathrm{ch}_M = \sum_{\chi \in P} \pi_M(\chi) e^\chi.$$

Theorem. *Set*

$$L' = \sum_{w \in W} (\det w) e^{w\rho}.$$

Then for A_λ, where $\lambda \in P^+$, we have

$$L' \mathrm{ch}_{A_\lambda} = \sum_{w \in W} (\det w) e^{w(\lambda+\rho)}.$$

Proof. The mapping $j : \mathcal{E}_c \longrightarrow F(\mathfrak{h})$ defined by the formula $j(u) = \sum_{\chi \in \mathfrak{h}^*} u(\chi) e^\chi$ is a ring homomorphism. Inserting $\lambda = \rho$ in formula (*) of 8.2, we obtain

$$\sum_{w \in W} \det w \cdot \delta_{w\rho} = L * \pi_{A_0} = L * \delta_0 = L.$$

Hence, $j(L) = L'$. The result now follows by applying j to formula (*) of 8.2 with $A_\lambda = L_{\lambda+\rho}$.   □

Remark. (1) When $\mathbb{K} = \mathbb{C}$, all the power series involved in Theorem 8.3 converge and define analytic functions on $\mathfrak{h}$. Theorem 8.3 claims the equality of two such functions.

(2) Let $\mathcal{G}$ be a complex semisimple Lie group with Lie algebra $\mathfrak{g}$ and $\mathcal{H} \subset \mathcal{G}$ the Cartan subgroup corresponding to the Lie subalgebra $\mathfrak{h}$. Consider the finite dimensional representation T of $\mathcal{G}$ corresponding to the $\mathfrak{g}$-module A_λ. Let $h \in \mathcal{H}$. Then $h = \exp(H)$, where $H \in \mathfrak{h}$. It is easy to derive from Theorem 8.3 that

$$\mathrm{Tr}\, T(h) = \frac{\sum\limits_{w \in W} \det w \cdot e^{(w(\lambda+\rho))}(H)}{\sum\limits_{w \in W} \det w \cdot e^{(w\rho)}(H)}.$$

This is the well-known H. Weyl's formula for characters of irreducible representations of complex semisimple Lie groups.

8.4 Weyl's Dimension Formula

Theorem. *Let* $\lambda \in P^+$. *Then*

$$\dim A_\lambda = \prod_{\gamma \in R^+} \frac{\langle \lambda + \rho, \gamma \rangle}{\langle \rho, \gamma \rangle}.$$

Proof. Set

$$F_\chi = \sum_{w \in W} \det w \cdot e^{w\chi} \text{ for any } \chi \in \mathfrak{h}^*.$$

Clearly, $F_\rho = L' = \prod_{\gamma \in R^+}(e^{\gamma/2} - e^{-\gamma/2})$. For any $\chi \in \mathfrak{h}^*$ and $H \in \mathfrak{h}$, we may consider $F_\chi(tH)$ as a formal power series in one variable t.

Let ρ' and λ' be elements of $\mathfrak{h}$ corresponding to ρ and λ, respectively, after the identification of $\mathfrak{h}$ with $\mathfrak{h}^*$ by means of the Killing form. Then

$$\dim A_\lambda = \mathrm{ch} A_\lambda(0) = \frac{F_{\lambda+\rho}(t\rho')}{F_\rho(t\rho')} \Big|_{t=0}.$$

Observe that

$$F_{\lambda+\rho}(t\rho') = \sum_{w \in W} \det w \cdot e^{t \langle \lambda+\rho, w^{-1}\rho \rangle} = F_\rho(t(\lambda' + \rho')).$$

Hence by the product formula we have

$$\dim A_\lambda = \frac{F_\rho(t(\lambda' + \rho'))}{F_\rho(t\rho')}\Big|_{t=0} = \prod_{\gamma \in R^+} \left(\frac{e^{t/2(\gamma(\lambda' + \rho'))} - e^{-t/2(\gamma(\lambda' + \rho'))}}{e^{t/2(\gamma(\rho'))} - e^{-t/2(\gamma(\rho'))}}\Big|_{t=0} \right)$$

The quantity on the right hand side is evaluated easily to be

$$\prod_{\gamma \in R^+} \frac{\langle \gamma, \lambda + \rho \rangle}{\langle \gamma, \rho \rangle}.$$

$\square$

8.5   *Summary of Results*

We collect here the results we have proven for finite dimensional representations of $\mathfrak{g}$.

1. For any weight $\lambda \in P^+$ we have constructed a finite dimensional irreducible $\mathfrak{g}$-module A_λ. All such modules are nonisomorphic. Any finite dimensional irreducible $\mathfrak{g}$-module is isomorphic to one of A_λ, where $\lambda \in P^+$.
2. **Complete reducibility**
   Any finite dimensional $\mathfrak{g}$-module M is isomorphic to a direct sum of A_λ.
3. The module A_λ is $\mathfrak{h}$-diagonalizable and has the unique (up to a factor) highest weight vector a_λ. The weight of a_λ is equal to λ. The module A_λ is called a **highest weight module** of highest weight λ.
4. **Harish–Chandra theorem on ideal.**
   The module A_λ is generated by the vector a_λ as $U(\mathfrak{n}_-)$-module (in particular, all the weights of A_λ are less than or equal to λ). The ideal of relations $I = \{X \in U(\mathfrak{n}_-) \mid X a_\lambda = 0\}$ is generated by the elements $E_{-\alpha}^{m_\alpha + 1}$, where $\alpha \in B$ and $m_\alpha = \lambda(H_\alpha)$.
5. The function π_{A_λ} is W-invariant.
6. If ψ is a weight of A_λ, then either $\lambda \sim \psi$ or $|\psi| < |\lambda|$.
7. A_λ has infinitesimal character $\theta_{\lambda + \rho}$. Explicitly, for any $a \in A_\lambda$ and $z \in \mathfrak{z}(\mathfrak{g})$ we have $za = \theta_{\lambda + \rho}(z)a$.
   If $\lambda_1, \lambda_2 \in P^+$ and $\lambda_1 \neq \lambda_2$, then homomorphisms $\theta_{\lambda_1 + \rho}$ and $\theta_{\lambda_2 + \rho}$ are distinct.
8. **Weyl character formula**

$$L \cdot \mathrm{ch}_{A_\lambda} = \sum_{w \in W} (\det w) e^{w(\lambda + \rho)}, \quad \text{where } L = \sum_{w \in W} (\det w) e^{w(\rho)}$$

9. **Kostant multiplicity formula.**

$$\pi_{A_\lambda}(\mu) = \sum_{w \in W} (\det w) K(\mu + \rho - w(\lambda + \rho)).$$

10. **Weyl dimension formula**

$$\dim A_\lambda = \prod_{\gamma \in R^+} \frac{\langle \gamma, \lambda + \rho \rangle}{\langle \gamma, \rho \rangle}.$$

11. For any finite dimensional $\mathfrak{g}$-module V, the module V is $\mathfrak{h}$-diagonalizable and its character π_V is W-invariant.

9   Proof of the Harish–Chandra Theorem

The proof we describe here will be obtained by first reducing Harish–Chandra's theorem to Chevalley's restriction theorem. The proof of Chevalley's theorem is obtained using characters of finite dimensional representations A_λ of $\mathfrak{g}$.

The proof we present uses implicitly a group action without defining the group that acts. The existence of the action should not be surprising in view of Corollary 2.5 that finite representations of the Lie algebra $\mathfrak{sl}(2)$ admits an action of the group $SL(2)$. A similar idea applies in general. Instead of providing a formal statement let us briefly explain how to obtain such a group.

Let G be the adjoint group of automorphisms of $\mathfrak{g}$. This is the group generated by groups $SL(2)_\gamma$ corresponding to all the roots γ. This group acts on $\mathfrak{g}$, on $U(\mathfrak{g})$, $S(\mathfrak{g})$ and preserves natural structures on all these spaces. On each of these spaces V the actions of $\mathfrak{g}$ and G are related as follows.

(*) Let $X \in \mathfrak{g}_\alpha$ and $g = \exp \ ad(X) \in G$. Then for any vector $v \in V$ we have

$$gv = \exp(X)v := \sum_k \frac{1}{k!} X^k v.$$

This expression makes sense since $X^k v = 0$ for large k.

In particular the invariants with respect to G and $\mathfrak{g}$ in each of these spaces are the same.

9.1   Reduction to Chevalley's Theorem

We constructed a morphism $j : \mathfrak{z}(\mathfrak{g}) \to Pol(\mathfrak{h}^*)^W = U(\mathfrak{h})^W$ and would like to show that it is an isomorphism. By construction, j is the restriction to $\mathfrak{z}(\mathfrak{g})$ of a linear map $j : U(\mathfrak{g}) \to U(\mathfrak{h})$ defined by Harish–Chandra projection (see 6.1).

Morphism j is compatible with natural filtrations on $\mathfrak{Z}(\mathfrak{g})$ and $U(\mathfrak{h})^W$ obtained by restrictions of standard filtrations on $U(\mathfrak{g})$ and $U(\mathfrak{h})$. So in order to show that j is an isomorphism it is enough to check that the associated graded morphism $\alpha := grj : gr\mathfrak{Z}(\mathfrak{g}) \rightarrow grU(\mathfrak{h})^W$ is an isomorphism. Let us identify these two spaces.

First of all notice that $\mathfrak{Z}(\mathfrak{g}) = U(\mathfrak{g})^{\mathfrak{g}}$ where we consider the adjoint action of $\mathfrak{g}$ on $U(\mathfrak{g})$, $ad(X)(u) = [X, u]$. Let us also consider the adjoint action of $\mathfrak{g}$ on the algebra $S(\mathfrak{g})$ such that $ad(X)$ is the derivation of the algebra $S(\mathfrak{g})$ satisfying $ad(X)(Y) = [X, Y]$ for $Y \in \mathfrak{g} \subset S(\mathfrak{g})$. Using the morphism *symm* discussed in Corollary 1.5. We see that the space gr $\mathfrak{Z}(\mathfrak{g})$ coincides with the space $S(\mathfrak{g})^{\mathfrak{g}}$ (this follows from the fact that *symm* is a morphism of $\mathfrak{g}$-modules). Similarly, $gr((U(\mathfrak{h})^W))$ coincides with the space $S(\mathfrak{h})^W$.

Consider the morphism $\beta : S(\mathfrak{g}) = S(\mathfrak{n}_-) \otimes S(\mathfrak{h}) \otimes S(\mathfrak{n}_-) \rightarrow S(\mathfrak{h})$ obtained by mapping $\mathfrak{n}_-$ and $\mathfrak{n}_+$ to 0. Analyzing the explicit description of the morphism α described above it is easy to see that it coincides with the restriction of β to $\mathfrak{g}$-invariant elements.

Using Killing form we will identify $\mathfrak{g}$ with $\mathfrak{g}^*$ and $\mathfrak{h}$ with $\mathfrak{h}^*$. In this way we interpret $S(\mathfrak{g})$ as the algebra $Pol(\mathfrak{g})$ of polynomial functions on $\mathfrak{g}$ and $S(\mathfrak{h})$ as the algebra $Pol(\mathfrak{h})$ of polynomial functions on $\mathfrak{h}$. Morphism β after this identification is just the restriction of polynomial functions on $\mathfrak{g}$ to $\mathfrak{h}$.

This shows that Harish–Chandra theorem follow from the following result

Theorem (The Chevalley's restriction theorem). Let $Pol(\mathfrak{g})$ and $Pol(\mathfrak{h})$ be algebras of polynomial functions on $\mathfrak{g}$ and $\mathfrak{h}$, respectively, and $\eta : Pol(\mathfrak{g}) \longrightarrow Pol(\mathfrak{h})$ the restriction homomorphism. Then $Pol(\mathfrak{g})^{\mathfrak{g}} \longrightarrow Pol(\mathfrak{h})^W$ is an isomorphism.

9.2 Proof of Injectivity in Chevalley's Theorem

Let us choose an ordering $\gamma_1, ..., \gamma_r$ of roots of the algebra $\mathfrak{g}$ and consider the algebraic variety $Y = \prod \mathfrak{g}_{\gamma_i} \times \mathfrak{h}$; in fact this is just an affine space isomorphic to $\mathfrak{g}$. Let us define a morphism of algebraic varieties $a : Y \rightarrow \mathfrak{g}$ by

$$a(X_1, ..., X_r, H) = \exp ad(X_1) \exp ad(X_2)... \exp ad(X_r)(H).$$

Clearly, any function $f \in Pol(\mathfrak{g})^{\mathfrak{g}}$ in the kernel of the morphism η will also lie in the kernel of morphism of algebras $a^* : Pol(\mathfrak{g}) \rightarrow Pol(Y)$ corresponding to the morphism a.

However, if we choose a regular element $H \in \mathfrak{h}$ (i.e., an element such that $\gamma(H) \neq 0$ for every root γ) and consider the point $y = (0, ..., 0, H) \in Y$, then easy computation shows that the differential da at this point is an isomorphism of linear spaces. This implies that the kernel of the homomorphism a^* is 0.

9.3   Proof of Surjectivity in Chevalley's Theorem

Fix a non-negative integer k. To every finite dimensional representation (ρ, V) of the Lie algebra $\mathfrak{g}$, we assign a polynomial function $P_{k,V}$ on the Lie algebra $\mathfrak{g}$ as follows $P_{k,V}(X) = \mathrm{tr}(\rho(X)^k)$. Clearly, this is a $\mathfrak{g}$-invariant polynomial function on $\mathfrak{g}$. The surjectivity of the morphism η follows from

Proposition.   *The collection of functions $P_{k,V}$ on $\mathfrak{h}$ spans $Pol(\mathfrak{h})^W$.*

Proof. Let us denote by $F(\mathfrak{h})$ the completion of the algebra $Pol(\mathfrak{h})$ at maximal ideal m corresponding to the point $0 \in \mathfrak{h}$. In other words, if (y_i) is a coordinate system on the linear space $\mathfrak{h}$, then $F(\mathfrak{h}) = \mathbb{K}[[y_1, ..., y_r]]$. Since polynomials $P_{k,V}$ are homogeneous in order to prove the proposition, it is enough to prove that the $\mathbb{K}$-linear span of the collection of polynomials $P_{k,V}$ is dense in the algebra $F(\mathfrak{h})^W$.

To see this we will consider a different model for the algebra $F(\mathfrak{h})$. Namely consider the category $\mathcal{R}(\mathfrak{h})$ of finite dimensional $\mathfrak{h}$-modules. We say that an object V of $\mathcal{R}(\mathfrak{h})$ is **integrable** if the action of $\mathfrak{h}$ is completely reducible and all coroots H_γ have integral spectrum. We denote by $\mathcal{R}$ the full subcategory of $\mathcal{R}(\mathfrak{h})$ of integrable objects. The Grothendieck group $K(\mathcal{R})$ of this category is naturally isomorphic to the group algebra $\mathbb{Z}(P)$ of the lattice P. Namely, a weight $\lambda \in P$ corresponds to a one-dimensional representation T_λ of the Lie algebra $\mathfrak{h}$ of weight λ.

Consider a homomorphism of algebras $\sigma : K(\mathcal{R}) \to F(\mathfrak{h})$ defined by

$$\sigma((\rho, V))(x) = tr_V(\exp(\rho(x))) = \sum_k \frac{1}{k!} tr_V(\rho(x)^k)$$

In particular, $\sigma(T_\lambda) = \exp(\lambda)$.

It is easy to see that the $\mathbb{K}$-span of the image of morphism σ is dense in $F(\mathfrak{h})$ (in fact $F(\mathfrak{h})$ can be realized as the completion of the algebra $\mathbb{K}(P) := \mathbb{Z}(P) \otimes_{\mathbb{Z}} \mathbb{K}$ at the maximal ideal corresponding to the homomorphism $\mathbb{K}(P) = \mathbb{K}(\mathfrak{R}) \otimes \mathbb{K} \to \mathbb{K}$ given by $V \mapsto \dim(V)$.

Now consider the category $\mathcal{R}(\mathfrak{g})$ of finite dimensional $\mathfrak{g}$-modules and the restriction functor $r : \mathcal{R}(\mathfrak{g}) \to \mathcal{R}(\mathfrak{h})$. Based on the $\mathfrak{sl}(2)$ theory we may view r as a functor $r : \mathcal{R}(\mathfrak{g}) \to \mathcal{R}$. Denote by π the corresponding morphism of Grothendieck groups $\pi : K(\mathcal{R}(\mathfrak{g})) \to K(\mathcal{R})$.

For every $V \in \mathcal{R}(\mathfrak{g})$ the element $\pi(V)$ considered as a function on P is just the character π_V of V, which was defined in Sect. 4.6.

Now, the image $\sigma(\pi(V)) \in F(\mathfrak{h})$ equals $\sum_k P_{k,V}/k!$. Thus, in order to show that polynomials $P_{k,V}$ span a dense subset of $F(\mathfrak{h})^W$, it is enough to prove the following.

Lemma.   *The image of morphism $\pi : K(\mathcal{R}(\mathfrak{g})) \to K(\mathcal{R})$ equals to the subgroup $K(\mathcal{R})^W \subset K(\mathcal{R})$ of W-invariant elements.*

The lemma easily follows from Theorem 5.3. Namely, if an element $u \in K(\mathcal{R}) \simeq \mathbb{Z}(P)$ is W-invariant, then induction on the maximal length of weights in the support of u implies that u can be written as a $\mathbb{Z}$-linear combination of $\pi(A_\lambda)$, where $\lambda \in P^+$. $\square$

A. Appendix: Eigenspaces Decomposition

In this section, we present the standard Eigen-space decomposition of linear algebra with few variations that are needed in the text.

A.1   Standard Eigenspace Decomposition

Let $\mathbb{K}$ be an algebraically closed field. Let T be an operator on a finite dimensional $\mathbb{K}$-vector space V.

We denote by $Spec(T, V)$ the set of $\lambda \in \mathbb{K}$ such that the operator $T - \lambda\mathbf{1}$ is not invertible. Since V is finite dimensional, the operator T satisfies some equation $P(T) = 0$ for some monic polynomial P that could be written as $\prod(T - \lambda_i \mathbf{1}) = 0$. This shows that if $V \neq \{0\}$ then the set $Spec(T, V)$ is not empty.

For any $\lambda \in \mathbb{K}$, we denote by $V(\lambda)$ the space of vectors $v \in V$ annihilated by some power of the operator $T - \lambda\mathbf{1}$. Vectors of the these spaces are called generalized eigenvectors. It is clear that $V(\lambda) \neq 0$ iff $\lambda \in Spec(T, V)$.

Denote by $V^{ss}(\lambda) = Ker(T - \lambda\mathbf{1}) \subset V(\lambda)$. Vectors of these spaces are called *eigenvectors*. We say that T is semisimple if V is spanned by eigenvectors of T.

Note that if S is an operator commuting with T then S preserves all the spaces $V^{ss}(\lambda), V(\lambda)$.

Proposition. $V = \oplus V(\lambda)$ *where the sum is taken over all* $\lambda \in \mathbb{K}$.

Proof. (a) We first prove linear independence. Otherwise, take the shortest dependence of the form $v_1 + \ldots + v_k = 0$, where each v_i is a generalized eigenvector with eigenvalues λ_i, and all eigenvalues are distinct. Clearly, $k \geq 2$. Applying $T - \lambda_1\mathbf{1}$ several times to the above identity, we get a shorter dependency.

(b) For every $\lambda \in \mathbb{K}$ consider the quotient space $Q_\lambda = V/V(\lambda)$. We claim that $Spec(T, Q_\lambda)$ does not contain λ. Indeed, let $V'(\lambda) \subset V$ be the preimage of the space $Q_\lambda(\lambda)$. Then some power of the operator $T - \lambda\mathbf{1}$ maps $V'(\lambda)$ to $V(\lambda)$ and hence some larger power maps it to 0. This implies that $V'(\lambda) = V(\lambda)$ and hence $Q_\lambda(\lambda) = 0$.

Consider now the space $Q = V/\sum_\lambda V(\lambda)$. Since this space is a quotient of all the spaces Q_λ, the set $Spec(T, Q) \subset \cap_\lambda Spec(T, Q_\lambda)$ is empty and hence $Q = 0$. $\square$

Corollary.   *If T is semisimple then $V = \oplus V^{ss}(\lambda)$.*

A.2   Eigenspace Decomposition for Commuting Families

Let now A be a commutative $\mathbb{K}$-algebra acting on a $\mathbb{K}$-vector space V. For each character $\chi : A \to \mathbb{K}$ we denote by $\mathfrak{m}_\chi = \ker(\chi)$ the corresponding maximal ideal of A. We denote by $V(\chi)$ the subspace of vectors in V that are annihilated by some power of $\mathfrak{m}$. They are called generalized eigenvectors corresponding to the character χ. We denote by $V^{ss}(\chi)$ the space of vectors annihilated by $\mathfrak{m}$. They are called eigenvectors.

We say that the action is *locally finite* if V is a union of finite dimensional A-submodules.

Proposition. *Let A be a commutative algebra and V be a locally finite A-module. We have:*

(1) $V = \oplus V(\chi)$ *where the sum is taken over all characters χ of A.*
(2) *If each $a \in A$ acts semisimply on V, then $V = \oplus V^{ss}(\chi)$.*

Proof. We first consider the case $\dim(V) < \infty$.

For (1) note that the linear independence of the spaces $V(\chi)$ follows from the previous proposition. To show that V is a direct sum we argue by induction on dimension of V. If each $a \in A$ has only one eigenvalue $\alpha(a)$, then α is a character and we are done. Otherwise, we can split V, using the previous proposition, as a sum of generalized eigenspaces for some $a \in A$. Since each of these spaces is invariant with respect to the algebra A, we can apply induction. The same proof gives the decomposition in the semi-simple case.

Now the locally finite case is an obvious formal consequence of the finite dimensional case.   □

Corollary. *Let A be a finite dimensional commutative algebra over $\mathbb{K}$ with unit. Then*

(1) *In A, there is a finite number of maximal ideals $\mathfrak{m}_i$, where $i = 1, \ldots, k$.*
(2) *There are elements $e_i \in A$, where $i = 1, \ldots, k$, such that*

$$e_i e_j = 0 \text{ for } i \neq j \text{ and } e_i^2 = e_i;$$

$$e_1 + e_2 + \ldots + e_k = 1;$$

$$e_i \in \mathfrak{m}_j \text{ for } i \neq j;$$

$$e_i \mathfrak{m}_i^n = 0 \text{ for } n > \dim A.$$

Proof. Let A act on itself by multiplication. By the previous proposition, we have a projection $P_\chi : A \to A(\chi)$ for each character χ of A. Write the identity operator as a sum $\mathbf{1} = \sum P_i$ where all $P_i = P_{\chi_i}$ are non zero.

If P is one of these projectors, then it is given by multiplication by an element $e = P(1) \in A$ (Indeed, $P(b) = P((b \cdot 1)) = b \cdot P(1) = b \cdot e$).

These elements $e_i = P_i(1)$ and the maximal ideals $\mathfrak{m}_i = \ker(\chi_i)$ satisfy the statement of the corollary.   □

Acknowledgements I am thankful to D. Leites for saving these lectures, translating them into English and preparing them for publication, and the Department of Mathematics of Stockholm University for financing a preprint of these lectures as part of Leites' *Seminar on Supermanifolds* proceedings.

I am thankful to E. Sayag for the great help in preparing the final draft of these lectures.

References

[BG] A. Braverman, D. Gaitsgory, *Poincare-Birkhoff-Witt theorem for quadratic algebras of Koszul type*. J. Algebra 181 (1996), 315-328.

[Bu] Bourbaki N., *Groupes et algébres de Lie*. Hermann, Paris, Ch. II–III: 1972; Ch. IV–VI: 1968; Ch. VII–VIII: 1974

[Di] Dixmier J., *Enveloping algebras*, Graduate study in Mathenatics, N. 11, AMS, 1996

[Dy] Dynkin E., The review of principal concepts and results of the theory of linear representations of semi-simple Lie algebras (Complement to the paper "Maximal subgroups of the classical groups") (Russian) Trudy Moskov. Mat. Obšč. 1, (1952). 39–166

[Ge] Gelfand, I. (ed.) *Representations of Lie groups and Lie algebras*. Acad. Kiado, Budapest, 1975

[Ke] Kempf, G.R. *Algebraic varieties*. London Mathematical Society Lecture Note Series, vol. 172, Cambridge University Press, Cambridge, 1993, pp. 163

[Ki] Kirillov, A. (ed.) *Representations of Lie groups and Lie algebras*. Acad. Kiado, Budapest, 1985

[OV] Onishchik A., Vinberg E., *Lie groups and Algebraic Groups*, Springer, 1987

[Pr] Prasolov V., *Problems and theorems in linear algebra*, AMS, 1991

[Se] Serre J.-P. *Lie algebras and Lie groups*. N.Y, W. A. Benjamin, 1965; *Algébres de Lie semisimples complexes*. N.Y, W. A. Benjamin, 1966

Stein–Sahi Complementary Series and Their Degenerations

Yuri A. Neretin

Abstract The paper is an introduction to the Stein–Sahi complementary series and to unipotent representations. We also discuss some open problems related to these objects. For the sake of simplicity, we consider only the groups $U(n, n)$.

Keywords Unitary representations • Complementary series • Symmetric spaces • Non-commutative harmonic analysis • Classical groups • Unitary group • Highest weight representations • Unipotent representations

Mathematics Subject Classification (2010): 42B35, 22D10

1 Introduction

This paper[1] is an attempt to present an introduction to the Stein–Sahi complementary series available for non-experts and beginners.

1.1 History of the Subject

The theory of infinite-dimensional representations of semi-simple groups was initiated in the pioneer works of I. M. Gelfand and M. A. Naimark (1946–1950), V. Bargmann [2] (1947), and K. O. Friedrichs [12] (1951–1953). The book by

---

[1]It is a strongly revised version of two sections of my preprint [30].

Y.A. Neretin (✉)
Mathematical Department, University of Vienna, Nordbergsrasse 15, Vienna, Austria

Institute for Theoretical and Experimental Physics, Moscow, Russia

MechMath Department, Moscow State University, Russia

I. M. Gelfand and M. A. Naimark [14] (1950) contains a well-developed theory
for the complex classical groups GL$(n, \mathbb{C})$, SO$(n, \mathbb{C})$, and Sp$(2n, \mathbb{C})$ (the parabolic
induction, complementary series, spherical functions, characters, Plancherel theo-
rems). However, this classic book[2] contains various statements and asseverations
that were not actually proved. In the modern terminology, some of the chapters were
"mathematical physics". Most of these statements were really proved by 1958–1962
in works of different authors (Harish-Chandra, F. A. Berezin, etc.).

In particular, I. M. Gelfand and M. A. Naimark (1950) claimed that they classi-
fied all unitary representations of GL$(n, \mathbb{C})$, SO$(n, \mathbb{C})$, and Sp$(2n, \mathbb{C})$. E. Stein [46]
compared Gelfand–Naimark constructions for the groups SL$(4, \mathbb{C}) \simeq$ SO$(6, \mathbb{C})$ and
observed that they are not equivalent. In 1967, E. Stein constructed "new" unitary
representations of SL$(2n, \mathbb{C})$.

D. Vogan [48] in 1986 obtained the classification of unitary representations of
groups GL$(2n)$ over real numbers $\mathbb{R}$ and quaternions $\mathbb{H}$. In particular, this work
contains an extension of Stein's construction to these groups. In the 1990s, the
Stein-type representations were a topic of interest of S. Sahi (see [40–42]), S. Sahi–
E. Stein [44], and A. Dvorsky–S. Sahi [8, 9]. In particular, Sahi extended the
construction to other series of classic groups, specifically to the groups SO$(2n, 2n)$,
U(n, n), Sp(n, n), Sp$(2n, \mathbb{R})$, SO$^*(4n)$, Sp$(4n, \mathbb{C})$, and SO$(2n, \mathbb{C})$.

1.2   Stein–Sahi Representations for U(n, n)

Denote by U(n) the group of unitary $n \times n$ matrices. Consider the *pseudo-unitary*
group U(n, n). We realize it as the group of $(n + n) \times (n + n)$-matrices $g = \begin{pmatrix} a & b \\ c & d \end{pmatrix}$
satisfying the condition

$$\begin{pmatrix} a & b \\ c & d \end{pmatrix} \begin{pmatrix} 1 & 0 \\ 0 & -1 \end{pmatrix} \begin{pmatrix} a & b \\ c & d \end{pmatrix}^* = \begin{pmatrix} 1 & 0 \\ 0 & -1 \end{pmatrix}.$$

Lemma 1.1. *The formula*

$$z \mapsto z^{[g]} := (a + zc)^{-1}(b + zd) \tag{1.1}$$

determines an action of the group U(n, n) *on the space* U(n).

The unitary group is equipped by the Haar measure $d\mu(z)$; hence, we can
determine the Jacobian of a transformation (1.1) by

$$J(g, z) = \frac{d\mu(z^{[g]})}{d\mu(z)}.$$

---

[2]Unfortunately, the book has been published only in Russian and German.

Lemma 1.2. *The Jacobian of the transformation $z \mapsto z^{[g]}$ on $U(n)$ is given by*

$$J(g,z) = |\det(a + zc)|^{-2n}.$$

Fix $\sigma, \tau \in \mathbb{C}$. For $g \in U(n,n)$, we define the following linear operator in the space $C^\infty(U(n))$:

$$\rho_{\sigma|\tau}(g)f(z) = f(z^{[g]})\det(a + zc)^{-n-\tau}\det\overline{(a + zc)}^{-n-\sigma}. \tag{1.2}$$

The formula includes powers of complex numbers, the precise definition of which is given below. In fact, $g \mapsto \rho_{\sigma|\tau}(g)$ is a well-defined operator-valued function on the universal covering group $U(n,n)^\sim$ of $U(n,n)$.

The chain rule for Jacobians,

$$J(g_1 g_2, z) = J(g_1, z)J(g_2, z^{[g_1]}), \tag{1.3}$$

implies

$$\rho_{\sigma|\tau}(g_1)\rho_{\sigma|\tau}(g_2) = \rho_{\sigma|\tau}(g_1 g_2).$$

In other words, $\rho_{\sigma|\tau}$ is a linear representation of the group $U(n,n)^\sim$.

Observation 1.3. *If $\operatorname{Re}\sigma + \operatorname{Re}\tau = -n$, $\operatorname{Im}\sigma = \operatorname{Im}\tau$, then a representation $\rho_{\sigma|\tau}$ is unitary in $L^2(U(n))$.*

This easily follows from the formula for the Jacobian.

Next, let σ, τ be real. We define the Hermitian form on $C^\infty(U(n))$ by the formula

$$\langle f_1, f_2 \rangle_{\sigma|\tau} := \int_{U(n)} \int_{U(n)} \det(1 - zu^*)^\sigma (1 - z^*u)^\tau f_1(z)\overline{f_2(u)}\,d\mu(z)\,d\mu(u). \tag{1.4}$$

Proposition 1.4. *The operators $\rho_{\sigma|\tau}(g)$ preserve the Hermitian form $\langle \cdot, \cdot \rangle_{\sigma|\tau}$.*

Theorem 1.5. *For σ, $\tau \notin \mathbb{Z}$, the Hermitian form $\langle \cdot, \cdot \rangle_{\sigma|\tau}$ is positive iff integer parts of numbers $-\sigma - n$ and τ are equal.*

In fact, the domain of positivity is the square $-1 < \tau < 0$, $-n < \sigma < -n + 1$ and its shifts by vectors $(-j, j)$, $j \in \mathbb{Z}$; see Fig. 5.

In particular, under this condition, a representation $\rho_{\sigma|\tau}$ is unitary.

For some values of (σ, τ) the form $\langle \cdot, \cdot \rangle_{\sigma|\tau}$ is positive semi-definite. The two most important such cases are:

1. For $\tau = 0$, we get the highest weight representations (or holomorphic representations). Thus, the Stein–Sahi representations are the nearest relatives of holomorphic representations.
2. For $\tau = 0$, $\sigma = 0, -1, -2, \ldots, -n$, we obtain some exotic "small" representations of $U(n,n)$.

1.3   The Structure of the Paper

We discuss only groups[3] $U(n, n)$.

In Sect. 2 we consider the case $n = 1$ and present the Pukanszky classification [37] of unitary representations of the universal covering group of $SL(2, \mathbb{R}) \simeq SU(1, 1)$.

In Sect. 3 we discuss Stein–Sahi representations of arbitrary $U(n, n)$. In Sect. 4 we explain the relationships of Stein–Sahi representations and holomorphic representations. In Sect. 5 we give explicit constructions of the Sahi "unipotent" representations.

In Sect. 6 we discuss some open problems of harmonic analysis.

1.4   Notation

Let $a, u, v \in \mathbb{C}$. Denote

$$a^{\{u|v\}} := a^u \overline{a}^v. \tag{1.5}$$

If $u - v \in \mathbb{Z}$, then this expression is well defined for all $a \neq 0$. However, the expression is well defined in many other situations, for instance, if $|1 - a| < 1$ and u, v are arbitrary (and even for $|1 - a| = 1, a \neq 1$).

The *norm* $\|z\|$ of an $n \times n$-matrix z is the usual norm of a linear operator in the standard Euclidean space $\mathbb{C}^n$.

We denote the Haar measure on the unitary group $U(n)$ by μ; assume that the complete measure of the group is 1.

---

[3]A comment for experts: Stein–Sahi representations of a semisimple Lie group G are complementary series induced from a maximal parabolic subgroup with an Abelian nilpotent radical.

The cases $G = U(n, n)$, $Sp(2n, \mathbb{R})$, and $G = SO^*(4n)$ (related to tube-type Hermitian symmetric spaces) are parallel. The only difficulty is Theorem 3.11 (the expansion of the integral kernel in characters); we choose $G = U(n, n)$ because this can be done by elementary tools. In the general Hermitian case, one can refer to the version of the Kadell integral [20] from [29] (the integrand is a product of a Jack polynomial and a Selberg-type factor).

For other series of groups, Stein–Sahi representations depend on one parameter, and picture is clear (in particular, inner products for degenerate ["unipotent"] representations can be written immediately). A BC analog of the Kadell integral is unknown (certainly, it must exist, and some special cases were evaluated in the literature; see, e.g., [30]). On the other hand, Stein–Sahi representations have multiplicity-free K-spectra. In such situation, there are a lot of ways to examine the of positivity of inner products; see, e.g., [5,41,42].

New elements of this paper are a "blow-up construction" for unipotent representations and (apparently) tame models for representations of universal coverings. The representations themselves were constructed in works of Sahi.

The *Pochhammer symbol* is given by

$$(a)_n := \frac{\Gamma(a+n)}{\Gamma(a)} = \begin{cases} a(a+1)\ldots(a+n-1), & \text{if } n \geqslant 0 \\ \frac{1}{(a-1)\ldots(a-n)}, & \text{if } n < 0. \end{cases} \tag{1.6}$$

2 Unitary Representations of SU(1, 1)

Denote by $SU(1,1)^\sim$ the universal covering group of $SU(1,1)$.

In this section, we present constructions of all irreducible unitary representations of $SU(1,1)^\sim$. According to the Bargmann–Pukanszky theorem, there are four types of such representations:

(1.) Unitary principal series
(2.) Complementary series
(3.) Highest-weight and lowest-weight representations
(4.) The one-dimensional representation

Models of these representations are given below.

The general Stein–Sahi representations are a strange "higher copy" of the $SU(1,1)$ picture.

References. The classification of unitary representations of $SL(2,\mathbb{R}) \simeq SU(1,1)$ was obtained by V. Bargmann [2]; it was extended to $SU(1,1)^\sim$ by L. Pukanszky [37]; see also P. Sally [45]. □

2.1 Preliminaries

2.1.1 Fourier Series and Distributions

By S^1 we denote the unit circle $|z| = 1$ in the complex plane $\mathbb{C}$. We parameterize S^1 by $z = e^{i\varphi}$.

By $C^\infty(S^1)$ we denote the space of smooth functions on S^1. Recall, that

$$f(\varphi) = \sum_{n=-\infty}^{\infty} a_n e^{in\varphi} \in C^\infty(S^1) \qquad \text{iff } |a_n| = o(|n|^{-L}) \quad \text{for all } L.$$

Recall that a distribution $h(\varphi)$ on the circle admits an expansion into a Fourier series:

$$h(\varphi) = \sum_{n=0}^{\infty} b_n e^{in\varphi}, \quad \text{where } |b_n| = O(|n|^L) \text{ for some } L.$$

For $s \in \mathbb{R}$, we define the *Sobolev space* $W^s(S^1)$ as the space of distributions

$$h(\varphi) = \sum_{n=0}^{\infty} b_n e^{in\varphi} \quad \text{such that} \sum |b_n|^2 (1 + |n|)^{2s} < \infty.$$

By definition, $W^0(S^1) = L^2(S^1)$. For a positive integer $s = k$, this condition is equivalent to $\frac{\partial^k}{\partial \varphi^k} h \in L^2(S^1)$. Evidently, $s < s'$ implies $W^s \supset W^{s'}$.

2.1.2   The Group SU(1, 1)

The group $SU(1, 1) \simeq SL(2, \mathbb{R})$ consists of all complex 2×2-matrices having the form

$$g = \begin{pmatrix} a & b \\ \bar{b} & \bar{a} \end{pmatrix}, \qquad \text{where } |a|^2 - |b|^2 = 1.$$

This group acts on the disk $|z| < 1$ and on the circle $|z| = 1$ by the *Möbius transformations*

$$z \mapsto (a + \bar{b}z)^{-1}(b + \bar{a}z).$$

2.1.3   A Model of the Universal Covering Group SU(1, 1)~

Recall that the fundamental group of $SU(1, 1)$ is $\mathbb{Z}$. A loop generating the fundamental group is

$$\mathfrak{R}(\varphi) = \begin{pmatrix} e^{i\varphi} & 0 \\ 0 & e^{-i\varphi} \end{pmatrix}, \qquad \mathfrak{R}(2\pi) = \mathfrak{R}(0) = 1. \tag{2.1}$$

Some examples of multi-valued continuous function on $SU(1, 1)$ are

$$\begin{pmatrix} a & b \\ \bar{b} & \bar{a} \end{pmatrix} \mapsto \ln a, \qquad \begin{pmatrix} a & b \\ \bar{b} & \bar{a} \end{pmatrix} \mapsto a^\lambda := a^{\lambda \ln a}.$$

We can realize $SU(1, 1)^\sim$ as a subset in $SU(1, 1) \times \mathbb{C}$ consisting of pairs

$$\left(\begin{pmatrix} a & b \\ \bar{b} & \bar{a} \end{pmatrix}, \sigma \right), \qquad \text{where } e^\sigma = a.$$

Thus, for a given matrix $\begin{pmatrix} a & b \\ \bar{b} & \bar{a} \end{pmatrix}$ the parameter σ ranges if the countable set $\sigma = \ln a + 2\pi k i$.

Define a multiplication in $SU(1, 1) \times \mathbb{C}$ by

$$(g_1, \sigma_1) \circ (g_2, \sigma_2) = (g_1 g_2, \sigma_1 + \sigma_2 + c(g_1, g_2)),$$

where $c(g_1, g_2)$ is the *Berezin–Guichardet* cocycle,

$$c(g_1, g_2) = \ln \frac{a_3}{a_1 a_2}.$$

Here a_3 is the matrix element of $g_3 = g_1 g_2$.

Theorem 2.1. (a) $\left| \frac{a_3}{a_1 a_2} - 1 \right| < 1$, *and therefore the logarithm is well defined.*
(b) *The operation* $\circ$ *determines the structure of a group on* $SU(1, 1) \times \mathbb{C}$.
(c) $SU(1, 1)^{\sim}$ *is a subgroup in the latter group.*

The proof is a simple and nice exercise.
Now we can define the single-valued function $\ln a$ on $SU(1, 1)$ by setting $\ln a := \sigma$.

2.2   Non-Unitary and Unitary Principal Series

2.2.1   Principal Series of Representations of SU(1, 1)

Fix $p, q \in \mathbb{C}$. For $g \in SU(1, 1)$, define the operator $T_{p|q}(g)$ in the space $C^\infty(S^1)$ by the formula

$$T_{p|q} \begin{pmatrix} a & b \\ \bar{b} & \bar{a} \end{pmatrix} f(z) = f \left(\frac{b + \bar{a}z}{a + \bar{b}z} \right) (a + \bar{b}z)^{\{-p|-q\}}. \tag{2.2}$$

We use the notation (1.5) for complex powers here.

Observation 2.2. (a) $T_{p|q}$ *is a well-defined operator-valued function on* $SU(1, 1)^{\sim}$.
(b) *It satisfies*

$$T_{p|q}(g_1) T_{p|q}(g_2) = T_{p|q}(g_1 g_2).$$

Proof. (a) First,

$$(a + \bar{b}z)^{-p} \overline{(a + \bar{b}z)}^{-q} = a^{-p} \cdot \bar{a}^{-q} (1 + a^{-1} \bar{b} z)^{-p} \overline{(1 + a^{-1} \bar{b} z)}^{-q}.$$

Since $|z| = 1$ and $|a| > |b|$, the last two factors are well defined. Next,

$$a^{-p} \bar{a}^{-q} := \exp\left\{ -p \ln a + q \overline{\ln a} \right\}$$

and $\ln a$ is a well-defined function on $SU(1, 1)^{\sim}$.

Proof of (b). One can verify this identity for g_1, g_2 near the unit and refer to the analytic continuation.   □

The representations $T_{p|q}(g)$ are called *representations of the principal (non-unitary) series*.

Remark. (a) A representation $T_{p|q}$ is a single-valued representation of SU(1, 1) iff $p - q$ is integer.

2.2.2 The Action of the Lie Algebra

The Lie algebra $\mathfrak{su}(1, 1)$ of SU(1, 1) consists of matrices

$$\begin{pmatrix} i\alpha & \beta \\ \overline{\beta} & -i\alpha \end{pmatrix}, \qquad \text{where } \alpha \in \mathbb{R}, \beta \in \mathbb{C}.$$

It is convenient to take the following basis in the complexification $\mathfrak{su}(1, 1)_{\mathbb{C}} = \mathfrak{sl}(2, \mathbb{C})$:

$$L_0 := \frac{1}{2}\begin{pmatrix} -1 & 0 \\ 0 & 1 \end{pmatrix}, \quad L_- := \begin{pmatrix} 0 & 1 \\ 0 & 0 \end{pmatrix}, \quad L_+ := \begin{pmatrix} 0 & 0 \\ -1 & 0 \end{pmatrix} \qquad (2.3)$$

These generators act in $C^\infty(S^1)$ by the following operators:

$$L_0 = z\frac{d}{dz} + \frac{1}{2}(p - q), \qquad L_- = \frac{d}{dz} - qz^{-1}, \qquad L_+ = z^2\frac{d}{dz} + pz. \qquad (2.4)$$

Equivalently,

$$L_0 z^n = \left(n + \frac{1}{2}(p - q)\right)z^n, \qquad L_- z^n = (n - q)z^{n-1}, \qquad L_+ z^n = (n + p)z^{n+1}. \tag{2.5}$$

2.2.3 Subrepresentations

Proposition 2.3. *A representation $T_{p|q}$ is irreducible iff $p, q \notin \mathbb{Z}$.*

Proof. Let $p, q \notin \mathbb{Z}$. Consider an L_0-eigenvector z^n. Then all vectors $(L_+)^k z^n$, $(L_-)^l z^n$ are nonzero. They span the whole space $C^\infty(S^1)$.   □

Observation 2.4. (a) *If $q \in \mathbb{Z}$, then z^q, z^{q+1}, ... span a subrepresentation in $T_{p|q}$.*
(b) *If $p \in \mathbb{Z}$, then z^{-p}, z^{-p-1}, z^{-p-2}, ... span a subrepresentation in $T_{p|q}$.*

Proof of (a). Clearly, our subspace is L^0-invariant and L^+-invariant. On the other hand, $L^- z^q = 0$, and we cannot leave our subspace.   □
   All possible positions of subrepresentations of $T_{p|q}$ are listed in Fig. 1.

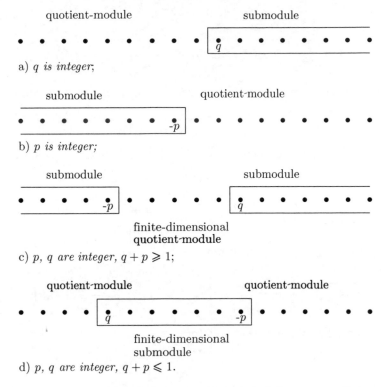

quotient-module   submodule

a) q *is integer;*

submodule   quotient-module

b) p *is integer;*

submodule   submodule

finite-dimensional
quotient-module

c) p, q *are integer,* $q + p \geqslant 1$;

quotient-module   quotient-module

finite-dimensional
submodule

d) p, q *are integer,* $q + p \leqslant 1$.

Fig. 1 Subrepresentations of the principal series. *Black circles* indicate vectors z^n. A representation $T_{p|q}$ is reducible iff $p \in \mathbb{Z}$ or $q \in \mathbb{Z}$

2.2.4   Shifts of Parameters

Observation 2.5. *If k is an integer, then $T_{p+k|q-k} \simeq T_{p|q}$. The intertwining operator is*

$$Af(z) = z^k f(z).$$

A verification is trivial.   □

2.2.5   Duality

Consider the bilinear map

$$\Pi : C^\infty(S^1) \times C^\infty(S^1) \to \mathbb{C}$$

given by

$$(f_1, f_2) \mapsto \frac{1}{2\pi} \int_0^{2\pi} f_1(e^{i\varphi}) f_2(e^{i\varphi}) \, d\varphi = \frac{1}{2\pi} \int_0^{2\pi} f_1(z) f_2(z) \frac{dz}{z}. \qquad (2.6)$$

Observation 2.6. *Representations* $T_{p|q}$ *and* $T_{1-p|1-q}$ *are dual with respect to* Π; *i.e.*,

$$\Pi\left(T_{p|q}(g)f_1, T_{1-p|1-q}(g)f_2\right) = \Pi(f_1, f_2). \qquad (2.7)$$

Proof. After simple cancellations, we get the following expression on the left-hand side of (2.7):

$$\frac{1}{2\pi i} \int_{|z|=1} f_1\left(\frac{b + \bar{a}z}{a + \bar{b}z}\right) f_2\left(\frac{b + \bar{a}z}{a + \bar{b}z}\right) \cdot (a + \bar{b}z)^{-1} \overline{(a + \bar{b}z)}^{-1} \frac{dz}{z}.$$

Keeping in mind $\bar{z} = z^{-1}$, we transform

$$(a + \bar{b}z)^{-1} \overline{(a + \bar{b}z)}^{-1} \frac{dz}{z} = (a + \bar{b}z)^{-1} (\bar{b} + az)^{-1} \, dz = \left(\frac{b + \bar{a}z}{a + \bar{b}z}\right)^{-1} d\left(\frac{b + \bar{a}z}{a + \bar{b}z}\right).$$

Now the integral comes into the desired form:

$$\frac{1}{2\pi i} \int_{|u|=1} f_1(u) \, f_2(u) \, \frac{du}{u}. \qquad \square$$

We also define a sesquilinear map

$$\Pi^* : C^\infty(S^1) \times C^\infty(S^1) \to \mathbb{C}$$

by

$$\Pi^*(f_1, f_2) := \Pi(f_1, \bar{f}_2) = \int_0^{2\pi} f_1(z) \overline{f_2(z)} \frac{dz}{z}. \qquad (2.8)$$

Observation 2.7. *Representations* $T_{p|q}$ *and* $T_{1-\bar{q}|1-\bar{p}}$ *are dual with respect to* Π^*.

The proof is the same. $\qquad \square$

2.2.6 Intertwining Operators

Consider the integral operator

$$I_{p|q} : C^\infty(S^1) \to C^\infty(S^1)$$

given by

$$I_{p|q} f(u) = \frac{1}{2\pi i \, \Gamma(p+q-1)} \int_{|z|=1} (1 - z\bar{u})^{\{p-1|q-1\}} f(z) \frac{dz}{z}, \qquad (2.9)$$

where the function $(1 - z\bar{u})^{\{p-1|q-1\}}$ is defined by

$$(1 - z\bar{u})^{\{p-1|q-1\}} := \lim_{t \to 1^-} (1 - tz\bar{u})^{\{p-1|q-1\}} \qquad (2.10)$$

The integral converges if $\mathrm{Re}(p+q) > -1$.

Theorem 2.8. *The map* $(p|q) \mapsto I_{p|q}$ *admits the analytic continuation to a holomorphic operator-valued function on* $\mathbb{C}^2$.

Theorem 2.9. *The operator* $I_{p|q}$ *intertwines* $T_{p|q}$ *and* $T_{1-q|1-p}$; *i.e.,*

$$T_{1-p|1-q}(g) \, I_{p|q} = I_{p|q} \, T_{p|q}(g).$$

Corollary 2.10. *If* $p \notin \mathbb{Z}$, $q \notin \mathbb{Z}$, *then the representations* $T_{p|q}$ *and* $T_{1-q|1-p}$ *are equivalent.*

2.2.7   Proof of Theorems 2.8 and 2.9

Lemma 2.11. *The expansion of the distribution* (2.10) *into the Fourier series is given by*

$$(1 - z\bar{u})^{p-1}(1 - \bar{z}u)^{q-1} = \frac{\Gamma(p+q-1)}{\Gamma(p)\Gamma(q)} \sum_{n=-\infty}^{\infty} \frac{(1-q)_n}{(p)_n} \left(\frac{z}{u}\right)^n \qquad (2.11)$$

$$= \Gamma(p+q-1) \sum_{n=-\infty}^{\infty} \frac{(-1)^n}{\Gamma(p+n)\Gamma(q-n)} \left(\frac{z}{u}\right)^n. \qquad (2.12)$$

Proof. Let $\mathrm{Re}\, p$, $\mathrm{Re}\, q$ be sufficiently large. Then we write

$$(1 - z\bar{u})^{p-1}(1 - \bar{z}u)^{q-1} = \left[\sum_{j \geq 0} \frac{(1-p)_j}{j!} \left(\frac{z}{u}\right)^j\right] \cdot \left[\sum_{l \geq 0} \frac{(1-q)_l}{l!} \left(\frac{u}{z}\right)^l\right] \qquad (2.13)$$

and open brackets in (2.13). For instance, the coefficient at $(z/u)^0$ is

$$\sum_{k \geq 0} \frac{(1-p)_k(1-q)_k}{k!k!} = {}_2F_1(1-p, 1-q; 1; 1),$$

where $_2F_1$ is the Gauss hypergeometric function. We evaluate the sum with the Gauss summation formula for $_2F_1(1)$; see [18], (2.1.14).   □

Proof of Theorem 2.8. Denote by

$$c_n := \frac{(-1)^n}{\Gamma(p+n)\Gamma(q-n)}$$

the Fourier coefficients in (2.12). Evidently, c_n admits a holomorphic continuation to the whole plane[4] $\mathbb{C}^2$.

By [18], (1.18.4),

$$\frac{\Gamma(n+a)}{\Gamma(n+b)} \sim |n|^{a-b} \qquad \text{as } n \to \pm\infty.$$

Keeping (2.11) in mind, we get

$$c_n \sim \text{const} \cdot |n|^{1-p-q} \qquad \text{as } n \to \pm\infty. \tag{2.14}$$

Then

$$I_{p|q} : z^n \mapsto c_{-n} z^n$$

and

$$I_{p|q} : \sum a_n z^n \mapsto \sum a_n c_{-n} z^n.$$

Obviously, this map sends smooth functions to smooth functions.   □

Proof of Corollary 2.10. In this case, all $c_n \neq 0$.   □

Proof of Theorem 2.9. The calculation is straightforward:

$$T_{1-q|1-p}(g) I_{p|q} f(u)$$

$$= \frac{1}{2\pi i} (a + \bar{b}u)^{\{q-1|p-1\}} \int_{|u|=1} \left(1 - \left(\frac{b + \bar{a}u}{a + \bar{b}u}\right)\bar{z}\right)^{\{q-1|p-1\}} f(z)\, \frac{dz}{z}.$$

Next, we observe

$$(a + \bar{b}u)\left(1 - \left(\frac{b + \bar{a}u}{a + \bar{b}u}\right)\bar{z}\right) = (a - b\bar{z})\left(1 - u\left(\frac{-\bar{b} + \bar{a}\bar{z}}{a - b\bar{z}}\right)\right)$$

---

[4]The Gamma function $\Gamma(z)$ has simple poles at $z = 0, -1, -2, \ldots$ and does not have zeros. Therefore, $1/(\Gamma(p+n)\Gamma(q-n))$ has zeros at $p = -n, -n-1, \ldots$ and at $q = n, n-1, \ldots$.

In particular, if both p, q are integers and $q < p$, when $I_{p|q} = 0$.

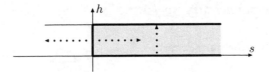

Fig. 2 The unitary principal series in coordinates
$h = (p - q + 1)/2, s = \frac{1}{i}(p + q - 1)/2$.
Equivalently,
$p = h + is, q = 1 - h + is$.
The shift $h \mapsto h + 1$ does not change a representation. Also, the symmetry $s \mapsto -s$ sends
a representation to an equivalent one. Therefore, representations of the principal series are
enumerated by the a semi-strip $0 \leqslant h < 1, s \geqslant 0$. It is more reasonable to think that representations
of the unitary principal series are enumerated by points of a semicylinder (s, h), where $s \geqslant 0$ and
h is defined modulo equivalence $h \sim h + k$, where $h \in \mathbb{Z}$

and come to

$$\frac{1}{2\pi i} \int_{|z|=1} \left(1 - u\left(\frac{-\overline{b} + \overline{a}z}{a - b\overline{z}}\right)\right)^{\{q-1|p-1\}} (a - b\overline{z})^{\{q-1|p-1\}} f(z) \frac{dz}{z}.$$

Now we change a variable again,

$$z = \frac{b + \overline{a}w}{a + \overline{b}w}, \qquad \overline{w} = \frac{-\overline{b} + \overline{a}\,\overline{z}}{a - b\overline{z}},$$

and come to the desired expression:

$$\frac{1}{2\pi i} \int_{|w|=1} (1 - u\overline{w})^{\{p-1|q-1\}} f\left(\frac{b + \overline{a}w}{a + \overline{b}w}\right) (a + \overline{b}w)^{\{-p|-q\}} \frac{dw}{w}.$$

2.2.8 The Unitary Principal Series

Observation 2.12. *A representation $T_{p|q}$ is unitary in $L^2(S^1)$ iff*

$$\operatorname{Im} p = \operatorname{Im} q, \qquad \operatorname{Re} p + \operatorname{Re} q = 1. \qquad (2.15)$$

The proof is straightforward; also, this follows from Observation 2.7.   □

2.3   The Complementary Series

2.3.1   The Complementary Series

Now let

$$0 < p < 1, \qquad 0 < q < 1. \tag{2.16}$$

Consider the Hermitian form on $C^\infty(S^1)$ given by

$$\langle f_1, f_2 \rangle_{p|q} = \frac{1}{(2\pi i)^2 \, \Gamma(p+q-1)} \int_{|z|=1} \int_{|u|=1} (1 - z\bar{u})^{\{p-1|q-1\}} f_1(z) \overline{f_2(u)} \, \frac{dz}{z} \frac{du}{u}. \tag{2.17}$$

By (2.12),

$$\langle z^n, z^m \rangle_{p|q} = \frac{1}{\Gamma(p)\Gamma(q)} \frac{(1-q)_n}{(p)_n} \cdot \delta_{m,n}. \tag{2.18}$$

Theorem 2.13. *If $0 < p < 1$, $0 < q < 1$, then the inner product (2.17) is positive definite.*

Proof. Indeed, in this case all coefficients

$$\frac{(1-q)_n}{(p)_n} = \frac{(1-p)_{-n}}{(q)_{-n}}$$

in (2.17) are positive. □

Theorem 2.14. *Let $0 < p < 1$, $0 < q < 1$. Then the representation $T_{p|q}$ is unitary with respect to the inner product $\langle \cdot, \cdot \rangle_{p|q}$; i.e.,*

$$\langle T_{p|q}(g) f_1, T_{p|q}(g) f_2 \rangle_{p|q} = \langle f_1, f_2 \rangle_{p|q}.$$

Proof. This follows from Theorem 2.9 and Observation 2.7. Indeed,

$$\langle f_1, f_2 \rangle_{p|q} = \Pi^*(I_{p|q} f_1, f_2)$$

and

$$\Pi^*(I_{p|q} T_{p|q}(g) f_1, T_{p|q}(g) f_2) = \Pi^*(T_{1-q|1-p}(g) I_{p|q} f_1, T_{p|q}(g) f_2)$$

$$= \Pi^*(I_{p|q} f_1, f_2) = \langle f_1, f_2 \rangle_{p|q}.$$

Fig. 3 The complementary
series. The diagonal is
contained in the principal
series (the segment of the axis
Oh in Fig. 2). The symmetry
with respect to the diagonal
sends a representation to an
equivalent representation

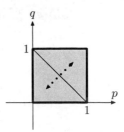

Keeping in mind our future purposes, we propose another (homotopic) proof
(Fig. 3). Substitute

$$z = \frac{b + \bar{a}z'}{a + \bar{b}z'}, \qquad u = \frac{b + \bar{a}u'}{a + \bar{b}u'}$$

into the integral in (2.17). Applying the identity

$$1 - \left(\frac{b + \bar{a}z'}{a + \bar{b}z'}\right)\overline{\left(\frac{b + \bar{a}u'}{a + \bar{b}u'}\right)} = (a + \bar{b}z')^{-1}(1 - z'\bar{u'})\overline{(a + \bar{b}u')}^{-1},$$

we get

$$\langle T_{p|q}(g)f_1, T_{p|q}(g)f_2\rangle_{p|q}. \qquad\qquad \square$$

2.3.2 Sobolev Spaces

Denote by $\mathcal{H}_{p|q}$ the completion of $C^\infty(S^1)$ with respect to the inner product of the
complementary series.

First, we observe that the principal series and the complementary series have an
intersection [see (2.15), (2.16)], namely, the interval

$$p + q = 1, \qquad 0 < p < 1.$$

In this case the inner product (2.18) is the L^2-inner product; i.e., $\mathcal{H}_{p|1-p} \simeq L^2(S^1)$.

Next consider arbitrary (p, q), where $0 < p < 1, 0 < q < 1$. By (2.14), the
space $\mathcal{H}_{p|q}$ consists of Fourier series $\sum a_n z^n$ such that

$$\sum_{n=-\infty}^{\infty} |a_n|^2 n^{1-p-q} < \infty.$$

Thus, $\mathcal{H}_{p,q}$ is the Sobolev space $W^{(1-p-q)/2}(S^1)$.

2.4 Holomorphic and Anti-Holomorphic Representations

Denote by D the disk $|z| < 1$ in $\mathbb{C}$.

2.4.1 Holomorphic (Highest-Weight) Representations

Set $q = 0$, and

$$T_{p|0} f(z) = f\left(\frac{b + \bar{a}z}{a + \bar{b}z}\right)(a + \bar{b}z)^{-p}.$$

Since $|a| > |b|$, the factor $(a + \bar{b}z)^{-p}$ is holomorphic in the disk D. Therefore, the space of holomorphic functions in D is $SU(1,1)^{\sim}$-invariant. Denote the representation of $SU(1,1)^{\sim}$ in the space of holomorphic functions by T_p^+.

Theorem 2.15. (a) For $p > 0$, the representation T_p^+ is unitary, and the invariant inner product in the space of holomorphic functions is

$$\left\langle \sum_{n \geqslant 0} a_n z^n, \sum_{n \geqslant 0} b_n z^n \right\rangle = \sum_{n > 0} \frac{n!}{(p)_n} a_n \bar{b}_n. \tag{2.19}$$

(b) For $p > 1$, the invariant inner product admits the following integral representation:

$$\langle f_1, f_2 \rangle = \frac{p-1}{\pi} \iint_{|z|<1} f_1(z) \overline{f_2(z)} \, (1 - |z|^2)^{p-2} d\lambda(z),$$

where $d\lambda(z)$ is the Lebesgue measure in the disk.

(c) For $p = 1$, the invariant inner product is

$$\langle f_1, f_2 \rangle = \frac{1}{2\pi} \int_0^{2\pi} f_1(e^{i\varphi}) \overline{f_2(e^{i\varphi})} \, d\varphi = \frac{1}{2\pi i} \int_{|z|=1} f_1(z) \overline{f_2(z)} \, \frac{dz}{z}. \tag{2.20}$$

We denote this Hilbert space of holomorphic functions by $\mathcal{H}_p^+$.

Proof. The invariance of inner products in (b) and (c) can be easily verified by straightforward calculations.

To prove (a), we note that weight vectors z^n must be pairwise orthogonal.

Next, operators of the Lie algebra $\mathfrak{su}(1,1)$ must be skew-self-adjoint. The generators of the Lie algebra must satisfy

$$(L_+)^* = L_-.$$

Therefore,

$$\langle L_+ z^n, z^{n+1} \rangle = \langle z^n, L_- z^{n+1} \rangle$$

or

$$(n + p) \langle z^{n+1}, z^{n+1} \rangle = (n + 1) \langle z^n, z^n \rangle .$$

This implies(a).

If $p = 1$, then $\langle z^n, z^n \rangle = 1$ for $n \geqslant 0$; i.e., we get the L^2-inner product. $\square$

The theorem does not provide us with an explicit integral formula for the inner product in $\mathcal{H}_p^+$ if $0 < p < 1$. There is another way to describe inner products in spaces of holomorphic functions.

2.4.2 Reproducing Kernels

Theorem 2.16. *For each $p > 0$, for any $f \in \mathcal{H}_p^+$, and for each $a \in D$,*

$$\langle f(z), (1 - z\overline{a})^{-p} \rangle = f(a) \qquad (\textit{the reproducing property}). \qquad (2.21)$$

Proof. Indeed,

$$\left\langle \sum a_n z^n, \sum \frac{(p)_n}{n!} z^n \overline{u}^n \right\rangle = \sum a_n \frac{(p)_n}{n!} u^n \langle z^n, z^n \rangle = \sum a_n u^n = f(u). \qquad \square$$

In fact, the identity (2.21) is an all-sufficient definition of the inner product. We will not discuss this (see [10, 31]), and prefer another way.

2.4.3 Realizations of Holomorphic Representations in Quotient Spaces

Consider the representation $T_{-1|-1-p}$ of the principal series,

$$T_{-1|-1-q} f(z) = f \left(\frac{b + \overline{a}z}{a + \overline{b}z} \right) (a + \overline{b}z)^{-1} \overline{(a + \overline{b}z)}^{-1-p} .$$

The corresponding invariant Hermitian form in $C^\infty(S^1)$, is

$$\langle f_1, f_2 \rangle_{-1|-1-p} = \frac{1}{(2\pi i)^2} \int_{|z|=1} \int_{|u|=1} (1 - \overline{z}u)^{-p} f_1(z) \overline{f_2(u)} \frac{dz}{z} \frac{du}{u}. \qquad (2.22)$$

[we write another pre-integral factor in comparison with (2.17)]. The integral diverges for $p > 1$. However, we can define the inner product by

$$\langle z^n, z^n \rangle = \begin{cases} \frac{(p)_n}{n!} & \text{if } n \geqslant 0 \\ 0 & \text{if } n < 0 \end{cases} ;$$

the latter definition is valid for all $p > 0$.

We denote by $L \subset C^\infty(S^1)$ the subspace consisting of the series $\sum_{n<0} a_n z^n$. This subspace is $\mathrm{SU}(1,1)$-invariant and our form is nondegenerate and positive definite on the quotient space $C^\infty(S^1)/L$.

Next, we consider the intertwining operator

$$\widetilde{I}_{-1|-1-p} : C^\infty(S^1) \to C^\infty(S^1)$$

as above (but we change a normalization of the integral):

$$\widetilde{I}_{-1|-1-p} f(u) = \frac{1}{2\pi i} \int_{|z|=1} (1 - \bar{z}u)^{-p} f(z) \frac{dz}{z}$$

The kernel of the operator is L and the image consists of holomorphic functions.

Observation 2.17. *(a) The operator $I_{-1|-1-p}$ is a unitary operator*

$$C^\infty(S^1)/L \to \mathcal{H}_p^+.$$

(b) The representation $T_{-1|-1-p}$ in $C^\infty(S^1)/L$ is equivalent to the highest-weight representation T_p^+.

2.4.4 Lowest-Weight Representations

Now set $p = 0$, $q > 0$. Then operators $T_{0|q}$ preserve the subspace consisting of "antiholomorphic" functions $\sum_{n \leq 0} a_n z^n$. Denote by T_q^- the corresponding representation in the space of antiholomorphic functions. These representations are unitary.

We omit further discussion because these representations are twins of highest-weight representations (Fig. 4).

2.5 The Blow-Up Trick

Here we discuss a trick that produces "unipotent" representations of $\mathrm{U}(n,n)$ for $n \geq 2$; see Sect. 5.2.

2.5.1 The Exotic Case $p = 1, q = 0$

In this case,

$$T_{1|0} = T_1^+ \oplus T_1^-.$$

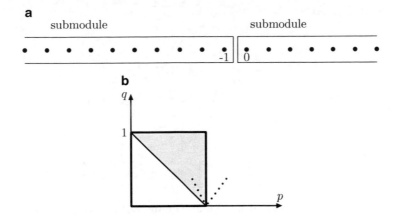

Fig. 4 (**a**) The structure of the representation $T_{1|0}$.
(**b**) Ways to reach $(p, q) = (1, 0)$ from different directions give origins to different invariant Hermitian forms on $T_{1|0}$. By our normalization, the inner product is positive definite in the *gray triangle* and negative definite in the *white triangle*. Therefore, coming to $(1, 0)$ from the *gray triangle*, we get a positive form

Let us discuss the behavior of the inner product of the complementary series near the point $(p|q) = (1|0)$:

$$\langle f_1, f_2 \rangle_{p|q} = \frac{1}{(2\pi i)^2} \int_{|z|=1} (1 - z\bar{u})^{\{p-1|q-1\}} f_1(z) \overline{f_2(z)} \, \frac{dz}{z}. \tag{2.23}$$

Consider the limit of this expression as $p \to 1$, $q \to 0$. The Fourier coefficients of the kernel are the following meromorphic functions:

$$c_n(p, q) = \frac{(-1)^n \Gamma(p + q - 1)}{\Gamma(q - n) \Gamma(p + n)}.$$

Note that

1. $c_n(p, q)$ has a pole at the line $p + q = 1$.
2. For $n \geqslant 0$, the function $c_n(p, q)$ has a zero on the line $q = 0$.
3. For $n < 0$, the function $c_n(p, q)$ has a zero at the line $p = 0$.

Thus, our point $(p, q) = (1, 0)$ lies on the intersection of a pole and of a zero of the function $c_n(p, q)$. Let us substitute

$$p = 1 + \varepsilon s \qquad q = \varepsilon t, \qquad \text{where } s + t \neq 0,$$

to $c_n(p, q)$ and pass to the limit as $\varepsilon \to 0$. Recall that

$$\Gamma(z) = \frac{(-1)^n}{n!\,(z + n)} + O(1), \qquad \text{as } z \to -n, \text{ where } n = 0, 1, 2, \ldots. \tag{2.24}$$

Therefore, we get

$$\lim_{\varepsilon \to 0} c_n(1 + \varepsilon s, \varepsilon t) = \begin{cases} \frac{t}{t+s} & \text{if } n \geqslant 0 \\ -\frac{s}{t+s} & \text{if } n < 0. \end{cases}$$

In particular, for $s = 0$ we get the T_1^+-inner product, and for $t = 0$ we get the T_1^--inner product. Generally,

$$\lim_{\varepsilon \to 0} \left\langle \sum_{n=-\infty}^{\infty} a_n z^n, \sum_{n=-\infty}^{\infty} b_n z^n \right\rangle_{1+\varepsilon s | \varepsilon t} = \frac{t}{t+s} \sum_{n=0}^{\infty} a_n \overline{b}_n - \frac{s}{t+s} \sum_{n=-\infty}^{-1} a_n \overline{b}_n.$$

Therefore, we get a one-parametric family of invariant inner products for $T_{1|0}$. However, all of them are linear combinations of two basis inner products mentioned above ($t = 0$ and $s = 0$).

3 Stein–Sahi Representations

Here we extend constructions of the previous section to the groups $G := \mathrm{U}(n,n)$. The analogy of the circle S^1 is the space $\mathrm{U}(n)$ of unitary matrices.

3.1 Construction of Representations

3.1.1 Distributions $\ell_{\sigma|\tau}$

Let z be an $n \times n$ matrix with norm < 1. For $\sigma \in \mathbb{C}$, we define the function $\det(1-z)^\sigma$ by

$$\det(1 - z)^\sigma := \det\left[1 - \sigma z + \frac{\sigma(\sigma - 1)}{2!} z^2 - \frac{\sigma(\sigma - 1)(\sigma - 2)}{3!} z^3 + \dots \right].$$

Extend this function to matrices z satisfying $\|z\| \leqslant 1$, $\det(1-z) \neq 0$ by

$$\det(1 - z)^\sigma := \lim_{u \to z, \, \|u\| < 1} \det(1 - u)^\sigma.$$

The expression $\det(1 - z)^\sigma$ is continuous in the domain $\|z\| \leqslant 1$ except for the surface $\det(1 - z) = 0$.

Denote by $\det(1 - z)^{\{\sigma|\tau\}}$ the function

$$\det(1 - z)^{\{\sigma|\tau\}} := \det(1 - z)^\sigma \det(1 - \overline{z})^\tau.$$

We define the function $\ell_{\sigma|\tau}(g)$ on the unitary group $U(n)$ by

$$\ell_{\sigma|\tau}(z) := 2^{-(\sigma+\tau)n} \det(1-z)^{\{\sigma|\tau\}}. \tag{3.1}$$

Obviously,

$$\ell_{\sigma|\tau}(h^{-1}zh) = \ell_{\sigma|\tau}(z) \qquad \text{for } z, h \in U(n). \tag{3.2}$$

Lemma 3.1. *Let* $e^{i\psi_1}, \ldots, e^{i\psi_n}$, *where* $0 \leqslant \psi_k < 2\pi$, *be the eigenvalues of* $z \in U(n)$. *Then*

$$\ell_{\sigma|\tau}(z) = \exp\left\{\frac{i}{2}(\sigma-\tau)\sum_k(\psi_k-\pi)\right\} \prod_{k=1}^n \sin^{\sigma+\tau}\frac{\psi_k}{2}. \tag{3.3}$$

Proof. It suffices to verify the statement for diagonal matrices; equivalently, we must check the identity

$$(1-e^{i\psi})^{\{\sigma|\tau\}} = \exp\left\{\frac{i}{2}(\sigma-\tau)(\psi-\pi)\right\} \sin^{\sigma+\tau}\frac{\psi}{2}.$$

We have

$$\frac{1}{2}(1-e^{i\psi}) = \exp\left\{\frac{i}{2}(\psi-\pi)\right\} \sin\frac{\psi}{2}.$$

Further, both the sides of the equality

$$2^{-\sigma}(1-e^{i\psi})^\sigma = \exp\left\{\frac{i}{2}\sigma(\psi-\pi)\right\} \sin^\sigma\frac{\psi}{2},$$

are real-analytic on $(0, 2\pi)$ and the substitution $\psi = \pi$ gives 1 on both sides. $\qquad\square$

3.1.2 Positivity

Let $\text{Re}(\sigma + \tau) < 1$. Consider the sesquilinear form on $C^\infty(U(n))$ given by

$$\langle f_1, f_2 \rangle_{\sigma|\tau} = \iint_{U(n)\times U(n)} \ell_{\sigma|\tau}(zu^{-1}) f_1(z)\overline{f_2(u)}\, d\mu(z)\, d\mu(u). \tag{3.4}$$

For $\sigma, \tau \in \mathbb{R}$ this form is Hermitian; i.e.,

$$\langle f_2, f_1 \rangle_{\sigma|\tau} = \overline{\langle f_1, f_2 \rangle_{\sigma|\tau}}.$$

Observation 3.2. *For fixed* $f_1, f_2 \in C^\infty(U(n))$, *this expression admits a mero-morphic continuation in* σ, τ *to the whole* $\mathbb{C}^2$.

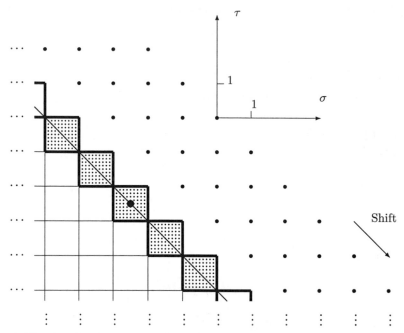

1. The dotted squares correspond to unitary representations $\rho_{\sigma|\tau}$.
2. Vertical and horizontal rays in the south-west of Figure correspond to nondegenerate highest weight and lowest weight representations. Fat points correspond to degenerated highest and lowest weight representations, and also to the unipotent representations. The point $(\sigma, \tau) = (0, 0)$ corresponds to the trivial one-dimensional representation.
3. In points of the thick segments, we have some exotic unitary sub-quotients.
4. The shift $(\sigma, \tau) \mapsto (\sigma + 1, \tau - 1)$ send a representation $\rho_{\sigma|\tau}$ of $\mathrm{SU}(n,n)^\sim$ to an equivalent representation.
5. The permutation of the axes $(\tau, \sigma) \mapsto (\sigma, \tau)$ gives a complex conjugate representation.
6. The symmetry with respect to the point $(-n/2, -n/2)$ (black circle) gives a dual representation (for odd n this point is a center of a dotted square; for even n this point is a common vertex of two dotted squares).
7. For $\sigma + \tau = n$ (the diagonal line) our Hermitian form is the standard L^2-product.
8. Linear (non-projective) representations of $\mathrm{U}(n,n)$ correspond to the family of parallel lines $\sigma - \tau \in \mathbb{Z}$.

Fig. 5 Unitarizability conditions for $\mathrm{U}(n, n)$. The case $n = 5$

This follows from general facts about distributions; however, this fact is a corollary of the expansion of the distributions $\ell_{\sigma|\tau}$ in characters; see Theorem 3.11. This expansion also implies the following theorem:

Theorem 3.3. *For $\sigma, \tau \in \mathbb{R} \setminus \mathbb{Z}$, the inner product (3.4) is positive definite (up to a sign) iff integer parts of $-\sigma - n$ and τ are equal.*

The domain of positivity is the union of the dotted squares in Fig. 5.

For σ, τ satisfying this theorem, denote by $\mathcal{H}_{\sigma|\tau}$ the completion of $C^\infty(\mathrm{U}(n))$ with respect to our inner product.

3.1.3 The Group $\mathrm{U}(n, n)$

Consider the linear space $\mathbb{C}^n \oplus \mathbb{C}^n$ equipped with the indefinite Hermitian form

$$\{v \oplus w, v' \oplus w'\} = \langle v, v'\rangle_{\mathbb{C}^n \oplus 0} - \langle w, w'\rangle_{0 \oplus \mathbb{C}^n}, \qquad (3.5)$$

where $\langle \cdot, \cdot \rangle$ is the standard inner product in $\mathbb{C}^n$. Denote by $\mathrm{U}(n, n)$ the group of linear operators in $\mathbb{C}^n \oplus \mathbb{C}^n$ preserving the form $\{\cdot, \cdot\}$. We write elements of this group as block $(n + n) \times (n + n)$ matrices $g := \begin{pmatrix} a & b \\ c & d \end{pmatrix}$. By definition, such matrices satisfy the condition

$$g \begin{pmatrix} 1 & 0 \\ 0 & -1 \end{pmatrix} g^* = \begin{pmatrix} 1 & 0 \\ 0 & -1 \end{pmatrix}. \qquad (3.6)$$

Lemma 3.4. *The following formula*

$$z \mapsto z^{[g]} := (a + zc)^{-1}(b + zd), \qquad z \in \mathrm{U}(n), \ g = \begin{pmatrix} a & b \\ c & d \end{pmatrix} \in \mathrm{U}(n, n), \quad (3.7)$$

determines an action of the group $\mathrm{U}(n, n)$ *on the* space $\mathrm{U}(n)$.

The proof is given in Sect. 3.3.3.

3.1.4 Representations $\rho_{\sigma|\tau}$ of $\mathrm{U}(n, n)$

Denote by $\mathrm{U}(n, n)^\sim$ the universal covering of the group $\mathrm{U}(n, n)$; for details, see Sect. 3.3.1. Fix σ, $\tau \in \mathbb{C}$. We define an action of $\mathrm{U}(n, n)^\sim$ in the space $C^\infty(\mathrm{U}(n))$ by the linear operators

$$\rho_{\sigma|\tau}(g) f(z) = f(z^{[g]}) \det^{\{-n-\tau|-n-\sigma\}}(a + zc). \qquad (3.8)$$

We must explain the meaning of the complex power in this formula. First,

$$a + zc = (1 + zca^{-1})a.$$

The defining (3.6) implies $\|ca^{-1}\| < 1$. Hence, for all matrices z satisfying $\|z\| \leqslant 1$, complex powers of $1 + zca^{-1}$ are well defined. Next,

$$\det(a)^{-n-\tau|-n-\sigma} := \exp\left\{-(n + \tau)\ln\det a - (n + \sigma)\overline{\ln\det a}\right\}$$

$$m_n \qquad\qquad\qquad m_3\ m_2 \qquad\qquad m_1$$

□ □ □ ■ □ □ □ □ □ □ □ □ ■ ■ □ □ □ ■ □

Fig. 6 A *Maya diagram* for signatures. We draw the integer "*line*" and fill the boxes $m_1, \ldots, m_n$ with *black*

It is a well-defined function on $U(n,n)^{\sim}$. We set

$$\det(a + zc)^{-n-\tau|-n-\sigma} := \det\left[(1 + zca^{-1})^{-n-\tau|-n-\sigma}\right] \det(a)^{-n-\tau|-n-\sigma}$$

3.1.5   The Stein–Sahi Representations

Proposition 3.5. *The operators $\rho_{\sigma|\tau}(g)$ preserve the form $\langle \cdot, \cdot \rangle_{\sigma|\tau}$.*

The proof is given in Sect. 3.3.6.

Corollary 3.6. *For σ, τ satisfying the positivity conditions of Theorem 3.3, the representation $\rho_{\sigma|\tau}$ is unitary in the Hilbert space $\mathcal{H}_{\sigma|\tau}$.*

3.1.6   The Degenerate Principal Series

Proposition 3.7. *Let $\mathrm{Re}(\rho + \sigma) = -n$, $\mathrm{Im}\,\sigma = \mathrm{Im}\,\tau$. Then the representation $\rho_{\sigma|\tau}$ is unitary in $L^2(U(n))$.*

Proposition 3.8.

$$\left.\frac{\langle f_1, f_2 \rangle_{\sigma|\tau}}{\prod_{j=1}^{n} \Gamma(\sigma + \tau + j)}\right|_{\sigma=-n-\tau} = \mathrm{const} \cdot \int_{U(n)} f_1(u)\overline{f_2(u)}\,\mathrm{d}\mu(u).$$

3.1.7   Shifts of Parameters

Proposition 3.9. *For integer k,*

$$\rho_{\sigma+k|\tau-k} \simeq (\det g)^k \cdot \rho_{\sigma|\tau}.$$

The intertwining operator is multiplication by the determinant

$$F(z) \mapsto F(z)\det(z)^k.$$

This operator also defines an isometry of the corresponding Hermitian forms (Fig. 6).

3.2   Expansions of Distributions $\ell_{\sigma|\tau}$ in Characters. Positivity

3.2.1   Characters of $U(n)$

See Weyl's book [49]. The set of finite-dimensional representations of $U(n)$ is parameterized by collections of integers (*signatures*)

$$\mathbf{m}: \quad m_1 > m_2 > \cdots > m_n.$$

The character $\chi_\mathbf{m}$ of the representation[5] $\pi_\mathbf{m}$ (a *Schur function*) corresponding to a signature $\mathbf{m}$ is given by

$$\chi_\mathbf{m}(z) = \frac{\det_{k,j=1,2,\ldots,n}\{e^{im_j\psi_k}\}}{\det_{k,j=1,2,\ldots,n}\{e^{i(j-1)\psi_k}\}}, \tag{3.9}$$

where $e^{i\psi_k}$ are the eigenvalues of z. Recall that the denominator admits the decomposition

$$\det_{k,j}\{e^{i(j-1)\psi_k}\} = \prod_{l<k}(e^{i\psi_l} - e^{i\psi_k}). \tag{3.10}$$

The dimension of π_m is

$$\dim \pi_\mathbf{m} = \chi_\mathbf{m}(1) = \frac{\prod_{0\leq\alpha<\beta\leq n}(m_\alpha - m_\beta)}{\prod_{j=1}^n j!}. \tag{3.11}$$

3.2.2   Central Functions

A function $F(z)$ on $U(n)$ is called *central* if

$$F(h^{-1}zh) = F(z) \qquad \text{for all } z, h \in U(n).$$

In particular, characters and $\ell_{\sigma|\tau}$ are central functions.
  For central functions F on $U(n)$, the following *Weyl integration formula* holds:

$$\int_{U(n)} F(z)\,d\mu(z) = \frac{1}{(2\pi)^n n!}\int_{0<\psi_1<2\pi}\cdots\int_{0<\psi_n<2\pi} F\left(\mathrm{diag}(e^{i\psi_1},\ldots,e^{i\psi_n})\right) \times$$

$$\times \left|\prod_{m<k}(e^{i\psi_m} - e^{i\psi_k})\right|^2 \prod_{k=1}^n d\varphi_k, \tag{3.12}$$

where $\mathrm{diag}(\cdot)$ is a diagonal matrix with given entries.

---

[5]Explicit constructions of representations of $U(n)$ are not used below.

Any central function $F \in L^2(U(n))$ admits an expansion in characters,

$$F(z) = \sum_{\mathbf{m}} c_{\mathbf{m}} \chi_{\mathbf{m}}(z),$$

where the summation is given over all signatures $\mathbf{m}$ and the coefficients $c_{\mathbf{m}}$ are L^2-inner products:

$$c_{\mathbf{m}} = \int_{U(n)} F(z) \overline{\chi_{\mathbf{m}}(z)} \, d\mu(z).$$

Note that $\overline{\chi}_{\mathbf{m}} = \chi_{\mathbf{m}^*}$, where

$$\mathbf{m}^* := (n-1-m_n, \ldots, n-1-m_2, n-1-m_1).$$

Applying formula (3.12), explicit expression (3.9) for characters, and formula (3.10) for the denominator, we obtain

$$c_{\mathbf{m}} = \frac{1}{(2\pi)^n n!} \int_{0 < \psi_1 < 2\pi} \cdots \int_{0 < \psi_n < 2\pi} F\left(\mathrm{diag}\{e^{i\psi_1}, \ldots, e^{i\psi_n}\}\right) \times$$

$$\times \det_{k,j=1,2,\ldots,n} \{e^{i(j-1)\psi_k}\} \det_{k,j=1,2,\ldots,n} \{e^{-im_j \psi_k}\} \prod_{k=1}^{n} d\varphi_k. \tag{3.13}$$

Let $F(z)$ be multiplicative with respect to eigenvalues,

$$F(z) = \prod_k f(e^{i\varphi_k})$$

[for, instance $F = \ell_{\sigma|\tau}$; see (3.3)]. Then we can apply the following simple lemma (see, e.g., [28]).

Lemma 3.10. *Let X be a set,*

$$\int_{X^n} \prod_{k=1}^{n} f(x_k) \det_{k,l=1,\ldots n} \{u_l(x_k)\} \det_{k,l=1,\ldots n} \{v_l(x_k)\} \prod_{j=1}^{n} dx_j =$$

$$= n! \det_{l,m=1,\ldots,n} \left\{ \int_X f(x) u_l(x) v_m(x) \, dx \right\}. \tag{3.14}$$

3.2.3 Lobachevsky Beta-Integrals

We wish to apply Lemma 3.10 to functions $\ell_{\sigma|\tau}$. For this purpose, we need for the following integral (see [15], 3.631,1, 3.631,8,)

$$\int_0^\pi \sin^{\mu-1}(\varphi)\, e^{ib\varphi}\, d\varphi = \frac{2^{1-\mu}\pi\Gamma(\mu)e^{ib\pi/2}}{\Gamma\big((\mu+b+1)/2\big)\Gamma\big((\mu-b+1)/2\big)}. \tag{3.15}$$

It is equivalent to the identity (2.12).

In a certain sense, the integral (3.21) is a multivariate analog of the Lobachevsky integral. On the other hand, (3.21) is a special case of the modified Kadell integral [29].

3.2.4 Expansion of the Function $\ell_{\sigma|\tau}$ in Character

Theorem 3.11. *Let* $\operatorname{Re}(\sigma+\tau) < 1$. *Then*

$$\ell_{\sigma|\tau}(g)$$

$$= \frac{(-1)^{n(n-1)/2}\sin^n(\pi\sigma)2^{-(\sigma+\tau)n}}{\pi^n}\prod_{j=1}^{n}\Gamma(\sigma+\tau+j)$$

$$\times \sum_{\mathbf{m}}\left\{\prod_{1\leqslant\alpha<\beta\leqslant n}(m_\alpha-m_\beta)\prod_{j=1}^{n}\frac{\Gamma(-\sigma+m_j-n+1)}{\Gamma(\tau+m_j+1)}\chi_{\mathbf{m}}(g)\right\} \tag{3.16}$$

$$= (-1)^{n(n-1)/2}2^{-(\sigma+\tau)n}\prod_{j=1}^{n}\Gamma(\sigma+\tau+j)$$

$$\times \sum_{\mathbf{m}}\left\{\frac{(-1)^{\sum m_j}\prod_{1\leqslant\alpha<\beta\leqslant n}(m_\alpha-m_\beta)}{\prod_{j=1}^{n}\Gamma(\sigma-m_j+n)\Gamma(\tau+m_j+1)}\chi_{\mathbf{m}}(g)\right\}. \tag{3.17}$$

The proof is contained in Sect. 3.2.6. For the calculation, we need Lemma 3.13 proved in the next subsection.

3.2.5 A Determinant Identity

Recall that the *Cauchy determinant* (see, e.g., [22]) is given by

$$\det_{kl}\left\{\frac{1}{x_k+y_l}\right\} = \frac{\prod_{1\leqslant k<l\leqslant n}(x_k-x_l)\cdot\prod_{1\leqslant k<l\leqslant n}(y_k-y_l)}{\prod_{1\leqslant k,l\leqslant n}(x_k+y_l)}. \tag{3.18}$$

The following version of the Cauchy determinant is also well known.

Lemma 3.12.

$$\det \begin{pmatrix} 1 & 1 & 1 & \cdots & 1 \\ \frac{1}{x_1+b_1} & \frac{1}{x_2+b_1} & \frac{1}{x_3+b_1} & \cdots & \frac{1}{x_n+b_1} \\ \frac{1}{x_1+b_2} & \frac{1}{x_2+b_2} & \frac{1}{x_3+b_2} & \cdots & \frac{1}{x_n+b_2} \\ \vdots & \vdots & \vdots & \ddots & \vdots \\ \frac{1}{x_1+b_{n-1}} & \frac{1}{x_2+b_{n-1}} & \frac{1}{x_3+b_{n-1}} & \cdots & \frac{1}{x_n+b_{n-1}} \end{pmatrix}$$

$$= \frac{\prod_{1\leqslant k<l\leqslant n}(x_k-x_l)\prod_{1\leqslant\alpha<\beta\leqslant n-1}(b_\alpha-b_\beta)}{\prod_{\substack{1\leqslant k\leqslant n \\ 1\leqslant\alpha\leqslant n-1}}(x_k+b_\alpha)}. \tag{3.19}$$

Proof. Let Δ be the Cauchy determinant (3.18). Then

$$y_1\Delta = \begin{pmatrix} \frac{y_1}{x_1+y_1} & \frac{y_1}{x_2+y_1} & \cdots & \frac{y_1}{x_1+y_1} \\ \frac{1}{x_1+y_2} & \frac{1}{x_2+y_2} & \cdots & \frac{1}{x_n+y_2} \\ \vdots & \vdots & \ddots & \vdots \\ \frac{1}{x_1+y_n} & \frac{1}{x_2+y_n} & \cdots & \frac{1}{x_n+y_n} \end{pmatrix}.$$

We take $\lim_{y_1\to\infty} y_1\Delta$ and substitute $y_{\alpha+1}=b_\alpha$. $\square$

The following determinant is a rephrasing of [22], Lemma 3.

Lemma 3.13.

$$\det \begin{pmatrix} 1 & 1 & 1 & \cdots & 1 \\ \frac{x_1+b_1}{x_1+a_1} & \frac{x_2+a_1}{x_2+b_1} & \frac{x_3+a_1}{x_3+b_1} & \cdots & \frac{x_n+a_1}{x_n+b_1} \\ \frac{(x_1+a_1)(x_1+a_2)}{(x_1+b_1)(x_1+b_2)} & \frac{(x_2+a_1)(x_2+a_2)}{(x_2+b_1)(x_2+b_2)} & \frac{(x_3+a_1)(x_3+a_2)}{(x_3+b_1)(x_3+b_2)} & \cdots & \frac{(x_n+a_1)(x_n+a_2)}{(x_n+b_1)(x_n+b_2)} \\ \vdots & \vdots & \vdots & \ddots & \vdots \\ \frac{\prod_{1\leqslant m\leqslant n-1}(x_1+a_m)}{\prod_{1\leqslant m\leqslant n-1}(x_1+b_m)} & \frac{\prod_{1\leqslant m\leqslant n-1}(x_2+a_m)}{\prod_{1\leqslant m\leqslant n-1}(x_2+b_m)} & \frac{\prod_{1\leqslant m\leqslant n-1}(x_3+a_m)}{\prod_{1\leqslant m\leqslant n-1}(x_3+b_m)} & \cdots & \frac{\prod_{m:1\leqslant m\leqslant n-1}(x_n+a_m)}{\prod_{m:1\leqslant m\leqslant n-1}(x_n+b_m)} \end{pmatrix}$$

$$= \frac{\prod_{1\leqslant k<l\leqslant n}(x_k-x_l)\prod_{1\leqslant\alpha\leqslant\beta\leqslant n-1}(a_\alpha-b_\beta)}{\prod_{1\leqslant k\leqslant n,1\leqslant\beta\leqslant n-1}(x_k+b_\beta)}. \tag{3.20}$$

Proof. Decomposing a matrix element into a sum of partial fractions, we obtain

$$\frac{(x_k + a_1)\ldots(x_k + a_\alpha)}{(x_k + b_1)\ldots(x_k + b_\alpha)} = 1 + \sum_{1 \leq \beta \leq \alpha} \frac{\prod_{j \leq \alpha}(a_j - b_\beta)}{\prod_{j \leq \alpha, j \neq \beta}(b_j - b_\beta)} \cdot \frac{1}{x_k + b_\beta}$$

Therefore, the $(\alpha + 1)$-th row is a linear combination of the following rows:

$$\begin{pmatrix} 1 & 1 & \ldots & 1 \end{pmatrix},$$

$$\begin{pmatrix} \frac{1}{x_1 + b_1} & \frac{1}{x_2 + b_1} & \cdots & \frac{1}{x_n + b_1} \end{pmatrix},$$

$$\cdots \quad \cdots \qquad \cdots \quad \cdots \quad \cdots \quad \cdots$$

$$\begin{pmatrix} \frac{1}{x_1 + b_\alpha} & \frac{1}{x_2 + b_\alpha} & \cdots & \frac{1}{x_n + b_\alpha} \end{pmatrix}.$$

Thus, our determinant equals

$$\prod_{\alpha=1}^{l-1} \frac{\prod_{j=1}^{\alpha}(a_j - b_\alpha)}{\prod_{j=1}^{\alpha-1}(b_j - b_\alpha)} \cdot \det \begin{pmatrix} 1 & 1 & \ldots & 1 \\ \frac{1}{x_1 + b_1} & \frac{1}{x_2 + b_1} & \cdots & \frac{1}{x_n + b_1} \\ \vdots & \vdots & \ddots & \vdots \\ \frac{1}{x_1 + b_\alpha} & \frac{1}{x_2 + b_\alpha} & \cdots & \frac{1}{x_n + b_\alpha} \end{pmatrix},$$

and we refer to Lemma 3.12.   □

3.2.6   Proof of Theorem 3.11

We must evaluate the inner product

$$\int_{U(n)} \ell_{\sigma|\tau}(g)\, \overline{\chi_{\mathbf{m}}(g)}\, d\mu(g).$$

Applying (3.13), we get

$$\frac{1}{(2\pi)^n\, n!} \int_{0 < \psi_k < 2\pi} \prod_{j=1}^{n} \left[\sin^{\sigma+\tau}(\psi_j/2) \cdot \exp\left\{ \frac{i}{2}(\sigma - \tau)(\psi_j - \pi) \right\} \right]$$

$$\times \det_{1 \leq k,l \leq n} \{e^{-im_k\psi_l}\} \cdot \det_{1 \leq k,l \leq n} \{e^{i(k-1)\psi_l}\} \prod_{l=1}^{n} d\psi_l. \tag{3.21}$$

By Lemma 3.10, we reduce this integral to

$$\frac{1}{(2\pi)^n} \det_{1\leq k,j \leq n} I(k,j),$$

where

$$I(k,j) = e^{-i(\sigma-\tau)\pi/2} \int_0^{2\pi} \sin^{\sigma-\tau}(\psi/2) \cdot \exp\{i((\sigma+\tau)/2 + k - 1 - m_j)\} \, d\psi.$$

We apply the Lobachevsky integral (3.15) and get

$$I(k,j) = \frac{2^{1-\sigma-\tau}\pi\Gamma(\sigma+\tau+1)(-1)^{k-1-m_j}}{\Gamma(\sigma+k-m_j)\Gamma(\tau-k+m_j+2)}$$

Applying standard formulas for the Γ-function, we come to

$$I(k,j) = 2^{1-\sigma-\tau}\Gamma(\sigma+\tau+1)\sin(-\sigma\pi) \cdot \frac{\Gamma(-\sigma+m_j-k+1)}{\Gamma(\tau+m_j-k+2)}$$

$$= 2^{1-\sigma-\tau}\Gamma(\sigma+\tau+1)\sin(-\sigma\pi) \cdot \frac{\Gamma(-\sigma+m_j-n+1)}{\Gamma(\tau+m_j-n+2)}$$

$$\cdot \boxed{\frac{(-\sigma+m_j-n+1)_{n-k}}{(\tau+m_j-n+2)_{n-k}}}$$

The factors outside the box do not depend on on k. Thus, we must evaluate the determinant

$$\det_{1\leq k,j \leq n} \frac{(-\sigma+m_j-n+1)_{n-k}}{(\tau+m_j-n+2)_{n-k}}.$$

Up to a permutation of rows, it is a determinant of the form described in Lemma 3.13 with

$$x_j = m_j, \qquad a_j = -\sigma - n + j, \qquad b = \tau - n + j + 1.$$

After a rearrangement of the factors, we obtain the required result.   □

3.2.7 Characters of Compact Groups. Preliminaries

First, recall some standard facts on characters of compact groups; for details, see, e.g., [21], 9.2, 11.1.

Let K be a compact Lie group equipped with the Haar measure μ, let $\mu(K) = 1$. Let $\pi_1, \pi_2, \ldots$ be the complete collection of pairwise distinct irreducible representations of K. Let $\chi_1, \chi_2, \ldots$ be their characters. Recall the orthogonality relations,

$$\langle \chi_k, \chi_l \rangle_{L^2(K)} = \int_K \chi_k(h) \overline{\chi_l(h)} \, d\mu(h) = \delta_{k,l} \tag{3.22}$$

and

$$\chi_k * \chi_l = \begin{cases} \frac{1}{\dim \pi_k} \chi_k & \text{if } k = l, \\ 0 & \text{if } k \neq l, \end{cases} \tag{3.23}$$

where $*$ denotes the convolution on the group,

$$u * v(g) = \int_K u(gh^{-1}) v(h) \, d\mu(h).$$

Consider the action of the group $K \times K$ in $L^2(K)$ by the left and right shifts

$$(k_1, k_2): \quad f(g) \mapsto f(k_1^{-1} g k_2).$$

The representation of $K \times K$ in $L^2(K)$ is a multiplicity-free direct sum of irreducible representations having the form $\pi_k \otimes \pi_k^*$, where π_k^* denotes the dual representation

$$L^2(K) \simeq \bigoplus_k \pi_k \otimes \pi_k^*. \tag{3.24}$$

Denote by $V_k \subset L^2(K)$ the space of representation $\pi_k \otimes \pi_k^*$. Each distribution f on K is a sum of "elementary harmonics",

$$f = \sum_k f^k, \qquad f_k \in V_k.$$

The projector to a subspace V_k is the convolution with the corresponding character,

$$f^k = \frac{1}{\dim \pi_k} f * \chi_k \tag{3.25}$$

(in particular, f^k is smooth).

Observation 3.14. *Let f be a function on $U(n)$, $f = \sum_{\mathbf{m}} a_{\mathbf{m}} f^{\mathbf{m}}$, where $f^{\mathbf{m}} \in V_{\mathbf{m}}$.*
(a) $f \in C^{\infty}(U(n))$ iff

$$\|f^{\mathbf{m}}\|_{L^2} = o \left(\sum m_j^2 \right)^{-L} \qquad \text{for all } L.$$

(b) f is a distribution on $U(n)$ iff there exists L such that

$$\|f^{\mathbf{m}}\|_{L^2} = o \left(\sum m_j^2 \right)^{L}.$$

Proof. Note that $f \in L^2(U(n))$ iff $\sum \|f^{\mathbf{m}}\|_{L^2}^2 < \infty$. Denote by Δ be the second-order invariant Laplace operator on $U(n)$. Then $\Delta f^{\mathbf{m}} = q(\mathbf{m}) f^{\mathbf{m}}$, where $q(\mathbf{m}) = \sum m_j^2 + \ldots$ is an explicit quadratic expression in $\mathbf{m}$. For $f \in C^{\infty}$ we have $\Delta^p f \in C^{\infty}$; this implies the first statement. Since $q(\mathbf{m})$ has a finite number of zeros (one), the second statement follows from (a) and the duality. □

3.2.8 Hermitian Forms Defined by Kernels

Let Ξ be a central distribution on K satisfying $\Xi(g^{-1}) = \overline{\Xi(g)}$. Consider the following Hermitian form on $C^{\infty}(K)$:

$$\langle f_1, f_2 \rangle = \iint_{K \times K} \Xi(gh^{-1}) f_1(h) \overline{f_2(g)} \, d\mu(h) \, d\mu(g). \qquad (3.26)$$

Consider the expansion of Ξ in characters

$$\Xi = \sum_k c_k \chi_k.$$

Lemma 3.15.

$$\langle f_1, f_2 \rangle = \sum_k \frac{c_k}{\dim \pi_k} \int_{U(n)} f_1^k(h) \overline{f_2^k(h)} \, d\mu(h). \qquad (3.27)$$

Proof. The Hermitian form (3.26) is $K \times K$-invariant. Therefore, the subspaces $V_k \simeq \pi_k \otimes \pi_k^*$ must be pairwise orthogonal. Since $\pi_k \otimes \pi_k^*$ is an irreducible representation of $K \times K$, it admits a unique up to a factor $K \times K$-invariant Hermitian form. Therefore, it is sufficient to find these factors.

Set $f_1 = f_2 = \chi_k$. We evaluate

$$\iint_{K \times K} \left(\sum_k c_k \chi_k(gh^{-1}) \right) \chi_k(h) \overline{\chi_k(g)} \, d\mu(g) \, d\mu(h) = \frac{c_k}{\dim \pi_k}$$

using (3.22) and (3.23).   □

3.2.9 Positivity

Let $\mathrm{Re}(\sigma + \tau) < 1$. Consider the sesquilinear form on $C^\infty(\mathrm{U}(n))$ given by

$$\langle f_1, f_2 \rangle_{\sigma|\tau} = \iint_{\mathrm{U}(n) \times \mathrm{U}(n)} \ell_{\sigma|\tau}(zu^{-1}) f_1(z) \overline{f_2(u)} \, d\mu(z) \, d\mu(u), \qquad (3.28)$$

where the distribution $\ell_{\sigma|\tau}$ is the same as above.

Observation 3.16. *For fixed f_1, $f_2 \in C^\infty(\mathrm{U}(n))$, the expression $\langle f_1, f_2 \rangle_{\sigma|\tau}$ admits a meromorphic continuation in σ, τ to the whole $\mathbb{C}^2$.*

Proof. Expanding f_1, f_2 in elementary harmonics

$$f_1(z) = \sum_{\mathbf{m}} f_1^{\mathbf{m}}(z), \qquad f_2(z) = \sum_{\mathbf{m}} f_2^{\mathbf{m}}(z),$$

we get (see Lemma 3.15)

$$\langle f_1, f_2 \rangle_{\sigma|\tau} = \sum_{\mathbf{m}} \frac{c_{\mathbf{m}}}{\dim \pi_{\mathbf{m}}} \int_{\mathrm{U}(n)} f_1^{\mathbf{m}}(z) \overline{f_2^{\mathbf{m}}(z)} \, d\mu(z),$$

where the meromorphic expressions for $c_{\mathbf{m}}$ were obtained in Theorem 3.11. The coefficients $c_{\mathbf{m}}$ have polynomial growth in $\mathbf{m}$. On the other hand, $\| f_j^{\mathbf{m}} \|$ rapidly decreases; see Observation 3.14. Therefore, the series converges.   □

Proof of positivity. Corollary 3.6. We look at expression (3.16). It suffices to examine the factor

$$\frac{\Gamma(-\sigma - n + m_j + 1)}{\Gamma(\tau + m_j + 1)}, \qquad (3.29)$$

because signs of all the remaining factors are independent on m_j. Let $n \in \mathbb{Z}$ and $\alpha \in (0, 1)$. Then

$$\mathrm{sign}\,\Gamma(n + \alpha) = \begin{cases} +1, & \text{if } n \geqslant 0, \\ (-1)^n, & \text{if } n < 0 \end{cases}.$$

Therefore, (3.29) is positive whenever integer parts of τ and $-\sigma - n$ are equal.   □

3.2.10   The L^2-limit. Proof of Proposition 3.8

Thus, let $\sigma + \tau = -n$. Then

$$\left(\prod_{j=1}^{n} \Gamma(\sigma + \tau + j)\right)^{-1} \ell_{\sigma|\tau} = \text{const} \cdot \sum (\dim \pi_{\mathbf{m}}) \chi_{\mathbf{m}}$$

Indeed, in this case Γ-factors in (3.16) cancel, and we use (3.11).
   Keeping in mind (3.27), we get Proposition 3.8.

3.3   Other Proofs

Here we prove that the operators $\rho_{\sigma|\tau}$ preserve the inner product determined by the distribution $\ell_{\sigma|\tau}$.

3.3.1   The Universal Covering of the Group $U(n, n)$

The fundamental group of $U(n, n)$ is[6]

$$\pi_1\big(U(n,n)\big) \simeq \mathbb{Z} \oplus \mathbb{Z}.$$

   The universal covering $U(n, n)^\sim$ of $U(n, n)$ can be identified with the set $\mathfrak{U}$ of triples

$$\left\{ \begin{pmatrix} a & b \\ c & d \end{pmatrix}, s, t \right\} \in U(n, n) \times \mathbb{C} \times \mathbb{C}$$

satisfying the conditions

$$\det(a) = e^s, \qquad \det(d) = e^t.$$

The multiplication of triples is given by the formula

$$(g_1, s_1, t_1) \circ (g_2, s_2, t_2) = \big(g_1 g_2, s_1 + s_2 + c^+(g_1, g_2), t_1 + t_2 + c^-(g_1, g_2)\big),$$

where the *Berezin cocycle* $c^\pm$ is given by

$$c^+(g_1, g_2) = \text{tr} \ln(a_1^{-1} a_3 a_2^{-1}), \qquad c^-(g_1, g_2) = \text{tr} \ln(d_1^{-1} d_3 d_2^{-1});$$

---

[6]By a general theorem, a real reductive Lie group G admits a deformation retraction to its maximal compact subgroup K. In our case, $K = U(n) \times U(n)$ and $\pi_1(U(n)) = \mathbb{Z}$.

here $g_3 = g_1 g_2$, and $g_j = \begin{pmatrix} a_j & b_j \\ c_j & d_j \end{pmatrix}$. It can be shown that $\|a_1^{-1} a_3 a_2^{-1} - 1\| < 1$; therefore, the logarithm is well defined. On the other hand,

$$e^{s_3} = e^{s_1 + s_2 + c^+(g_1, g_2)} = \det(a_1) \det(a_2) \det(a_1^{-1} a_3 a_2^{-1}) = \det(a_3).$$

This shows that the $\mathfrak{U}$ is closed with respect to multiplication.

For details, see [31].

In particular, $\det(a)$ is a well-defined single-valued function on $\mathrm{U}(n, n)^{\sim}$. In our notation, it is given by

$$(g, s, t) \mapsto s.$$

3.3.2 Another Model of U(n, n)

We can realize $\mathrm{U}(n, n)$ as the group of $(n + n) \times (n + n)$-matrices $g = \begin{pmatrix} \alpha & \beta \\ \gamma & \delta \end{pmatrix}$ satisfying the condition

$$g \begin{pmatrix} 0 & i \\ -i & 0 \end{pmatrix} g^* = \begin{pmatrix} 0 & i \\ -i & 0 \end{pmatrix} \tag{3.30}$$

3.3.3 Action of U(n, n) on the Space U(n). Proof of Lemma 3.4

We must show that for

$$z \in \mathrm{U}(n) \quad \text{and} \quad g = \begin{pmatrix} a & b \\ c & d \end{pmatrix} \in \mathrm{U}(n, n),$$

we have

$$z^{[g]} := (a + zc)^{-1}(b + zd) \in \mathrm{U}(n, n). \tag{3.31}$$

For $z \in \mathrm{U}(n)$, consider its graph $graph(z) \subset \mathbb{C}^n \oplus \mathbb{C}^n$. It is an n-dimensional linear subspace, consisting of all vectors $v \oplus vz$, where a vector-row v ranges in $\mathbb{C}^n$. Since $z \in \mathrm{U}(n)$, the subspace $graph(z)$ is isotropic[7] with respect to the Hermitian form $\begin{pmatrix} 1 & 0 \\ 0 & -1 \end{pmatrix}$. Conversely, any n-dimensional isotropic subspace in $\mathbb{C}^n \oplus \mathbb{C}^n$ is a graph of a unitary operator $z \in \mathrm{U}(n)$.

Thus, we get a one-to-one correspondence between the group $\mathrm{U}(n)$ and the Grassmannian of n-dimensional isotropic subspaces in $\mathbb{C}^n \oplus \mathbb{C}^n$.

---

[7]A subspace V in a linear space is *isotropic* with respect to Hermitian form Q if Q equals 0 on V.

The group $U(n,n)$ acts on the Grassmannian, and therefore $U(n,n)$ acts on the space $U(n)$. Then (3.31) is the explicit expression for the latter action. Indeed,

$$\left(v \oplus vz\right)\begin{pmatrix} a & b \\ c & d \end{pmatrix} = v(a + zc) \oplus v(b + zd).$$

We denote $\xi := v(a + zc)$ and get

$$\xi \oplus \xi(a + zc)^{-1}(b + zd),$$

and this completes the proof of Lemma 3.4.   □

Thus, $U(n)$ is a $U(n,n)$-homogeneous space. We describe without proof (it is a simple exercise) the stabilizer of a point $z = 1$. It is a maximal parabolic subgroup.

In the model (3.30) it can be realized as the subgroup of matrices having the structure

$$\begin{pmatrix} \alpha & 0 \\ \beta & \alpha^{*-1} \end{pmatrix}$$

It is a semidirect product of $GL(n, \mathbb{C})$ and the Abelian group $\mathbb{R}^{n^2}$.

In our basic model the stabilizer of $z = 1$ is the semi-direct product of two subgroups

$$\frac{1}{2}\begin{pmatrix} \alpha + \alpha^{*-1} & \alpha - \alpha^{*-1} \\ \alpha - \alpha^{*-1} & \alpha + \alpha^{*-1}, \end{pmatrix} \quad \text{where } g \in GL(n, \mathbb{C}), \tag{3.32}$$

and

$$\begin{pmatrix} 1 + iT & iT \\ -iT & 1 - iT \end{pmatrix}, \quad \text{where } T = T^*. \tag{3.33}$$

3.3.4   The Jacobian

Lemma 3.17. *For the Haar measure $\mu(z)$ on $U(n)$, we have*

$$\mu\left(z^{[g]}\right) = |\det^{-2n}(a + zc)| \cdot \mu(z). \tag{3.34}$$

Proof. A verification of this formula is straightforward; we only outline the main steps. First, $J(g, z) := |\det^{-2n}(a + zc)|$ satisfies the chain rule (1.3). Next, the formula (3.34) is valid for $g \in U(n,n)$ having the form $\begin{pmatrix} a & 0 \\ 0 & d \end{pmatrix}$, where $u, v \in U(n)$. Indeed, the corresponding transformation of $u \mapsto u^{[h]}$ is $u \mapsto a^{-1}ud$ and its Jacobian is 1.

Therefore, we can set $z = 1$, $z^{[g]} = 1$. Now we must evaluate the determinants of the differentials of maps $z \mapsto z^{[g]}$ at $z = 1$ for g given by (3.32) and (3.33). In the second case the differential is the identity map; in the first case the differential is $dz \mapsto \alpha^*(dz)\alpha$. We represent α as $p\Delta q$, where Δ is the diagonal with real eigenvalues and p, q are unitary. Now the statement becomes obvious. $\square$

3.3.5 The Degenerate Principal Series. Proof of Proposition 3.7

Thus, let $\mathrm{Re}(\sigma + \tau) = -n$, $\mathrm{Im}(\sigma) = \mathrm{Im}(\tau) = s$. Then

$$\det(a + uc)^{-n-\sigma|-n-\tau} = |\det(a + uc)|^{-n-2is} e^{i(\tau-\sigma)\mathrm{Arg}\det(a+uc)},$$

where $\mathrm{Arg}(\cdot)$ is the argument of a complex number. Therefore,

$$\langle T_{\sigma|\tau}(g)f_1, T_{\sigma|\tau}(g)f_2 \rangle_{L^2(\mathrm{U}(n))}$$
$$= \int_{\mathrm{U}(n)} f_1(u^{[g]})\overline{f_2(u^{[g]})} \left|\det(a + uc)^{-n-\sigma|-n-\tau}\right|^2 d\mu(u)$$
$$= \int_{\mathrm{U}(n)} f_1(u^{[g]})\overline{f_2(u^{[g]})}|\det(a + uc)|^{-2n} d\mu(u),$$

and we change the variable $z = u^{[g]}$, keeping Lemma 3.17 in mind. $\square$

3.3.6 The Invariance of the Kernel. Proof of Proposition 3.5

Lemma 3.18. *The distribution $\ell_{\sigma|\tau}$ satisfies the identity*

$$\ell_{\sigma|\tau}(u^{[g]}(v^{[g]})^*) = \ell_{\sigma|\tau}(uv^*)\det(a + uc)^{\{-\tau|-\sigma\}}\det(a + vc)^{\{-\sigma|-\tau\}}. \quad (3.35)$$

Proof. This follows from the identity

$$1 - u^{[g]}(v^{[g]})^* = (a + uc)^{-1}(1 - uv^*)(a + vc)^{*-1}, \qquad \text{where } g \in \mathrm{U}(n,n),$$

which can be easily verified by a straightforward calculation (see, e.g., [31]). $\square$

Proof of Proposition 3.5. First, let $\mathrm{Re}(\sigma + \tau) < 1$. Substitute $h_1 = u_1^{[g]}$, $h_2 = u_2^{[g]}$ in the integral

$$\langle f_1, f_2 \rangle_{\sigma|\tau} = \iint_{\mathrm{U}(n)\times\mathrm{U}(n)} \ell_{\sigma|\tau}(h_1 h_2^*) f_1(h_1)\overline{f_2(h_2)} d\mu(h_1) d\mu(h_2).$$

By the lemma, we obtain

$$\iint_{U(n) \times U(n)} \ell_{\sigma|\tau}(u_1 u_2^*) \det(a + u_1 c)^{\{-\tau|-\sigma\}} |\det(a + u_2 c)|^{-\sigma|-\tau\}}$$

$$\times f_1(u_1) \overline{f_2(u_2)} |\det(a + u_1 c)|^{-2n} |\det(a + u_2 c)|^{-2n} \, d\mu(u_1) \, d\mu(u_2)$$

$$= \langle \rho_{\sigma|\tau}(g) f_1, \rho_{\sigma|\tau}(g) f_2 \rangle_{\sigma|\tau}.$$

Thus, our operators preserve the form $\langle \cdot, \cdot \rangle_{\sigma|\tau}$.

For general $\sigma, \tau \in \mathbb{C}$, we consider the analytic continuation.   □

3.3.7   Shift of parameters Proof of Proposition 3.9

First, we recall *Cartan decomposition.* For $t_1 \geqslant \ldots \geqslant t_n$ denote

$$CH(t) := \begin{pmatrix} \cosh(t_1) & 0 & \cdots \\ 0 & \cosh(t_2) & \cdots \\ \vdots & \vdots & \ddots \end{pmatrix}, \quad SH(t) := \begin{pmatrix} \sinh(t_1) & 0 & \cdots \\ 0 & \sinh(t_2) & \cdots \\ \vdots & \vdots & \ddots \end{pmatrix}.$$

The following statement is well known.

Proposition 3.19. *Each element $g \in U(n, n)$ can be represented in the form*

$$g = \begin{pmatrix} u_1 & 0 \\ 0 & v_1 \end{pmatrix} \begin{pmatrix} CH(t) & SH(t) \\ SH(t) & CH(t) \end{pmatrix} \begin{pmatrix} u_2 & 0 \\ 0 & v_2 \end{pmatrix} \tag{3.36}$$

for some (uniquely determined) t and some $u_1, u_2, v_1, v_2 \in U(n)$.

Now we must show that the operator $f(z) \mapsto \det(z) f(z)$ intertwines $\rho_{\sigma|\tau}$ and $\rho_{\sigma+1|\tau-1}$. A straightforward calculation reduces this to the identity

$$\frac{\det(a + zc)}{\det \overline{(a + zc)}} = \frac{\det(z^{[g]})}{\det(z)},$$

which becomes obvious after the substitution (3.36).

Also,

$$\ell_{\sigma+1|\tau-1}(z) = -\ell_{\sigma|\tau}(z) \det z,$$

and this easily implies the second statement of Proposition 3.9.

4   Hilbert Spaces of Holomorphic Functions

Theorem 3.3 exhausts the cases when the form $\langle \cdot, \cdot \rangle_{\sigma|\tau}$ is positive definite on $C^\infty(U(n))$. However, there are cases of positive semi-definiteness. They are discussed in the next two sections (Fig. 7).

Fig. 7 Conditions of positivity of holomorphic representations ξ_σ (the *"Berezin–Wallach set"*)

Set $\tau = 0$. In this case, our construction produces holomorphic representations[8] of $U(n, n)$. Holomorphic representations were discovered by Harish–Chandra (holomorphic discrete series, [17]) and Berezin (analytic continuations of holomorphic discrete series, [3]). They are discussed in numerous texts (for partial expositions and further references, see, e.g., [10, 31]); our aim is to show a link with our considerations.

4.1 The case $\tau = 0$

Substituting $\tau = 0$, we get the action

$$\rho_{\sigma|0}(g)\, f(z) = f(z^{[g]})\, \det(a + zc)^{-n}\, \overline{\det(a + zc)}^{\,-n-\sigma}.$$

The Hermitian form is

$$\langle f_1, f_2 \rangle_{\sigma|0} = \int_{U(n)} \int_{U(n)} \det(1 - z^* u)^\sigma\, f_1(z)\, \overline{f_2(u)}\, d\mu(z)\, d\mu(u).$$

Theorem 4.1. *The form* $\langle f_1, f_2 \rangle_{\sigma|0}$ *is positive semi-definite iff* σ *is contained in the set*

$$\sigma = 0, -1, \ldots, -(n-1), \text{ or } \sigma < -(n-1).$$

This means that all coefficients $c_\mathbf{m}$ in the formula (3.27) are non-negative, but some coefficients vanish. In fact, the proof (see below) is the examination of these coefficients.

Under the conditions of the theorem we get a structure of a pre-Hilbert space in $C^\infty(U(n))$. Denote by $\mathcal{H}_\sigma$ the corresponding Hilbert space.

Next, consider the action of the subgroup $U(n) \times U(n)$ in $\mathcal{H}_\sigma$. We must get an orthogonal direct sum

$$\bigoplus_{\mathbf{m} \in \Omega_\sigma} \pi_\mathbf{m} \oplus \pi_\mathbf{m}^*$$

Some of summands of (3.24) disappear, when we pass to the quotient space; actually, the summation is taken over a proper subset Ω_σ of the set of all representations. The next theorem is the description of the set Ω_σ.

---

[8]Or highest-weight representations.

Theorem 4.2. *(a) If $\sigma < -(n-1)$, then*

$$\Omega_\sigma := \Big\{\mathbf{m} : m_n \geq 0\Big\}.$$

(b) If $\sigma = -n + \alpha$, where $\alpha = 1, 2, \ldots, n-1, n$, then

$$\Omega_\sigma = \Big\{\mathbf{m} : m_n = 0, \, m_{n-1} = 1, \ldots, m_{n-\alpha+1} = \alpha - 1\Big\}.$$

Proof. [9] Substitute $\tau = 0$ in (3.17),

$$c_{\mathbf{m}} = (-1)^{n(n-1)/2} 2^{-\sigma n} \prod_{j=1}^{n} \Gamma(\sigma + j)$$

$$\times \sum_{\mathbf{m}} \left\{ \frac{(-1)^{\sum m_j} \prod_{1 \leq \alpha < \beta \leq n} (m_\alpha - m_\beta)}{\prod_{j=1}^{n} \Gamma(\sigma - m_j + n) \boxed{\Gamma(m_j + 1)}} \chi_{\mathbf{m}}(g) \right\} \qquad (4.1)$$

$$= \frac{(-1)^{n(n-1)/2} \sin^n(\pi\sigma) 2^{-(\sigma)n}}{\pi^n} \prod_{j=1}^{n} \Gamma(\sigma + j) \qquad (4.2)$$

$$\times \sum_{\mathbf{m}} \left\{ \prod_{1 \leq \alpha < \beta \leq n} (m_\alpha - m_\beta) \prod_{j=1}^{n} \frac{\Gamma(-\sigma + m_j - n + 1)}{\boxed{\Gamma(m_j + 1)}} \chi_{\mathbf{m}}(g) \right\} \qquad (4.3)$$

We have $\Gamma(m_j + 1) = \infty$ for $m_j < 0$. Therefore, the corresponding fractions in (4.3) are zero, and the expansion of $\ell_{\sigma|0}$ has the form

$$\ell_{\sigma|0} = \sum_{\mathbf{m}: \, m_n \geq 0} c_{\mathbf{m}} \chi_{\mathbf{m}}. \qquad (4.4)$$

Let us list possible cases.

Case 1. If $\sigma < -n - 1$, then all coefficients $c_{\mathbf{m}}$ are positive, [see (4.3)]; in the line (4.2), poles of the Gamma functions cancel with zeros of sines.

Case 2. If $\sigma \geq -n - 1$ is non-integer, then all the coefficients $c_{\mathbf{m}}$ are non-zero, but they have different signs.

Case 3. Let σ be integer, $\sigma \geq -n + 1$. Consider a small perturbation of σ,

$$\sigma = -n + \alpha + \varepsilon.$$

---

[9]This is the original Berezin's proof; he started from explicit expansions of reproducing kernels (4.6).

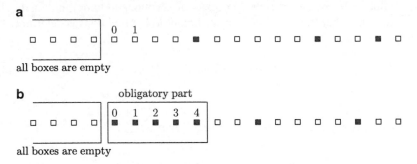

Fig. 8 "Maya diagrams" for signatures of harmonics in holomorphic representations.
(a) A general case, $\sigma < n - 1$.
(b) Degenerate case. Here $\sigma = -(n - 1) + 5$

In this case we get an uncertainty in the expression (4.1):

$$\frac{\prod_{j=1}^{n} \Gamma(-n + \alpha + \varepsilon + j)}{\prod_{j=1}^{n} \Gamma(\alpha - m_j + \varepsilon)}, \qquad \varepsilon \to 0.$$

The order of the pole of the numerator is $n - \alpha$. However, order of a pole in the denominator ranges between $n - \alpha$ and n according to $\mathbf{m}$. If the last order $> n - \alpha$, then the ratio is zero (Fig. 8). The only possibility to get the order of a pole $= n - \alpha$ is to set

$$m_n = 0, \quad m_{n-1} = 1, \quad \ldots, \quad m_{n-\alpha+1} = 0. \tag{4.5}$$

Thus, the coefficients $c_{\mathbf{m}}$ are nonzero only for signatures satisfying (4.5); they are positive.

We omit a discussion of positive integer σ (the invariant inner product is not positive). □

4.2 Intertwining Operators

Denote by B_n the space of complex $n \times n$ matrices with norm < 1.
   Consider the integral operator

$$I_\sigma f(z) = \int_{U(n)} \det(1 - zh^*)^\sigma f(h) \, d\mu(h), \qquad z \in B_n.$$

It intertwines $\rho_{\sigma|0}$ with the representation $\rho_{-n|-n-\sigma}$. Denote the last representation by ξ_σ:

$$\xi_\sigma(g) f(z) = f(z^{[g]}) \det(a + zc)^\sigma.$$

The I_σ-image $\mathcal{H}_\sigma^\circ$ of the space $\mathcal{H}_\sigma$ consists of functions holomorphic in B_n. The structure of a Hilbert space in the space of holomorphic functions is determined by the reproducing kernel

$$K_\alpha(z, \overline{u}) = \det(1 - zu^*)^\sigma. \tag{4.6}$$

4.3 Concluding Remarks (Without Proofs)

(a) For $\sigma < -(2n - 1)$, the inner product in $\mathcal{H}_\sigma^\circ$ can be written as an integral

$$\langle f_1, f_2 \rangle = \text{const} \int_{B_n} f_1(z)\, \overline{f_2(z)}\, \det(1 - zz^*)^{-\sigma - 2n}\, dz\, \overline{dz}.$$

(b) For $\sigma < n - 1$, the space $\mathcal{H}_\sigma^\circ$ contains all polynomials.
(c) Let $\sigma = 0, -1, \ldots, -(n - 1)$. Consider the matrix

$$\Delta = \begin{pmatrix} \dfrac{\partial}{\partial z_{11}} & \cdots & \dfrac{\partial}{\partial z_{1n}} \\ \vdots & \ddots & \vdots \\ \dfrac{\partial}{\partial z_{n1}} & \cdots & \dfrac{\partial}{\partial z_{nn}} \end{pmatrix}.$$

The space $\mathcal{H}_{-(n-1)}^\circ$ consists of functions f satisfying the partial differential equation

$$(\det \Delta) f(z) = 0.$$

The space $\mathcal{H}_\sigma^\circ$, where $\sigma = 0, -1, \ldots, -(n - 1)$, consists of functions that are annihilated by all $(-\sigma + 1) \times (-\sigma + 1)$ minors of the matrix Δ. Also, $\mathcal{H}_\sigma^\circ$ contains all polynomials satisfying this system of equations.

In particular, the space $\mathcal{H}_0^\circ$ is one-dimensional.

5 Unipotent Representations

Here we propose models for "unipotent representations" of Sahi [43] and Dvorsky–Sahi [8, 9] (Fig. 9).

5.1 Quotients of $\rho_{\sigma|\tau}$ at Integer Points

Set

$$\tau = 0, \qquad \sigma = -n + \alpha, \qquad \text{where } \alpha = 0, 1, \ldots, n - 1. \tag{5.1}$$

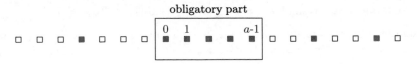

Fig. 9 Maya diagram for signatures $\in Z_j$; here j is the number of *black boxes* to the left of the "obligatory part."

Fig. 10 The case $n = 2$.
(a) $\alpha = 0$. The decomposition of $L^2(U(2))$ into a direct sum.
(b) $\alpha = 1$. White circles correspond to the big subrepresentation W_{tail}. The quotient is a direct sum of two subrepresentations.
(c) $\alpha = 2$. The quotient is one-dimensional.
(d) $0 < \tau < 1, \sigma = -n$. The invariant filtration. The subquotients are unitary

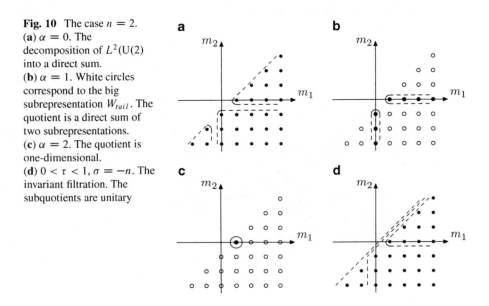

For $j = 0, 1, \ldots, n - \alpha$, denote by Z_j the set of all signatures **m** of the form (Fig. 10).

$$\mathbf{m} = (m_1, \ldots, m_{n-\alpha-j}, \alpha - 1, \alpha - 2, \ldots, 0, m_{n-j+1}, \ldots m_n).$$

Denote by $V_\mathbf{m}$ the $U(n) \times U(n)$ subrepresentation in $C^\infty(U(n))$ corresponding a signature **m**; see Sect. 3.2.7.

Theorem 5.1. *(a) The subspace*

$$W_{tail} := \bigoplus_{\mathbf{m} \notin \cup Z_j} V_\mathbf{m} \subset C^\infty(U(n)),$$

is $U(n, n)$-invariant.
(b) The quotient $C^\infty(U(n))/W_{tail}$ is a sum $n - \alpha + 1$ subrepresentations

$$W_j = \oplus_{\mathbf{m} \in Z_j} V_\mathbf{m}.$$

The representation of $U(n, n)$ in each W_j is unitary.

We formulate the result for $\alpha = 0$ separately. In this case $W_{\text{tail}} = 0$.

Theorem 5.2. *The representation $\rho_{-n|0}$ is a direct sum of $n + 1$ unitary representations W_j, where $0 \leqslant j \leqslant n$. We have $V_{\mathbf{m}} \subset W_j$ if the number of negative labels m_k is j.*

In particular, *we get a canonical decomposition of $L^2(\mathrm{U}(n))$ into a direct sum of* $(n + 1)$ *subspaces.*

The proof is given in the next subsection.

5.2   The Blow-Up Construction

[10] The distribution $\ell_{\sigma|\tau}$ depends meromorphically on two complex variables σ, τ. Its poles and zeros are located at $\sigma \in \mathbb{Z}$ and in $\tau \in \mathbb{Z}$. For this reason, values of $\ell_{\sigma|\tau}$ at points $(\sigma, \tau) \in \mathbb{Z}^2$ generally are not uniquely defined. Passing to such points from different directions, we get different limits.[11]

Thus, set

$$\sigma = -n + \alpha + s\varepsilon, \quad \tau = t\varepsilon \quad \text{where } (s, t) \neq (0, 0). \tag{5.2}$$

Substituting this in (3.17), we get

$$\ell_{-n+\alpha+\varepsilon s|\varepsilon t} = (-1)^{n(n-1)/2} 2^{-(\sigma+\tau)n} \prod_{k=1}^{n} \Gamma\big(-n + \alpha\varepsilon(s + t) + k\big)$$

$$\times \sum_{\mathbf{m}} \left\{ \frac{(-1)^{\sum m_j} \prod_{1 \leqslant a < b \leqslant n}(m_a - m_b)}{\prod_{k=1}^{n} \Gamma(\alpha + \varepsilon s - m_k)\Gamma(\varepsilon t + m_k + 1)} \chi_{\mathbf{m}}(g) \right\}. \tag{5.3}$$

Theorem 5.3. (a) *Let* $s \neq -t$. *Then there exists a limit in the sense of distributions:*

$$\ell^{s:t}(z) := \lim_{\varepsilon \to 0} \ell_{-n+\alpha+\varepsilon s|\varepsilon t}(z). \tag{5.4}$$

In other words, the function $(\sigma|\tau) \mapsto \ell_{\sigma|\tau}$ *has a removable singularity at* $\varepsilon = 0$ *on the line*

$$\sigma = -n + \alpha + \varepsilon s, \quad \tau = \varepsilon t, \quad \text{where } \varepsilon \in \mathbb{C}.$$

(b) *Denote by* $c_{\mathbf{m}}(s : t)$ *the Fourier coefficients of* $\ell^{s:t}$. *If* $\mathbf{m}$ *is in the "tail," i.e.,* $\mathbf{m} \notin \cup Z_j$, *then* $c_{\mathbf{m}}(s : t) = 0$.

---

[10]The case $\mathrm{U}(1, 1)$ was considered above in Sect. 2.5.1.

[11]A remark for an expert in algebraic geometry: We consider blow-up of the plane $\mathbb{C}^2$ at the point $(-n + \alpha, 0)$.

(c) Moreover, $\ell^{s:t}$ admits a decomposition

$$\ell^{s:t} = \sum_{j=0}^{n-\alpha} \frac{t^j s^{n-\alpha-j}}{(s+t)^{n-\alpha}} \mathfrak{L}_j, \tag{5.5}$$

where $\mathfrak{L}_j$ is of the form

$$\mathfrak{L}_j = \sum_{\mathbf{m}\in Z_j} a_{\mathbf{m}} \chi_{\mathbf{m}}, \tag{5.6}$$

where the $a_{\mathbf{m}}$ do not depend on s, t.
(d) For each j, all coefficients $a_{\mathbf{m}}^j$ in (5.6) are either positive or negative.

Proof. For the numerator of (5.3), we have the asymptotic

$$\prod_{k=1}^{n} \Gamma\big(-n + \alpha\varepsilon(s+t) + k\big) = C\,\varepsilon^{-n+\alpha}(s+t)^{-n+\alpha} + O(\varepsilon^{-n+\alpha+1}), \qquad \varepsilon \to 0.$$

Next, examine factors of the denominator,

$$\Gamma(\alpha + \varepsilon s - m_k)\Gamma(\varepsilon t + m_k + 1) \sim \begin{cases} A_1(m_k)(\varepsilon t)^{-1} & \text{if } m_k < 0 \\ A_2(m_k) & \text{if } 0 \leqslant m_k < \alpha, \\ A_3(m_k)(\varepsilon s)^{-1} & \text{if } m_k \geqslant \alpha, \end{cases} \qquad \varepsilon \to 0$$

where A_1, A_2, A_3 do not depend on s, t. Therefore, the order of the pole of denominator $\prod_k$ of (5.3) is

number of m_j outside the segment $[0, \alpha - 1]$.

The minimal possible order of a pole of the denominator is $n - \alpha$. In this case, $c_{\mathbf{m}}$ has a finite nonzero limit, of the form

$$c_{\mathbf{m}}(s:t) = A(\mathbf{m}) \cdot \frac{s^{\text{number of } m_k \geqslant \alpha} \cdot t^{\text{number of } m_k < 0.}}{(s+t)^{n-\alpha}}$$

If an order of pole in the denominator is $> n - \alpha$, then $c_{\mathbf{m}}(s:t) = 0$. This corresponds to the tail.

We omit the need to watch the positivity of $c_{\mathbf{m}}(s:t)$.

Formally, it is necessary to watch the growth of $c_{\mathbf{m}}(s:t)$ as $\mathbf{m} \to \infty$ and the growth of

$$\frac{\partial}{\partial\varepsilon} c_{\mathbf{m}}(-n + \alpha + \varepsilon s, \varepsilon t)$$

to be sure that (5.4) is a limit in the sense of distributions. This is a more or less trivial exercise on the Gamma function. □

There are many ways to express $\mathfrak{L}^j$ in the terms of $\ell^{s;t}$. One of variants is given in the following obvious proposition.

Proposition 5.4. *The distribution $\mathfrak{L}_j$ is given by the formula*

$$\mathfrak{L}_j(z) = \frac{1}{j!} \frac{\partial^j}{\partial t^j} (1+t)^{n-\alpha} \ell^{1;t}(z)\Big|_{t=0} \qquad (5.7)$$

5.3 The Family of Invariant Hermitian Forms

Thus, for $(\sigma, \tau) = (-n + \alpha, 0)$, we obtained the following families of $\rho_{-n+\alpha|0}$-invariant Hermitian forms:

$$R^{s;t}(f_1, f_2) := \iint_{U(n) \times U(n)} \ell^{s;t}(zu^*) f_1(z) \overline{f_2(u)} \, d\mu(z) \, d\mu(u) \qquad (5.8)$$

and

$$Q_j(f_1, f_2) := \iint_{U(n) \times U(n)} \mathfrak{L}_j(zu^*) f_1(z) \overline{f_2(u)} \, d\mu(z) \, d\mu(u). \qquad (5.9)$$

They are related as

$$R^{s;t}(f_1, f_2) = \sum_{j=0}^{n-\alpha} \frac{t^j s^{n-\alpha-j}}{(s+t)^{n-\alpha}} \mathfrak{L}_j(f_1, f_2).$$

A form $\mathfrak{L}_j$ is zero on

$$Y_j := W_{tail} \oplus (\oplus_{i \neq j} W_i)$$

and determines an inner product on $W_j \simeq C^\infty(U(n))/Y_j$.

6 Some Problems of Harmonic Analysis

6.1 Tensor Products $\rho_{\sigma|\tau} \otimes \rho_{\sigma'|\tau'}$

Nowadays the problem of decomposition of a tensor product of two arbitrary unitary representations does not seem interesting. We propose several informal arguments for the reasonableness of the problem in our case.

(a) For $n = 1$, it is precisely the well-known problem of the decomposition of tensor products of unitary representations of $\mathrm{SL}(2,\mathbb{R})^\sim \simeq \mathrm{SU}(1,1)^\sim$; see [16, 23, 32, 36, 39].

(b) Decomposition of tensor products $\rho_{\sigma,0} \otimes \rho_{\sigma',0}$ of holomorphic representations is a well-known combinatorial problem; see [19].

(c) Tensor products $\rho_{\sigma,0} \otimes \rho_{0|\tau'}$ are Berezin representations; see [4, 27, 47].

(d) All of the problems (a)–(c) have interesting links with the theory of special functions.

(e) There is a canonical isomorphism: [12]

$$\rho_{-n/2|-n/2} \otimes \rho_{-n/2|-n/2} \simeq L^2\big(\mathrm{U}(n,n)/\mathrm{GL}(n,\mathbb{C})\big). \qquad (6.1)$$

Thus, we again come to a classical problem, i.e., the problem of decomposition of L^2 on a pseudo-Riemannian symmetric space G/H; see [11, 35].[13] General tensor products $\rho_{\sigma|\tau} \otimes \rho_{\sigma'|\tau'}$ can be regarded as deformations of the space $L^2\big(\mathrm{U}(n,n)/\mathrm{GL}(n,\mathbb{C})\big)$.

6.2 Restriction Problems

1. Consider the group $G^* := \mathrm{U}(n,n)$ and its subgroup $G := \mathrm{O}(n,n)$. The group G has an open dense orbit on the space $\mathrm{U}(n)$, namely,

$$G/H := \mathrm{O}(n,n)/\mathrm{O}(n,\mathbb{C}).$$

The restriction of the representation $\rho_{-n/2|-n/2}$ to G is equivalent to the representation of G in $L^2(G/H)$. Restrictions of other $\rho_{\sigma|\tau}$ can be regarded as deformations of $L^2(G/H)$.

The same argument produces deformations of L^2 on some other pseudo-Riemannian symmetric spaces. In particular, we have the following variants:

2. $G^* = \mathrm{U}(2n,2n)$, $G/H = \mathrm{Sp}(n,n)/\mathrm{Sp}(2n,\mathbb{C})$.

3. $G^* = \mathrm{U}(n,n)$, $G/H = \mathrm{SO}^*(2n)/\mathrm{O}(n,\mathbb{C})$.

4. $G^* = \mathrm{U}(2n,2n)$, $G/H = \mathrm{Sp}(4n,\mathbb{R})/\mathrm{Sp}(2n,\mathbb{C})$.

5. $G^* = \mathrm{U}(p+q,p+q)$, $G/H = \mathrm{U}(p,q) \times \mathrm{U}(p,q)/\mathrm{U}(p,q)$. In this case, $G/H \simeq \mathrm{U}(p,q)$.

6. $G^* = \mathrm{U}(n,n)$, $G = \mathrm{GL}(n,\mathbb{C})$. In this case, we have $(n+1)$ open orbits $G/H_p = \mathrm{GL}(n,\mathbb{C})/\mathrm{U}(p,n-p)$.

7. $G^* = \mathrm{U}(n,n) \times \mathrm{U}(n,n)$, $G = \mathrm{U}(n,n)$. This is the problem about tensor products discussed above.

---

[12]Indeed, $\mathrm{U}(n) \simeq \mathrm{U}(n,n)/P$, where P is a maximal parabolic subgroup in $\mathrm{U}(n,n)$. The group $\mathrm{U}(n,n)$ has an open orbit on $\mathrm{U}(n,n)/P \times \mathrm{U}(n,n)/P$, and the stabilizer of a point is $\simeq \mathrm{GL}(n,\mathbb{C})$.

[13]In a certain sense, the Plancherel formula for $L^2(G/H)$ was obtained in [1, 6]. However, no Plancherel measure, nor spectra are known. The corresponding problems remain open.

6.3 The Gelfand–Gindikin Program

Recall the statement of the problem; see [13,34]. Let G/H be a pseudo-Riemannian symmetric space. The natural representation of G in $L^2(G/H)$ has several pieces of spectrum. Therefore, $L^2(G/H)$ admits a natural orthogonal decomposition into direct summands having uniform spectra. The problem is: *to describe explicitly the corresponding subspaces or corresponding projectors.*

In Sect. 5.1 we obtained a natural decomposition of $L^2(\mathrm{U}(n))$ into $(n+1)$ direct summands. Therefore, *in the cases listed in Sect. 6.2, we have a natural orthogonal decompositions of $L^2(G/H)$.*

In any case, for the one-sheet hyperboloid $\mathrm{U}(1,1)/\mathbb{C}^*$ we get the desired construction (see Molchanov [24,25]).

6.4 Matrix Sobolev Spaces?

Our inner product $\langle \cdot, \cdot \rangle_{\sigma|\tau}$ seems to be similar to Sobolev-type inner products discussed in Sect. 2.3.2. However, it is not a Sobolev inner product, because the kernel $\det(1-zu^*)^{\{\sigma|\tau\}}$ has a non-diagonal singularity.

Denote

$$s = -\sigma - \tau + n.$$

Let F be a distribution on $\mathrm{U}(n)$, and let $F = \sum F_{\mathbf{m}}$ be its expansion in a series of elementary harmonics. We have

$$F \in \mathcal{H}_{\sigma|\tau} \iff \sum_{\mathbf{m}} \frac{c_{\mathbf{m}}}{\dim \pi_{\mathbf{m}}} \|F_{\mathbf{m}}\|_{L^2}^2 < \infty$$

$$\iff \sum_{\mathbf{m}} \left\{ \|F_{\mathbf{m}}\|_{L^2}^2 \prod_{j=1}^{n} (1 + |m_j|)^s \right\} < \infty, \qquad (6.2)$$

where $\|F_{\mathbf{m}}\|_{L^2}$ denotes

$$\|F_{\mathbf{m}}\|_{L^2} := \left(\int_{\mathrm{U}(n)} |F_{\mathbf{m}}(h)|^2 \, \mathrm{d}\mu(h) \right)^{1/2}$$

Our Hermitian form defines a norm only in the case $|s| < 1$, but (6.2) makes sense for arbitrary real s. Thus, *we can define a Sobolev space H_s on $\mathrm{U}(n)$ of arbitrary order.*

The author does not know specific applications of this remark, but it seems that it can be useful in the following situation.

First, a reasonable harmonic analysis related to semisimple Lie groups is the analysis of unitary representations. But around 1980, Molchanov observed that many identities with special function admit interpretations on a "physical level of rigor" as formulas of non-unitary harmonic analysis. Until now, there have been no reasonable interpretations of this phenomenon [but formulas exist; see, e.g., [7], see also [27], Sect. 1.32, and formulas (2.6)–(2.15)]. In particular, we do not know reasonable functional spaces that can be the scene of action of such an analysis. It seems that our spaces H_s can be possible candidates.

Acknowledgement Supported by the grant FWF, project P19064, P22122, Russian Federal Agency for Nuclear Energy, Dutch grant NWO.047.017.015, and grant JSPS-RFBR-07.01.91209.

References

1. van den Ban, E.P., Schlichtkrull, H., *The most continuous part of the Plancherel decomposition for a reductive symmetric space.* Ann. Math., 145 (1997), 267–364
2. Bargmann, V. *Irreducible unitary representations of the Lorentz group.* Ann. Math, 48 (1947), 568–640
3. Berezin, F.A., *Quantization in complex symmetric spaces.* Izv. Akad. Nauk SSSR, Ser. Math., 39, 2, 1362–1402 (1975); English translation: Math USSR Izv. 9 (1976), No 2, 341–379(1976)
4. Berezin, F. A. *The connection between covariant and contravariant symbols of operators on classical complex symmetric spaces.* Sov. Math. Dokl. 19 (1978), 786–789
5. Branson, Th., Olafsson, G., Orsted, B. *Spectrum generating operators and intertwining operators for representations induced from a maximal parabolic subgroup,* J. Funct. Anal., 135, 163–205.
6. Delorme, P. *Formule de Plancherel pour les espaces symmétrique reductifs.* Ann. Math., 147 (1998), 417–452
7. van Dijk, G., Molchanov, V.F. *The Berezin form for rank one para-Hermitian symmetric spaces.* J. Math. Pure. Appl., 78 (1999), 99–119.
8. Dvorsky, A., Sahi, S. *Explicit Hilbert spaces for certain unipotent representations. II.* Invent. Math. 138 (1999), no. 1, 203–224.
9. Dvorsky, A., Sahi, S. *Explicit Hilbert spaces for certain unipotent representations. III.* J. Funct. Anal. 201(2003), no. 2, 430–456.
10. Faraut, J., Koranyi, A., *Analysis in symmetric cones.* Oxford Univ.Press, (1994)
11. Flensted-Jensen, M. *Discrete series for semisimple symmetric spaces.* Ann. of Math. (2) 111 (1980), no. 2, 253–311.
12. Friedrichs, K. O. *Mathematical aspects of the quantum theory of fields.* Interscience Publishers, Inc., New York, 1953.
13. Gelfand, I. M., Gindikin, S. G. *Complex manifolds whose skeletons are semisimple Lie groups and analytic discrete series of representations.* Funct. Anal. Appl., 11 (1978), 258–265
14. Gelfand, I.M., Naimark, M.I., *Unitary representations of classical groups. Unitary representations of classical groups.* Trudy MIAN., t.36 (1950); German translation: Gelfand I.N., Neumark M.A., *Unitare Darstellungen der klassischen gruppen.*, Akademie-Verlag, Berlin, 1957.
15. Gradshtein, I.S., Ryzhik, I.M. *Tables of integrals, sums and products.* Fizmatgiz, 1963; English translation: Acad. Press, NY, 1965
16. Groenevelt, W., Koelink, E., Rosengren, H. *Continuous Hahn polynomials and Clebsch–Gordan coefficients.* Theory and applications of special functions, 221–284, Dev. Math., 13, Springer, New York, 2005.

17. Harish-Chandra, *Representations of semisimple Lie groups IV,* Amer. J. Math., 743–777 (1955). Reprinted in Harish-Chandra *Collected papers,* v.2.

18. *Higher transcendental functions,* v.1., McGraw-Hill book company, 1953

19. Jakobsen, H.P., Vergne, M., *Restrictions and expansions of holomorphic representations.* J. Funct. Anal., 34 (1979), 29–53.

20. Kadell, K. *The Selberg–Jack symmetric functions.* Adv. Math., 130 (1997), 33-102

21. Kirillov, A.A. *Elements of representation theory,* Moscow, Nauka, 1972; English transl.: Springer, 1976.

22. Krattenthaler, C. *Advanced determinant calculus.* The Andrews Festschrift (Maratea, 1998). Sem. Lothar. Combin. 42 (1999), Art. B42q, 67 pp. (electronic).

23. Molchanov, V. F.*Tensor products of unitary representations of the three-dimensional Lorentz group.* Izv. Akad. Nauk SSSR Ser. Mat. 43 (1979), no. 4, 860–891, 967. English transl. in Izvestia.

24. Molchanov, V. F. *Quantization on the imaginary Lobachevsky plane.* Funct. Anal. Appl., 14 (1980), 162–144

25. Neretin, Yu. A. *The restriction of functions holomorphic in a domain to a curve lying in the boundary, and discrete* $SL_2(\mathbb{R})$*-spectra.* Izvestia: Mathematics, 62:3(1998), 493–513

26. Neretin, Yu.A., *Matrix analogs of B-function and Plancherel formula for Berezin kernel representations,* Mat. Sbornik, 191, No.5 (2000), 67–100;

27. Neretin, Yu.A., *Plancherel formula for Berezin deformation of* L^2 *on Riemannian symmetric space,* J. Funct. Anal. (2002), 189(2002), 336–408.

28. Neretin, Yu.A. *Matrix balls, radial analysis of Berezin kernels, and hypergeometric determinants,* Moscow Math. J., v.1 (2001), 157–221.

29. Neretin, Yu.A. *Notes Sahi–Stein representations and some problems of non-*L^2 *harmonic analysis.,* J. Math. Sci., New York, 141 (2007), 1452–1478

30. Neretin, Yu. A. *Notes on matrix analogs of Sobolev spaces and Stein–Sahi representations.* Preprint, http://arxiv.org/abs/math/0411419

31. Neretin, Yu. A. *Lectures on Gaussian integral operators and classical groups,* to appear.

32. Neretin, Yu. A. *Some continuous analogs of expansion in Jacobi polynomials and vector-valued orthogonal bases.* Funct. Anal. Appl., 39 (2005), 31–46.

33. Neretin, Yu.A., Olshanskii, G.I., *Boundary values of holomorphic functions , singular unitary representations of groups* $O(p,q)$ *and their limits as* $q \rightarrow \infty$. Zapiski nauchn. semin. POMI RAN 223, 9–91(1995); English translation: J.Math.Sci., New York, 87, 6 (1997), 3983–4035.

34. Olshanskij, G.I., *Complex Lie semigroups, Hardy spaces, and Gelfand–Gindikin program.* Deff. Geom. Appl., 1 (1991), 235–246

35. Oshima, T. *A calculation of c-functions for semisimple symmetric spaces. Lie groups and symmetric spaces,* 307–330, Amer. Math. Soc. Transl. Ser. 2, 210, Amer. Math. Soc., Providence, RI, 2003.

36. Pukanszky, L., *On the Kronecker products of irreducible unitary representations of the* 2×2 *real unimodular group.* Trans. Amer. Math. Soc., 100 (1961), 116–152

37. Pukanzsky, L. *Plancherel formula for universal covering group of* $SL(2,\mathbb{R})$. Math. Ann., 156 (1964), 96–143

38. Ricci, F., Stein, E. M. *Homogeneous distributions on spaces of Hermitean matrices.* J. Reine Angew. Math. 368 (1986), 142–164.

39. Rosengren, H. *Multilinear Hankel forms of higher order and orthogonal polynomials.* Math. Scand., 82 (1998), 53-88

40. Sahi, S. *A simple construction of Stein's complementary series representations.* Proc. Amer. Math. Soc. 108 (1990), no. 1, 257–266.

41. Sahi, S., *Unitary representations on the Shilov boundary of a symmetric tube domain,* Contemp. Math. 145 (1993) 275–286.

42. Sahi, S. *Jordan algebras and degenerate principal series,* J. Reine Angew.Math. 462 (1995) 1–18.

43. Sahi, S. *Explicit Hilbert spaces for certain unipotent representations.* Invent. Math. 110 (1992), no. 2, 409–418.

44. Sahi, S., Stein, E. M. *Analysis in matrix space and Speh's representations.* Invent. Math. 101 (1990), no. 2, 379–393.
45. Sally, P. J., *Analytic continuations of irreducible unitary representations of the universal covering group of* SL(2, ℝ). Amer. Math. Soc., Providence, 1967
46. Stein, E. M. *Analysis in matrix spaces and some new representations of* SL(N, C). Ann. of Math. (2) 86 1967 461–490.
47. Unterberger, A., Upmeier, H., *The Berezin transform and invariant differential operators.* Comm.Math.Phys.,164, 563–597(1994)
48. Vogan, D. A., *The unitary dual of* GL(n) *over an Archimedean field.* Invent. Math. 83 (1986), no. 3, 449–505.
49. Weyl, H. *The Classical Groups. Their Invariants and Representations.* Princeton University Press, Princeton, N.J., 1939.

The Special Symplectic Structure
of Binary Cubics

Marcus J. Slupinski and Robert J. Stanton

Abstract We present a thorough investigation of binary cubics over a field of
characteristic not 2 or 3 using equivariant symplectic methods. The primary
symplectic tools are the moment map and its norm, as well as the symplectic
gradient of the norm. Among the results obtained are a symplectic stratification
of the space of binary cubics, the identification of a group structure on generic
orbits, a symplectic derivation of the Cardano-Tartaglia formula, and a symplectic
formulation and proof of the Eisenstein syzygy.

Keywords Moment map • Binary cubic polynomials • Special symplectic
structure • Prehomogeneous vector space

Mathematics Subject Classification (2010): 11S90, 53D20

1 Introduction

Binary cubic polynomials have been studied since the nineteenth century, being the
natural setting for a possible extension of the rich theory of binary quadratic forms.
An historical summary of progress on this subject can be found in [5], especially
concerning results related to integral coefficients. While for a fixed binary cubic

M.J. Slupinski (✉)
IRMA, Université de Strasbourg (Strasbourg), 7 rue René Descartes,
F-67084 Strasbourg Cedex, France
e-mail: slupins@math.u-strasbg.fr

R.J. Stanton
Department of Mathematics, Ohio State University, 231 West 18th Avenue,
Columbus OH 43210-1174, USA
e-mail: stanton@math.ohio-state.edu

B. Krötz et al. (eds.), *Representation Theory, Complex Analysis, and Integral Geometry*,
DOI 10.1007/978-0-8176-4817-6_8, © Springer Science+Business Media, LLC 2012

interesting questions remain open, e.g. its range in the integers, the number of solutions, etc., it is the structure of the space of all binary cubics that is the topic of this paper.

The space of binary cubics, we will take coefficients in a field, is an example of a prehomogeneous vector space under $Gl(2, k)$, and from this point of view has been thoroughly investigated. Beginning with the fundamental paper by Shintani [14], recast adelically in [16], an analysis of this pv sufficient to obtain the properties of the Sato–Shintani zeta function was done. Subsequently, several descriptions of the orbit structure were obtained, in particular relating them to extensions of the coefficient field. A feature of this space, and some other prehomogeneous spaces, apparently never exploited is the existence of a symplectic structure which is preserved by the natural action of $Sl(2, k)$.

The purpose of this paper is to expose the rich structure of the space of binary cubics when viewed as a symplectic module using the standard tools of equivariant symplectic geometry, viz. the moment map, its norm square, and its symplectic gradient i.e. the natural Hamiltonian vector field. The advantages are several: somewhat surprisingly, the techniques are universally applicable, with the only hypothesis that the fields not be of characteristic 2 or 3; there are explicit symplectic parameters for each orbit type (including the singular ones not studied previously) that are easily computed for any specific field; the computations are natural; we obtain new results for the space of binary cubics, e.g. a group structure on orbits; we obtain ancient results for cubics, namely a symplectic derivation of the Cardano–Tartaglia formula for a root.

This paper arose as a test case to see the extent that we might push a more general project [15] on Heisenberg graded Lie algebras. A symplectic module can be associated with every such graded Lie algebra and in the case of the split Lie algebra G_2, this symplectic module turns out to be isomorphic to the space of binary cubics with the $Sl(2, k)$ action mentioned above. Although our approach to binary cubics is inspired by the general situation, in order to give an accessible and elementary presentation, we have made this paper essentially self-contained with only one or two results quoted without proof from [15].

The symplectic technology consists of the following. The moment map, μ, maps the space of binary cubics $S^3(k^{2^*})$ to the Lie algebra $\mathfrak{sl}(2, k)$ of $Sl(2, k)$. By means of the Killing form on $\mathfrak{sl}(2, k)$, one obtains a scalar valued function Q on $S^3(k^{2^*})$, the norm square of μ. Using the symplectic structure, one constructs Ψ, the symplectic gradient of Q, as the remaining piece of symplectic machinery. This symplectic module appears to be "special" in several ways, e.g. a consequence of our analysis is that all the $Sl(2, k)$ orbits in $S^3(k^{2^*})$ are co-isotropic (see [15] for the general case). Let us recall that over the real numbers it has been shown that there is also a very strong link between special symplectic connections (see [3]) and Heisenberg graded Lie algebras (called 2-graded in [3]).

Here is a more detailed overview of the paper. We will analyze each of the symplectic objects μ, Q, Ψ and determine for each of them their image, their fiber, the $\mathrm{Sl}(2, k)$ orbits in each fiber, and explicit parameters and isotropy for each orbit type. This is all done with symplectic methods, so that furthermore we identify the symplectic geometric meaning of these fibers. For example, we show that the null space, Z, of μ is the set of multiples of cubes of linear forms. As $\mathrm{Sl}(2, k)$ preserves the null space, we obtain a decomposition into a collection of isomorphic Lagrangian orbits which we show are parameterized by k^*/k^{*3}. Binary cubics whose moment lies in the nonzero nilpotent cone of $\mathfrak{sl}(2, k)$ turn out to be those which contain a factor that is the square of a linear form. For these there is only one orbit, whose image under μ we characterize. The pullback by means of μ of the natural symplectic structure on the image and the restriction of the symplectic form on $S^3(k^{2^*})$ essentially coincide. The generic case is when the image of the moment map lies in the semisimple orbits of $\mathfrak{sl}(2, k)$. In this case, the $\mathrm{Sl}(2, k)$ orbits are different from the $\mathrm{Gl}(2, k)$ orbits, in contrast to the earlier cases. Here, each of the values of Q in k^* determine a collection of $\mathrm{Sl}(2, k)$ orbits for which we give symplectic parameters using a 'sum of cubes' theorem. As a consequence, we show that the orbits for a fixed nonzero value of Q form a group (over $\mathbb{Z}$ see [1]) which we explicitly identify. Interestingly, a binary cubic is in the orbit corresponding to the identity of this group if and only if it is reducible. The set of binary cubics corresponding to a fixed nonzero value of Q is not stable under $\mathrm{Gl}(2, k)$. However, the set of binary cubics for which the value of Q belongs to a fixed nonzero square class of k is stable under $\mathrm{Gl}(2, k)$ and we obtain an explicit parametrization of all $\mathrm{Gl}(2, k)$ orbits on this set.

If the field of coefficients is specialized to say $\mathbb{C}$, then several of the results herein are known. For example, that the zero set of Q is the tangent variety to Z, or that the generic orbit is the secant variety of Z can be found in the complex algebraic geometric literature. For some other fields, other results are in the literature. However, the use of symplectic methods is new to all these cases and gives a unifying approach that seems to make transparent many classic results. For example, a careful analysis of μ and Ψ in the generic case leads to a proof of the Cardano–Tartaglia formula for a root of a cubic. As another application we conclude the paper with a symplectic generalization of the classical Eisenstein syzygy for the covariants (compare to [12],[10]) of a binary cubic. This is interesting because there is an analogue of this form of the Eisenstein syzygy for the symplectic module associated with any Heisenberg graded Lie algebra ([15]). Finally, we remark that the symplectic methodology used in this paper could be used to understand binary cubics over the integers or more general rings.

We are very pleased to acknowledge the support of our respective institutions that made possible extended visits. To the gracious faculty of the Université Louis Pasteur goes a sincere merci beaucoup from RJS. In addition, RJS wants to acknowledge the support of Max Planck Institut, Bonn, for an extended stay during which some of this research was done.

2   Binary Cubics as a Symplectic Space

Let k be a field such that $char(k) \neq 2, 3$. The vector space k^{2^*} has a symplectic
structure

$$\Omega(ax + by, a'x + b'y) = ab' - ba'.$$

Functorially, one obtains a symplectic structure on the set of binary cubics

$$S^3(k^{2^*}) = \{ax^3 + 3bx^2y + 3cxy^2 + dy^3 : a, b, c, d \in k\}.$$

Explicitly, if $P = ax^3 + 3bx^2y + 3cxy^2 + dy^3$ and $P' = a'x^3 + 3b'x^2y + 3c'$
$xy^2 + d'y^3$,

$$\omega(P, P') = ad' - da' - 3bc' + 3cb'. \tag{1}$$

In particular, we have

$$\omega(P, (ex + fy)^3) = P(f, -e). \tag{2}$$

Hence for $ex + fy \neq 0$,

$$(ex + fy) \mid P \iff \omega(P, (ex + fy)^3) = 0. \tag{3}$$

This indicates that one can use the symplectic form ω to study purely algebraic
properties of the space of binary cubics. More generally, the interplay of symplectic
methods and the algebra of binary cubics will be the primary theme of this paper.

The group

$$Sl(2, k) = \left\{ \begin{pmatrix} \alpha & \beta \\ \gamma & \delta \end{pmatrix} : \alpha\delta - \beta\gamma = 1 \right\}$$

acts on k^{2^*} via the transpose inverse:

$$\begin{pmatrix} \alpha & \beta \\ \gamma & \delta \end{pmatrix} \cdot x = \delta x - \beta y, \quad \begin{pmatrix} \alpha & \beta \\ \gamma & \delta \end{pmatrix} \cdot y = -\gamma x + \alpha y, \tag{4}$$

and this action identifies $Sl(2, k)$ with the group of transformations of k^{2^*} that
preserve the symplectic form Ω, i.e. $Sp(k^{2^*}, \Omega)$. It follows that the functorial action
of $Sl(2, k)$ on $S^3(k^{2^*})$ preserves the symplectic form ω. There is no kernel of this
action thus $Sl(2, k) \hookrightarrow Sp(S^3(k^{2^*}), \omega)$.

The Lie algebra $\mathfrak{sl}(2, k)$ acts on k^{2^*} via the negative transpose:

$$\begin{pmatrix} \alpha & \beta \\ \gamma & -\alpha \end{pmatrix} \cdot x = -\alpha x - \beta y, \quad \begin{pmatrix} \alpha & \beta \\ \gamma & -\alpha \end{pmatrix} \cdot y = -\gamma x + \alpha y, \tag{5}$$

which in terms of differential operators acting on polynomial functions on k^2 corresponds to the action

$$\begin{pmatrix} \alpha & \beta \\ \gamma & -\alpha \end{pmatrix} \cdot f = \alpha(-x\partial_x f + y\partial_y f) - \beta y \partial_x f - \gamma x \partial_y f. \tag{6}$$

In particular, this gives the following action of $\mathfrak{sl}(2, k)$ on cubics:

$$x^3 \mapsto -3\alpha x^3 - 3\beta x^2 y$$

$$x^2 y \mapsto -\gamma x^3 - \alpha x^2 y - 2\beta xy^2$$

$$xy^2 \mapsto -2\gamma x^2 y + \alpha xy^2 - \beta y^3$$

$$y^3 \mapsto -3\gamma xy^2 + 3\alpha y^3.$$

2.1 Symplectic Covariants

Among the basic tools of equivariant symplectic geometry are the moment map (μ), its norm square (Q) and the symplectic gradient of Q (Ψ). The symplectic structure on $S^3(k^{2^*})$ is not generic as it is consistent with one inherited from an ambient Heisenberg graded Lie algebra, hence the description "special". In [15] in the setting of Heisenberg graded Lie algebras, we derive the fundamental properties of the basic symplectic objects as well as give explanations for normalizing constants, and identify characteristic features of these special symplectic structures. For the purposes of this paper, the explicit formulae will suffice.

Definition 2.1. (i) The moment map $\mu : S^3(k^{2^*}) \to \mathfrak{sl}(2, k)$ here is

$$\mu(x^3 + 3bx^2 y + 3cxy^2 + dy^3) = \begin{pmatrix} ad - bc & 2(bd - c^2) \\ 2(b^2 - ac) & -(ad - bc) \end{pmatrix}. \tag{7}$$

(ii) The cubic covariant $\Psi : S^3(k^{2^*}) \to S^3(k^{2^*})$ is given by

$$\Psi(P) = \mu(P) \cdot P = (-3a\alpha - 3b\gamma)x^3 + (-3a\beta - 3b\alpha - 6c\gamma)x^2 y$$

$$+ (-6b\beta + 3c\alpha - 3d\gamma)xy^2 + (-3c\beta + 3d\alpha)y^3, \tag{8}$$

where $P = ax^3 + 3bx^2 y + 3cxy^2 + dy^3$ and

$$\begin{pmatrix} \alpha & \beta \\ \gamma & -\alpha \end{pmatrix} = \begin{pmatrix} ad - bc & 2(bd - c^2) \\ 2(b^2 - ac) & -(ad - bc) \end{pmatrix}.$$

(iii) The normalized quartic invariant $Q_n : S^3(k^{2*}) \to k$ is

$$Q_n(P) = -\det\mu(P) = (a^2d^2 - 3b^2c^2 - 6abcd + 4b^3d + 4ac^3). \quad (9)$$

Notice that $Q_n(P)$ is a multiple (−1) of the classic discriminant of the polynomial P.

Remark 2.2. The symmetric role of the coordinates x and y is implemented by

$$J = \begin{pmatrix} 0 & -1 \\ 1 & 0 \end{pmatrix},$$

which satisfies $J \cdot x = y, J \cdot y = -x$ and

$$J \cdot \left(ax^3 + 3bx^2y + 3cxy^2 + dy^3\right) = -dx^3 + 3cx^2y - 3bxy^2 + ay^3.$$

From (7) it follows that $\mu(J \cdot P)$ is the cofactor matrix of $\mu(P)$.

Remark 2.3. The set of symplectic covariants $\omega, \mu, \Psi, Q, Q_n$ defined above is not the only choice possible for the purposes of this article. One could just as well use

$$\omega_\lambda = \lambda\omega, \quad \mu_\lambda = \lambda\mu, \quad \Psi_\lambda = \lambda\Psi, \quad Q_\lambda = \lambda^2 Q,$$

where $\lambda \in k^*$.

The moment map is characterized by the identity

$$Tr(\mu(P)\xi) = -\frac{1}{3}\omega(\xi \cdot P, P) \qquad \forall P \in S^3(k^{2*}), \forall \xi \in \mathfrak{sl}(2,k), \quad (10)$$

which specialized to $\xi = \mu(P)$ gives a characterization of Ψ

$$Q(P) = 8\omega(P, \Psi(P)). \quad (11)$$

From (10), one gets that μ is $Sl(2,k)$-equivariant:

$$\mu(g \cdot P) = g\mu(P)g^{-1} \qquad \forall P \in S^3(k^{2*}), \forall g \in Sl(2,k),$$

and $\mathfrak{sl}(2,k)$-equivariant:

$$d\mu_P(\xi \cdot P) = [\xi, \mu(P)] \qquad \forall P \in S^3(k^{2*}), \forall \xi \in \mathfrak{sl}(2,k).$$

Here, $d\mu_P(Q) = 2B_\mu(P, Q)$ where $B_\mu : S^3(k^{2*}) \times S^3(k^{2*}) \to \mathfrak{sl}(2,k)$ is the unique symmetric bilinear map such that $\mu(P) = B_\mu(P, P)$.

From the $\mathrm{Sl}(2,k)$ and $\mathfrak{sl}(2,k)$ equivariance of μ one obtains the $\mathrm{Sl}(2,k)$ and $\mathfrak{sl}(2,k)$ equivariance of Ψ, Q and Q_n. Several useful relations among μ, Ψ and Q are derived in [15]. The following involves a relation between vanishing sets of symplectic covariants.

Proposition 2.4. *Let P be a binary cubic. Then*

$$\mu(P) = 0 \Rightarrow \Psi(P) = 0 \Rightarrow Q(P) = 0.$$

Proof. Since $\Psi(P) = \mu(P) \cdot P$, it is obvious that $\mu(P) = 0 \Rightarrow \Psi(P) = 0$. Suppose that $\Psi(P) = 0$. Then by equation (10)

$$Tr(\mu(P)^2) = -\frac{1}{3}\omega(\Psi(P), P) = 0.$$

But $\mu(P)^2 + \det\mu(P)Id = 0$ by the Cayley–Hamilton theorem, so $\det\mu(P) = 0$ and hence $Q(P) = 0$. $\qquad\square$

From the invariant theory point of view, a covariant is an $\mathrm{Sl}(2,k)$ invariant in $S^*(S^3(k^{2^*})) \otimes S^*(k^{2^*})$. Concerning completeness of the symplectic invariants one has the classic syzygy of Eisenstein for $k = \mathbb{C}$ [8].

Proposition 2.5. *(i) μ, Ψ, Q and the identity generate the $\mathrm{Sl}(2,k)$ invariants in $S^3(k^{2^*}) \otimes S^*(k^{2^*})$.*
(ii) The only relation among them viewed as functions on k^2 is

$$\Psi(P)(\cdot)^2 - 9Q_n(P)P(\cdot)^2 = -\frac{9}{2}\Omega_{k^2}(\mu(P)\cdot, \cdot)^3,$$

here Ω is extended by duality to $k^2 \times k^2$.

Proof. We shall give a symplectic proof of the relation (ii) for k in §3. $\qquad\square$

Remark 2.6. There are two interesting results related by a simple scaling to the Eisenstein syzygy. Fix $P \in S^3(k^{2^*})$ with $Q_n(P) \neq 0$. One can associate with P a type of Clifford algebra, $Cliff_P$, and in [9] it is shown that the center of $Cliff_P$ is the coordinate algebra of the genus one curve $X^2 - 27Q_n(P) = Z^3$. The other result arises from the observation that we could work over, say, $\mathbb{Z}$ instead of k. Then in [11] Mordell showed that all integral solutions (X, Y, Z) to $X^2 + kY^2 = Z^3$ with $(X, Z) = 1$ are obtained from some $P \in S^3(\mathbb{Q}^{2^*})$ with $Q_n(P) = -4k$ and evaluating *(ii)* at a lattice point in $\mathbb{Q}^2$. We will not use these results in this paper but we will give a symplectic proof at another time.

Remark 2.7. The Proposition gives a complete description of binary cubics from the point of view of $\mathrm{Sl}(2, \mathbb{C})$ invariant theory. From the symplectic theory point of view, in [15] we give characterizations of $\mathrm{Sl}(2, k)$ as the subgroup of $Sp(S^3(k^{2^*}), \omega)$ that preserves $Q(\cdot)$ and as the subgroup of $Sp(S^3(k^{2^*}), \omega)$ that commutes with Ψ.

2.2 The Image of the Moment Map

As $\mu : S^3(k^{2*}) \to \mathfrak{sl}(2,k)$ is equivariant, the image of μ is a union of $\mathrm{Sl}(2,k)$ invariant sets. Of course, the invariant functions on $\mathfrak{sl}(2,k)$ are generated by det. The following description of the orbits of $\mathrm{Sl}(2,k)$ acting on level sets of det uses the symplectic structure on k^{2*}. Lacking any reference for this probably known result we include a proof. Subsequently, Paul Ponomarev brought to our attention the material in [2] p.158–159 from which an alternate albeit non-symplectic proof can be extracted.

Proposition 2.8. *Let $\Delta \in k$ and set*

$$\mathfrak{sl}(2,k)_\Delta = \{X \in \mathfrak{sl}(2,k) \setminus \{0\} : \det X = \Delta\},$$

$$k_\Delta^* = \{x \in k^* : \exists a, b \in k \text{ such that } x = a^2 + b^2 \Delta\}.$$

Then the orbits of $\mathrm{Sl}(2,k)$ acting on $\mathfrak{sl}(2,k)_\Delta$ are in bijection with k^*/k_Δ^* under the map $\nu_\Delta : \mathfrak{sl}(2,k)_\Delta \to k^*/k_\Delta^*$ defined by

$$\nu_\Delta(X) = [\Omega(v, X \cdot v)], \tag{12}$$

where v is any element in k^{2*}, which is not an eigenvector of X.

Proof. We make some preliminary remarks before proving the result. First we observe that the definition of $\nu_\Delta(X)$ is independent of choice of v. Indeed, given v which is not an eigenvector of X, then $\{v, X \cdot v\}$ is a basis of k^{2*}. Given w any other vector which is not an eigenvector then $w = av + bX \cdot v$, and using Cayley–Hamilton we obtain that $[\Omega(v, X \cdot v)] = [\Omega(w, X \cdot w)]$.

Next, note that if $X \in \mathfrak{sl}(2,k)$ there exists $g \in \mathrm{Sl}(2,k)$ and $\beta, \gamma \in k$ such that

$$gXg^{-1} = \begin{pmatrix} 0 & \beta \\ \gamma & 0 \end{pmatrix}.$$

So to prove the result, we need only consider matrices in $\mathfrak{sl}(2,k)_\Delta$ of the form $X = \begin{pmatrix} 0 & \beta \\ \gamma & 0 \end{pmatrix}$ with either β or γ nonzero. Since

$$\begin{pmatrix} 0 & 1 \\ -1 & 0 \end{pmatrix} \begin{pmatrix} 0 & \beta \\ \gamma & 0 \end{pmatrix} \begin{pmatrix} 0 & 1 \\ -1 & 0 \end{pmatrix}^{-1} = \begin{pmatrix} 0 & -\gamma \\ -\beta & 0 \end{pmatrix},$$

we can further suppose that $\gamma \neq 0$. Then x is not an eigenvector of X and $\nu_{\det X}(X) = [\Omega(x, X \cdot x)] = [\Omega(x, \gamma x)] = [\gamma]$.

Suppose $\begin{pmatrix} 0 & \beta \\ \gamma & 0 \end{pmatrix}$ and $\begin{pmatrix} 0 & \beta' \\ \gamma' & 0 \end{pmatrix}$ in $\mathfrak{sl}(2,k)_\Delta$ have the same value of ν_Δ, i.e., $\beta\gamma = -\Delta = \beta'\gamma'$ and $[\gamma] = [\gamma']$.

Then there exist p, q in k such that $\gamma' = (p^2 + q^2 \det X)\gamma$. Take as *Ansatz*

$$\begin{pmatrix} a & b \\ c & d \end{pmatrix} = \begin{pmatrix} p & -q\frac{\Delta}{\gamma'} \\ \gamma q & p\frac{\gamma}{\gamma'} \end{pmatrix}.$$

Then

$$\det \begin{pmatrix} a & b \\ c & d \end{pmatrix} = p^2 \frac{\gamma}{\gamma'} + q^2 \Delta \frac{\gamma}{\gamma'}$$

$$= \frac{\gamma}{\gamma'}(p^2 + q^2 \Delta)$$

$$= 1.$$

A routine computation shows that

$$\begin{pmatrix} a & b \\ c & d \end{pmatrix} \begin{pmatrix} 0 & \beta' \\ \gamma' & 0 \end{pmatrix} \begin{pmatrix} d & -b \\ -c & a \end{pmatrix} = \begin{pmatrix} 0 & \beta \\ \gamma & 0 \end{pmatrix},$$

and so ν_Δ separates orbits.

To show that given $\alpha \neq 0$, there is an X with $\det X = \Delta$ and $\nu_\Delta(X) = [\alpha]$, take

$$X = \begin{pmatrix} 0 & -\frac{\Delta}{\alpha} \\ \alpha & 0 \end{pmatrix}.$$

Then $\det X = \Delta$ and $\nu_\Delta(X) = [\alpha]$. Finally, $\mathrm{Sl}(2, k)$ invariance of ν_Δ follows from the definition of ν_Δ. $\square$

Remark 2.9. We make some elementary observations concerning the $\mathrm{Sl}(2, k)$ adjoint orbits. If $-\Delta \in k^{*2}$, then $k_\Delta^* = k^*$ and there is only one orbit. If $\Delta = 0$ then $k_\Delta^* = k^{*2}$ and there is one nilpotent orbit for every element of k^*/k^{*2}. If $-\Delta \notin k^{*2}$ is nonzero, then k_Δ^* is the set of values in k^* taken by the norm function associated to the quadratic extension $k(\sqrt{-\Delta})$ or, equivalently, by the anisotropic quadratic form $x^2 + \Delta y^2$ on k^2. It is well known that this is a proper subgroup of k^*, at least in characteristic 0 (with thanks to P. Ponomarev for a discussion on characteristic p) and so in characteristic zero there are at least two orbits.

Remark 2.10. Since k^*/k_Δ^* is a group, the Proposition puts a natural group structure on the set of orbits of $\mathrm{Sl}(2, k)$ acting on trace free matrices of fixed determinant. Alternatively, $\mathfrak{sl}(2, k)$ can be $\mathrm{Sl}(2, k)$-equivariantly identified with $S^2(k^{2*})$, the space of binary quadratic forms, by

$$X \quad \longleftrightarrow \quad q_X(v) = \Omega(v, X \cdot v).$$

By transport of structure, the Proposition then puts a natural group structure on the set of orbits of $\mathrm{Sl}(2, k)$ acting on binary quadratic forms of fixed discriminant. One can check that this is Gauss composition. In Theorems 3.35 and 3.47, we will put a natural group structure on orbits of binary cubics with fixed nonzero discriminant.

The image of the moment map can be characterized as follows.

Theorem 2.11. *Let* $X \in \mathfrak{sl}(2, k) \setminus \{0\}$. *Then*

$$X \in \operatorname{Im} \mu \quad \Longleftrightarrow \quad \nu_{\det X}(X) = [2].$$

Proof. As before, we can suppose without loss of generality that $X = \begin{pmatrix} 0 & \beta \\ \gamma & 0 \end{pmatrix}$ with say β nonzero.

($\Rightarrow$) : If $X = \mu(P)$ and $P = ax^3 + 3bx^2y + 3cxy^2 + dy^3$, we have

$$ad - bc = 0$$
$$2(bd - c^2) = \beta$$
$$2(b^2 - ac) = \gamma.$$

Hence, $b\beta = d\gamma$ and

$$-\beta = 2\left(c^2 - d^2\frac{\gamma}{\beta}\right) = 2\left(c^2 + \left(\frac{d}{\beta}\right)^2(-\beta\gamma)\right) = 2\left(c^2 + \left(\frac{d}{\beta}\right)^2\det X\right)$$

so that $\nu_{\det X}(X) = [-\beta] = [2]$.

($\Leftarrow$): Since $\nu_{\det X}(X) = [-\beta]$ and by hypothesis $\nu_{\det X}(X) = [2]$, there exist p, q in k such that

$$-\beta = 2(p^2 + q^2\det X) = 2(p^2 - q^2\beta\gamma).$$

If we set

$$c = p, \quad a = \frac{\gamma}{\beta}p, \quad d = \beta q, \quad b = \gamma q$$

and

$$P = ax^3 + 3bx^2y + 3cxy^2 + dy^3,$$

it is easily checked that

$$\mu(P) = \begin{pmatrix} ad - bc & 2(bd - c^2) \\ 2(b^2 - ac) & -(ad - bc) \end{pmatrix} = \begin{pmatrix} 0 & \beta \\ \gamma & 0 \end{pmatrix} = X. \qquad \square$$

Remark 2.12. This result is a weak form of the Eisenstein identity. Indeed, if one cubes both sides of $\nu_{\det X}(X) = [2]$ and uses Gauss composition, one obtains the Eisenstein identity evaluated at a particular vector.

Remark 2.13. Varying the symplectic structure to $\omega_\lambda, \lambda \in k^*$ one can sweep out the other orbits with a moment map.

Remark 2.14. The image of the linearized moment map, $B_\mu(\cdot, \cdot)$, cannot be specified. Indeed, the convex hull of the image will contain all $\mathfrak{sl}(2, k)$.

Corollary 2.15. *Let P, P' be nonzero binary cubics such that $Q_n(P) = Q_n(P')$ and such that $\mu(P)$ and $\mu(P')$ are nonzero. Then there exists $g \in Sl(2, k)$ such that $g \cdot \mu(P) = \mu(P')$.*

Proof. Since $Q_n(P) = Q_n(P')$, we have $\det \mu(P) = \det \mu(P')$. By the previous theorem,

$$\nu_{\det \mu(P)}(\mu(P)) = \nu_{\det \mu(P')}(\mu(P'))$$

and the result follows from Proposition 2.8. $\square$

2.3 The Image and Fibers of Ψ

Proposition 2.16. $P \in S^3(k^{2*})$ *with* $Q_n(P) \neq 0$ *is in the image of Ψ if and only if* $9Q_n(P)$ *is a cube in* k^*.

Proof. ($\Rightarrow$) : Suppose that $P = \Psi(B)$. The key to the argument is a result from [15] that is special to Heisenberg graded Lie algebras, namely a formula for Ψ^2. From this result, one obtains $\Psi^2(B) = -(9Q_n(B))^2 B$. On the other hand, we have $\Psi^2(B) = \Psi(P)$. Hence, $B = -(9Q_n(B))^{-2}\Psi(P)$. Applying Ψ again and using that Ψ is cubic we obtain $P = \Psi(B) = -\eta^3(9Q_n(P))^2 P$, where $\eta = -(9Q_n(B))^{-2}$. So $(-\eta)^3 = (9Q_n(P))^{-2}$. Now $(-\eta(9Q_n(B))^2)^3 = 1$ so $(9Q_n(B))^6 = (-\eta)^{-3} = (9Q_n(P))^2$. Thus, we obtain $9Q_n(P) = (\pm 9Q_n(B))^3$. ($\Leftarrow$) : Suppose $9Q_n(P) = \lambda^3$. Set $B = -\frac{1}{\lambda^2}\Psi(P)$. Then as above, $\Psi(B) = P$. $\square$

Corollary 2.17. *For* $P \in S^3(k^{2*})$ *with* $9Q_n(P) \in k^{*3}$, *the fiber $\Psi^{-1}(P)$ consists of one element.*

Proof. From the previous proof, if $P = \Psi(B)$ then $B = -(9Q_n(B))^{-2}\Psi(P)$. $\square$

Remark 2.18. We will see later that a nonzero $P \in S^3(k^{2*})$ with $Q_n(P) = 0$ is in the image of Ψ if and only if $\mu(P) = 0$ and $I_T(P) = [6]$ (cf Proposition 3.19). The fiber of Ψ is then given by Proposition 3.23.

3 Orbits and Fibers

3.1 Symplectic Covariants and Triple Roots

One has the natural 'algebraic' condition

Definition 3.1. $T - \{P \in S^3(k^{2*}) : P \neq 0 \text{ and } P \text{ has a triple root}\}$,

and the natural 'symplectic' condition

Definition 3.2. $Z_\mu = \{P \in S^3(k^{2^*}) : P \neq 0 \text{ and } \mu(P) = 0\}$.

The next proposition shows that the symplectic quantity μ detects the purely algebraic property of whether or not a binary cubic has a triple root.

Proposition 3.3. $T = Z_\mu$.

Proof. Let $P = ax^3 + 3bx^2y + 3cxy^2 + dy^3$. Then $P \in Z_\mu$ iff $\mu(P) = 0$ iff $ad = bc, bd = c^2$ and $b^2 = ac$.

If $bc = 0$, then $cbd = c^3 = 0$ and $b^3 = acb = 0$. Hence $b = c = 0$ and either $a = 0$ or $d = 0$. In the first case $P = dy^3$ and in the second $P = ax^3$.

If $bc \neq 0$, then $a = \frac{b^2}{c}$ and $d = \frac{c^2}{b}$ which means $P = \frac{1}{bc}(bx + cy)^3$.   □

In order to determine the $Sl(2,k)$ orbit structure in the level set $Z_\mu = \mu^{-1}(0) \setminus \{0\}$ we need to construct an invariant that separates the orbits. We begin with the observation that the factorization of $P \in T$ is not unique.

Lemma 3.4. *Let* $\lambda, \mu \in k^*$ *and* $\phi, \psi \in k^{2^*}$ *be such that* $\lambda\phi^3 = \mu\psi^3$. *Then* $\frac{\lambda}{\mu}$ *is a cube and* ϕ *and* ψ *are proportional.*

Proof. Unique factorization.   □

This means the following (algebraic) definition makes sense.

Definition 3.5. Define $I_T : T \to k^*/k^{*3}$ by

$$I_T(P) = [\lambda]_{k^*/k^{*3}},$$

where $P = \lambda\phi^3, \lambda \in k^*$ and $\phi \in k^{2^*}$.

One can formulate the definition using symplectic methods. Given a nonzero $\phi \in k^{2^*}$ there is a $g \in Sl(2,k)$ with $\Omega(\phi, g \cdot \phi) = 1$. If $P = \lambda\phi^3$, then

$$\omega(P, (g \cdot \phi)^3) = \lambda\omega(\phi^3, (g \cdot \phi)^3) = \lambda\Omega(\phi, g \cdot \phi)^3 = \lambda. \tag{13}$$

Thus, $I_T(P) = [\omega(P, (g \cdot \phi)^3)]$.

Proposition 3.6. *(i) Let* $P_1, P_2 \in T$. *Then*

$$Sl(2,k) \cdot P_1 = Sl(2,k) \cdot P_2 \iff I_T(P_1) = I_T(P_2). \tag{14}$$

(ii) The map I_T *induces a bijection of the space of orbits*

$$Z_\mu/Sl(2,k) \longleftrightarrow k^*/k^{*3}. \tag{15}$$

(iii) Let $P \in T$ *and let* $G_P = \{g \in Sl(2,k) : g \cdot P = P\}$ *be the isotropy subgroup of* P. *Then*

$$G_P = \{g \in Sl(2,k) : \exists \mu \in k^* \text{ s.t. } g \cdot \phi = \mu\phi \text{ and } \mu^3 = 1\},$$

where $P = \lambda\phi^3, \lambda \in k^*$ *and* $\phi \in k^{2^*}$.

Proof. (i): Suppose that $P_1 = \lambda \phi^3$ and that there exists $g \in \mathrm{Sl}(2, k)$ such that $g \cdot P_1 = P_2$. Then $P_2 = g \cdot (\lambda \phi^3) = \lambda(g \cdot \phi)^3$ and $I_T(P_2) = [\lambda] = I_T(P_1)$.

Conversely, suppose $P_1 = \lambda_1 \phi_1^3$, $P_2 = \lambda_2 \phi_2^3$ and $I_T(P_1) = I_T(P_2)$. The action of $\mathrm{Sl}(2, k)$ on nonzero vectors of k^{2*} is transitive, so we can find $g \in \mathrm{Sl}(2, k)$ such that $g \cdot \phi_1 = \phi_2$ and hence such that

$$g \cdot P_1 = \lambda_1 \phi_2^3.$$

Since $I_T(P_1) = I_T(P_2)$, there exists $\rho \in k$ such that $\lambda_1 = \rho^3 \lambda_2$ and

$$g \cdot P_1 = \lambda_2 (\rho \phi_2)^3.$$

Choosing $h \in \mathrm{Sl}(2, k)$ such that $h \cdot (\rho \phi_2) = \phi_2$, we have $(hg) \cdot P_1 = P_2$.

(ii): By (i), the map I_T induces an injection of the space of orbits of $\mathrm{Sl}(2, k)$ acting on T into k^* / k^{*3}. This is in fact a surjection since if $\lambda \in k^*$, $I_T(\lambda x^3) = [\lambda]$.

(iii): This follows from unique factorization.   □

Remark 3.7. Extending ϕ to a basis of k^{2*}, we have the isomorphism

$$G_P \cong \left\{ \begin{pmatrix} \mu & a \\ 0 & \frac{1}{\mu} \end{pmatrix} : \mu \in k^*, \mu^3 = 1 \text{ and } a \in k \right\}.$$

Consequently, all the $\mathrm{Sl}(2, k)$ orbits in Z_μ are isomorphic. Hence, Z_μ is a smooth variety, and in [15] we show that it is Lagrangian.

As the center of $\mathrm{Gl}(2, k)$ acts on Z_μ by "cubes" it preserves I_T, and thus the $\mathrm{Sl}(2, k)$ orbits in Z_μ are the same as the $\mathrm{Gl}(2, k)$ orbits. From the point of view of algebraic groups, the result by Demazure [4] characterizes $\mathrm{Sl}(2, k)$ as the subgroup of the automorphisms of $S^3(k^{2*})$ that preserves Z_μ.

3.2 Symplectic Covariants and Double Roots

In a similar way, next we consider the 'algebraic' condition

Definition 3.8. $D = \{P \in S^3(k^{2*}) : P \neq 0 \text{ and } P \text{ has a double root}\}$,

and the 'symplectic' condition

Definition 3.9. $N_\mu = \{P \in S^3(k^{2*}) : P \neq 0 \text{ and } \mu(P) \text{ is nonzero nilpotent}\}$.

Again it turns out that the symplectic quantity μ detects the purely algebraic property of whether or not a binary cubic has a double root.

Theorem 3.10. $D = N_\mu$.

Proof. The inclusion $D \subseteq N_\mu$ follows from the

Lemma 3.11. *Let* $P \in D$ *and write* $P = (ex + fy)^2(rx + sy)$ *with* $ex + fy$ *and* $rx + sy$ *independent. Then*

$$\mu(P) = \frac{2}{9}(es - fr)^2 \begin{pmatrix} -ef & -f^2 \\ e^2 & ef \end{pmatrix}.$$

In particular, $\operatorname{Ker} \mu(P)$ *is spanned by the double root* $ex + fy$.

Proof. Straightforward calculation.   □

To prove the inclusion $N_\mu \subseteq D$, suppose $\mu(P)$ is a nonzero nilpotent. Then $\operatorname{Ker} \mu(P)$ is one-dimensional, spanned by, say, $v \in k^{2^*}$. Since $\operatorname{Sl}(2, k)$ acts transitively on nonzero vectors in k^{2^*}, there exists $g \in \operatorname{Sl}(2, k)$ such that $g \cdot v = x$. Then $\mu(g \cdot P) = g\mu(P)g^{-1}$ is nonzero nilpotent with kernel spanned by x. Let $g \cdot P = ax^3 + 3bx^2y + 3cxy^2 + dy^3$. Then by the formulae (6) and (7), the condition $\mu(g \cdot P) \cdot x = 0$ is equivalent to the system

$$ad - bc = 0$$

$$bd - c^2 = 0.$$

If $(c, d) \neq (0, 0)$, this implies there exists $\lambda, \nu \in k$ such that $(a, b) = \lambda(c, d)$ and $(b, c) = \nu(c, d)$. Hence, $c = \nu d, b = \nu^2 d, a = \nu^3 d$ and $\mu(g \cdot P) = 0$ which is a contradiction. Thus, $c = d = 0$ and $g \cdot P = ax^3 + 3bx^2y = x^2(ax + 3by)$. We have $b \neq 0$ (otherwise $\mu(P) = 0$) so x and $ax + by$ form a basis of k^2. Applying g^{-1} to $g \cdot P = x^2(ax + 3by)$ completes the proof.   □

Again, in order to obtain parameters for the orbit structure of N_μ we need standard representatives. The factorization of $P \in N_\mu$ given by Theorem 3.10 is not unique. However, we can use the symplectic form Ω on k^{2^*} to get a canonical form for P.

Lemma 3.12. *Let* $P \in N_\mu$. *There exists a unique basis* $\{\phi, \xi\}$ *of* k^{2^*} *such that* $P = \phi^2\xi$ *and* $\Omega(\phi, \xi) = 1$.

Proof. If $P \in N_\mu$ then P has a double root by Theorem 3.10. Fix a factorization $P = \phi_1^2\xi_1$. By unique factorization, any other factorization is of the form $P = \phi^2\xi$, where

$$\phi = \lambda\phi_1, \quad \xi = \frac{1}{\lambda^2}\xi_1$$

for some $\lambda \in k^*$. Then $\Omega(\phi, \xi) = 1$ iff $\lambda = \Omega(\phi_1, \xi_1)$ and this proves the claim.   □

Proposition 3.13. *The group $Sl(2,k)$ acts simply transitively on N_μ. Consequently, $Gl(2,k)$ has one orbit on N_μ.*

Proof. Let $P, Q \in N_\mu$ and write $P = \phi^2 \xi$ and $Q = \phi'^2 \xi'$ with $\Omega(\phi, \xi) = \Omega(\phi', \xi') = 1$. The element g of $GL(2,k)$ defined by $g \cdot \phi = \phi'$ and $g \cdot \xi = \xi'$ is clearly in $Sl(2,k)$, satisfies $g \cdot P = Q$ and is the unique element of $Sl(2,k)$ sending P to Q. □

Remark 3.14. In [15] when char $k = 0$ we show that N_μ is the tangent variety to Z_μ.

Remark 3.15. From Proposition 3.3, Theorem 3.10 and (9) we see that $Q_n(P) = 0$ iff P has a multiple root, which is consistent with the classic discriminant interpretation. Also, the open subset of double roots is isomorphic to $Sl(2,k)$. Consequently the variety $Q_n(P) = 0$ is not smooth, but has singular set which is a union over k^*/k^{*3} of isomorphic Lagrangian $Sl(2,k)$-orbits.

Image and Fibers of $\mu : N_\mu \to \mathfrak{sl}(2,k)$

The image of the moment map on N_μ is given by Theorem 2.11:

Corollary 3.16. $\mu(N_\mu) = \{X \in \mathfrak{sl}(2,k) \setminus \{0\} : \det X = 0 \text{ and } v_0(X) = [2]\}$.

Now we give two descriptions of the fibers of $\mu : N_\mu \to \mathfrak{sl}(2,k)$: the first symplectic, the second algebraic. Note that the fibers of the moment map are symplectic objects so it is not a priori clear that they have a purely algebraic description.

Proposition 3.17. *Let $P \in N_\mu$ and let $\phi \in k^{2*}$ be a square factor of P.*

(a) $\mu^{-1}(\mu(P)) = \{P + a\Psi(P) : a \in k\} \cup \{-P + b\Psi(P) : b \in k\}$.
(b) $\mu^{-1}(\mu(P)) = \{P + a\phi^3 : a \in k\} \cup \{-P + b\phi^3 : b \in k\}$.
(c) The affine lines in (a) and (b) are disjoint.

Proof. Since $Sl(2,k)$ acts transitively on N_μ we can assume without loss of generality that $P = 3x^2 y$. Then by (7) and (8),

$$\mu(3x^2 y) = \begin{pmatrix} 0 & 0 \\ 2 & 0 \end{pmatrix}, \quad \Psi(3x^2 y) = -6x^3.$$

We want to find all $Q \in S^3(k^{2*})$ such that

$$\mu(Q) = \begin{pmatrix} 0 & 0 \\ 2 & 0 \end{pmatrix}. \tag{16}$$

By Theorem 3.10, a solution of this equation is of the form $Q = (ex + fy)^2(rx + sy)$ with $es - fr \neq 0$. Substituting back in (16), we get

$$\frac{2}{9}(es - fr)^2 \begin{pmatrix} -ef & -f^2 \\ e^2 & ef \end{pmatrix} = \begin{pmatrix} 0 & 0 \\ 2 & 0 \end{pmatrix}$$

from which it follows that the set of solutions of equation (16) is:

$$\{x^2(e^2rx+3y) : e \in k^*, r \in k\} \cup \{x^2(e^2rx-3y) : e \in k^*, r \in k\}.$$

Since $P = 3x^2y$ and $\Psi(P) = -6x^3$, this proves (a), (b) and (c). $\qquad\square$

The fiber of μ at $\mu(P)$ is also the orbit through P of the isotropy group of $\mu(P)$.

Corollary 3.18. *Let* $P \in N_\mu$ *and let* $G_{\mu(P)} = \{g \in Sl(2,k) : g\mu(P)g^{-1} = \mu(P)\}$. *Then* $\mu^{-1}(\mu(P)) = G_{\mu(P)} \cdot P$.

Proof. Since $\mu(P)$ is nilpotent nonzero, a simple calculation shows that

$$G_{\mu(P)} = \{Id + a\mu(P) : a \in k\} \cup \{-Id + b\mu(P) : b \in k\}$$

and the result follows from Proposition 3.17. $\qquad\square$

It appears that N_μ is a regular contact variety. If one endows the nilpotent variety $\mathcal{N}$ in $\mathfrak{sl}(2,k)$ with the KKS symplectic structure, then $\mu : N_\mu \to \mathcal{N}$ is a prequantization of the image of μ.

Image and Fibers of $\Psi:N_\mu \to Z_\mu$

We begin with some properties of Ψ.

Proposition 3.19. *Let* $P = \phi^2\xi$ *with* $\phi, \xi \in k^{2^*}$. *Then:*

(i) $\mu(\Psi(P)) = 0$;
(ii) ϕ^3 *divides* $\Psi(P)$;
(iii) $\Psi(P) = 0$ *iff* $\mu(P) = 0$;
(iv) $\Psi(P) \neq 0 \Rightarrow I_T(\Psi(P)) = [6]_{k^*/k^{*3}}$.

Proof. Set $\phi = ex + fy$ and $\xi = rx + sy$. Then calculation gives

$$\mu(P) = \frac{2}{9}(es-fr)^2 \begin{pmatrix} -ef & -f^2 \\ e^2 & ef \end{pmatrix},$$

$$\Psi(P) = -\frac{2}{9}(es-fr)^3(ex+fy)^3 \qquad (17)$$

and all parts of the proposition follow immediately from these formulae. $\qquad\square$

Corollary 3.20. *The image of* Ψ *on* N_μ *is* $Z_\mu[6]$.

Proof. According to Proposition 3.19(iv), if $P \in N_\mu$ then $\Psi(P) \in Z_\mu$ and $I_T(\Psi(P)) = [6]_{k^*/k^{*3}}$. Since Ψ is $Sl(2,k)$-equivariant and $Sl(2,k)$ acts transitively on both N_μ and $Z_\mu[6]$, it is clear that Ψ maps N_μ onto $Z_\mu[6]$. $\qquad\square$

To describe the fibers, we need a symplectic characterization of the double root of a $P \in Z_\mu$. Recall that $ex + fy \neq 0$ is a root of P iff $\omega(P, (ex+fy)^3) = 0$. Analogous to this result we have

Proposition 3.21. *Let P be a binary cubic and* $(ex + fy) \in k^{2*}$ *be nonzero.*

$$(ex + fy)^2 \mid P \iff B_\mu(P, (ex + fy)^3) = 0. \tag{18}$$

Proof. We begin with two remarks. First, since $\mathrm{Sl}(2, k)$ acts transitively on nonzero elements of k^{2*} and since B_μ and Ψ are $\mathrm{Sl}(2, k)$-equivariant, we can assume without loss of generality that $ex + fy = x$. Second, the formula for B_μ obtained by polarizing (7) is

$$B_\mu(P, P') = \begin{pmatrix} \frac{1}{2}(ad' + da' - bc' - cb') & (bd' + db') - 2cc' \\ 2bb' - (ac' + ca') & -\frac{1}{2}(ad' + da' - bc' - cb') \end{pmatrix} \tag{19}$$

if $P = ax^3 + 3bx^2y + 3cxy^2 + dy^3$ and $P' = a'x^3 + 3b'x^2y + 3c'xy^2 + d'y^3$. Let $P = ax^3 + 3bx^2y + 3cxy^2 + dy^3$. Then

$$B_\mu(P, x^3) = \begin{pmatrix} \frac{1}{2}d & 0 \\ -c & -\frac{1}{2}d \end{pmatrix}$$

and hence x^2 divides P iff $c = d = 0$ iff $B_\mu(P, x^3) = 0$. □

Now since Ψ maps D to T we expect a criterion involving Ψ for $ex + fy \neq 0$ to be a double root of P.

Proposition 3.22. *Let P be a binary cubic and* $(ex + fy) \in k^{2*}$ *be nonzero.*

(i) If $(ex + fy)^2$ *divides P, then* $\Psi(P)$ *is proportional to* $(ex + fy)^3$.
(ii) If $\Psi(P)$ *is a nonzero multiple of* $(ex + fy)^3$, *then* $(ex + fy)^2$ *divides P.*
(iii) $\{P \in S^3(k^{2*}) : B_\mu(P, (ex + fy)^3) = 0\}$ *is a Lagrangian subspace of* $S^3(k^{2*})$.

Proof. (i): If x^2 divides P, then taking $e = 1$ and $f = 0$ in the formulae (17) we get $\Psi(P) = -\frac{2}{9}d^3x^3$.
(ii): If there exists $\lambda \in k^*$ such that $(ex + fy)^3 = \frac{1}{\lambda}\Psi(P)$, we have

$$B_\mu(P, (ex + fy)^3) = \frac{1}{\lambda}B_\mu(P, \mu(P) \cdot P).$$

But $B_\mu(P, \mu(P) \cdot P) + B_\mu(\mu(P) \cdot P, P) = [\mu(P), \mu(P)] = 0$ since B_μ is $\mathfrak{sl}(2, k)$-equivariant. Hence, $B_\mu(P, \mu(P) \cdot P) = 0$ and $B_\mu(P, (ex + fy)^3) = 0$, which implies by the previous result that $(ex + fy)^2$ divides P.
(iii): Let $L = \{P \in S^3(k^{2*}) : B_\mu(P, (ex + fy)^3) = 0\}$. As we saw in the proof above, the binary cubic $ax^3 + 3bx^2y + 3cxy^2 + dy^3$ is in L iff $c = d = 0$ and hence L is of dimension two. It follows from (1) that $\omega(P, P') = 0$ if $P, P' \in L$ and hence L is Lagrangian. □

We can now give two descriptions of the fibers of $\Psi : N_\mu \to Z_\mu[6]$, the first symplectic, the second algebraic. Again, as the fibers of Ψ are symplectic objects it is not a priori clear that they have a purely algebraic description.

Proposition 3.23. *Let* $P \in N_\mu$ *and let* $\phi \in k^{2^*}$ *be a square factor of* P.

(i) $\Psi^{-1}(\Psi(k^* P)) = \{aP + b\Psi(P) : a \in k^*, b \in k\}$.
(ii) $\Psi^{-1}(\Psi(k^* P)) = \{Q \in N_\mu : \phi^2$ *divides* $Q\}$.

Explicit factorization of P when $Q_n(P) = 0$

From what has been done thus far we obtain readily

Proposition 3.24. *Let* $P = ax^3 + 3bx^2y + 3cxy^2 + dy^3$ *be a nonzero binary cubic over a field* k *such that* $\mathrm{char}(k) \neq 2, 3$.

(i) *If* $\mu(P) = 0$, *then* $Q_n(P) = 0$ *and*

$$P = \begin{cases} ax^3 \text{ or } dy^3 & \text{if } bc = 0, \\ \frac{1}{bc}(bx + cy)^3 & \text{if } bc \neq 0. \end{cases}$$

(ii) *If* $\mu(P) \neq 0$ *and* $Q_n(P) = 0$, *then*

$$P = \begin{cases} x^2(ax + 3by) \text{ or } (3cx + d)y^2 & \text{if } ad - bc = 0, \\ \left(-(b^2 - ac)x + \frac{1}{2}(ad - bc)y\right)^2 \left(\frac{a}{(b^2-ac)^2}x + \frac{4d}{(ad-bc)^2}y\right) & \text{if } ad - bc \neq 0. \end{cases}$$

3.3 Symplectic Covariants and Sums of Coprime Cubes

We have seen that a P with multiple roots corresponds to $Q_n(P) = 0$. So we begin the study of P with $Q_n(P) \neq 0$, in which case the $\mathrm{Sl}(2, k)$ orbits are not the same as the $\mathrm{Gl}(2, k)$ orbits. The values of the symplectic invariant $Q_n(P)$ will have much to say about the roots of P. We begin with the 'natural' condition

Definition 3.25. $\mathcal{O}_{[1]} = \{P \in S^3(k^{2^*}) : Q_n(P) \text{ is a square in } k^*\}$.

The relevant 'algebraic' definition turns out to be

Definition 3.26. $S = \{P \in S^3(k^{2^*}) : \exists T_1, T_2 \in T \text{ s.t } P = T_1 + T_2 \text{ with } T_1, T_2$ coprime$\}$.

Specializing to the space of binary cubics a general theorem valid for the symplectic covariants of the $\mathfrak{g}_1$ of any Heisenberg graded Lie algebra $\mathfrak{g}$, we get the

Theorem 3.27. (i) *Let* $P \in S$ *and let* $P = T_1 + T_2$ *with* $T_1, T_2 \in T$ *coprime. Then* T_1, T_2 *are unique up to permutation.*
(ii) *Let* $P = T_1 + T_2$ *with* $T_1, T_2 \in T$. *Then*

$$Q_n(P) = \omega(T_1, T_2)^2. \tag{20}$$

(iii) Let $P \in \mathcal{O}_{[1]}$ and suppose $Q_n(P) = q^2$ with $q \in k^*$. Then

$$T_1 = \frac{1}{2}\left(P + \frac{1}{3q}\Psi(P)\right), \quad T_2 = \frac{1}{2}\left(P - \frac{1}{3q}\Psi(P)\right)$$

are coprime elements of T such that $P = T_1 + T_2$.

Proof. For k algebraically closed an argument that P is a sum of cubes can be found in [6, 17–18]. The fact that $Q_n(P) = \omega(T_1, T_2)^2$ and (i) and (iii) are proved for general k and for Heisenberg graded Lie algebras in [15]. □

Corollary 3.28. $S = \mathcal{O}_{[1]}$.

Remark 3.29. There is a natural bi-Lagrangian foliation of $\mathcal{O}_{[1]}$ obtained by means of the decomposition $P = T_1 + T_2$. Modulo some technicalities, if one fixes T_2 and varies over T such that $\omega(T, T_2) = \omega(T_1, T_2)$ mod k^{*2}, then does the same with T_1, one obtains a pair of foliations that are transverse and Lagrangian, for details see [15].

Recall that elements of T are, up to a scalar factor, cubes of linear forms. Hence a binary cubic P is in S iff there exist a basis $\{\phi_1, \phi_2\}$ of k^{2*} and $\lambda_1, \lambda_2 \in k^*$ such that

$$P = \lambda_1 \phi_1^3 + \lambda_2 \phi_2^3. \tag{21}$$

The λ_i and ϕ_i in this equation are not unique but the direct sum decomposition

$$k^{2*} = <\phi_1> \oplus <\phi_2>$$

is canonically associated with P as is described in the next result.

Corollary 3.30. *(i) $P \in \mathcal{O}_{[1]}$ iff $\mu(P) \neq 0$ is diagonalizable over k, hence $\mu(P)$ is contained in a semisimple orbit.*

(ii) Let $P \in \mathcal{O}_{[1]}$ and let $\{\phi_1, \phi_2\}$ be a basis of k^{2*}. The following are equivalent:

(a) There exist $\lambda_1, \lambda_2 \in k^*$ such that $P = \lambda_1 \phi_1^3 + \lambda_2 \phi_2^3$.
(b) $\{\phi_1, \phi_2\}$ is a basis of eigenvectors of $\mu(P)$.

(iii) Let $P \in \mathcal{O}_{[1]}$ and suppose $P = \lambda_1 \phi_1^3 + \lambda_2 \phi_2^3$, where $\lambda_1, \lambda_2 \in k^*$ and $\{\phi_1, \phi_2\}$ is a basis of k^{2*}. Then if q is the square root $\lambda_1 \lambda_2 \Omega(\phi_1, \phi_2)^3$ of $Q_n(P)$,

$$\mu(P) \cdot \phi_1 = -q\phi_1,$$
$$\mu(P) \cdot \phi_2 = q\phi_2.$$

Proof. (i): By Cayley–Hamilton and equation (9),

$$0 = \mu(P)^2 + \det\mu(P)\mathrm{Id} = \mu(P)^2 - Q_n(P)\mathrm{Id}.$$

Hence, $\mu(P)$ is diagonalizable over k iff $Q_n(P)$ is a square in k.

(ii): Since there exists $g \in Sl(2, k)$ with $< g \cdot \phi_1 >=< x >$ and $< g \cdot \phi_2 >=< y >$, we can assume without loss of generality that $\phi_1 = x$ and $\phi_2 = y$. Setting $P = ax^3 + 3bx^2 y + 3cxy^2 + dy^3$, we have: $\{x^3, y^3\}$ is a basis of eigenvectors of $\mu(P)$ iff $\mu(P)$ is diagonal iff (by equation (7))

$$bd - c^2 = b^2 - ac = 0.$$

This equation implies $b(ad - bc) = 0$ and hence, since $Q_n(P) \neq 0$, that $b = 0$ and $c^2 = bd = 0$. It follows that $\{x^3, y^3\}$ is a basis of eigenvectors of $\mu(P)$ iff $b = c = 0$ iff $P = ax^3 + dy^3$.

(iii): As above, we can suppose without loss of generality that $P = ax^3 + dy^3$ and then

$$\mu(P) = \begin{pmatrix} ad & 0 \\ 0 & -ad \end{pmatrix},$$

which implies $\mu(P) \cdot x = -adx$ and $\mu(P) \cdot y = ady$. This proves (iii) since $\Omega(x, y) = 1$. □

Corollary 3.31 (Fibers of μ on $\mathcal{O}_{[1]}$). *Let $X \in \mathfrak{sl}(2, k)$ be diagonalizable over k, let $\pm q$ be its eigenvalues and let ϕ_+ and ϕ_- be corresponding eigenvectors in k^{2*}. Then*

$$\mu^{-1}(X) = \left\{ a\phi_-^3 + \frac{q}{a\Omega(\phi_-, \phi_+)^3} \phi_+^3 : a \in k^* \right\}.$$

Proof. This follows from Corollary 3.30(ii) and (iii). □

Orbit parameters for $\mathcal{O}_{[1]}$

For generic k, there will be many $Sl(2, k)$ orbits on $\mathcal{O}_{[1]}$. So the first task is to obtain parameters for the orbits. For this the symplectic result Theorem 3.27 leads to a new and effective method. Let $P \in \mathcal{O}_{[1]}$. Then as we have seen, there exist a unique *unordered* pair of elements T_1, T_2 in T such that

$$P = T_1 + T_2,$$

$$Q_n(P) = \omega(T_1, T_2)^2. \tag{22}$$

Hence, the map $I_{\mathcal{O}_{[1]}} : \mathcal{O}_{[1]} \rightarrow k^* \times_{Z_2} k^*/k^{*3}$

$$I_{\mathcal{O}_{[1]}}(P) = [\omega(T_1, T_2), I_T(T_1)I_T(T_2)^{-1}] \tag{23}$$

is well defined where $k^* \times_{Z_2} k^*/k^{*3}$ denotes the quotient of $k^* \times k^*/k^{*3}$ by the Z_2-action

$$-1 \cdot (\lambda, \alpha) = (-\lambda, \alpha^{-1}).$$

Remark 3.32. The invariant $I_{\mathcal{O}_{[1]}}(\cdot)$ is symplectic not algebraic since its definition requires the symplectic form. We have not found this invariant for binary cubics in the literature.

The next result shows that the image of $I_{\mathcal{O}_{[1]}}$ is constrained.

Proposition 3.33. *Let $P \in \mathcal{O}_{[1]}$, let $q \in k^*$ be a square root of $Q_n(P)$ and let $(\tau_1(q), \tau_2(q)) \in T \times T$ be defined by (22). Choose a basis $\{\phi_1, \phi_2\}$ of k^{2^*} and $\lambda_1, \lambda_2 \in k^*$ such that $\tau_1(q) = \lambda_1 \phi_1^3$ and $\tau_2(q) = \lambda_2 \phi_2^3$. Then*

$$\frac{q}{\lambda_1 \lambda_2} = \Omega(\phi_1, \phi_2)^3, \tag{24}$$

$$(q\lambda_1)(q\lambda_2) = \left(\frac{q}{\Omega(\phi_1, \phi_2)}\right)^3. \tag{25}$$

Proof. The two equations are equivalent and follow immediately from

$$q = \omega(\tau_1(q), \tau_2(q)) = \lambda_1 \lambda_2 \omega(\phi_1^3, \phi_2^3) = \lambda_1 \lambda_2 \Omega(\phi_1, \phi_2)^3. \qquad \square$$

Theorem 3.34. *Let $I_{\mathcal{O}_{[1]}} : \mathcal{O}_{[1]} \to k^* \times_{Z_2} k^*/k^{*3}$ be defined by (23) above.*

(i) Let $P, P' \in \mathcal{O}_{[1]}$. Then

$$Sl(2, k) \cdot P' = Sl(2, k) \cdot P \iff I_{\mathcal{O}_{[1]}}(P') = I_{\mathcal{O}_{[1]}}(P).$$

(ii) The map $I_{\mathcal{O}_{[1]}}$ induces a bijection

$$\mathcal{O}_{[1]}/Sl(2, k) \longleftrightarrow k^* \times_{Z_2} k^*/k^{*3}.$$

(iii) Let $P \in \mathcal{O}_{[1]}$ and suppose $P = \lambda_1 \phi_1^3 + \lambda_2 \phi_2^3$, where $\lambda_1, \lambda_2 \in k^*$ and $\{\phi_1, \phi_2\}$ is a basis of k^{2^*}. Let $G_P = \{g \in Sl(2, k) : g \cdot P = P\}$. Then

$$G_P = \left\{ g \in Sl(2, k) : \exists \mu \in k^* \ s.t. \ g \cdot \phi_1 = \mu \phi_1, g \cdot \phi_2 = \frac{1}{\mu} \phi_2 \ and \ \mu^3 = 1 \right\}.$$

Proof. (i): Since ω and I_T are $Sl(2, k)$-invariant, it is clear from (23) that the map $I_{\mathcal{O}_{[1]}} : \mathcal{O}_{[1]} \to k^* \times_{Z_2} k^*/k^{*3}$ factors through the action of $Sl(2, k)$. To show that the induced map on orbit space is injective, suppose that P and P' are binary cubics such that $I_{\mathcal{O}_{[1]}}(P') = I_{\mathcal{O}_{[1]}}(P)$. First choose $g, g' \in Sl(2, k)$ such that

$$g \cdot P = ax^3 + by^3,$$
$$g' \cdot P' = a'x^3 + b'y^3. \tag{26}$$

From equations (1) and (9), we have

$$\omega(x^3, y^3) = 1, \quad Q_n(P) = a^2 b^2, \quad Q_n(P') = a'^2 b'^2.$$

Hence, $I_{\mathcal{O}_{[1]}}(P') = I_{\mathcal{O}_{[1]}}(P)$ implies

$$[ab, [a][b]^{-1}] = [a'b', [a'][b']^{-1}]$$

in $k^* \times_{Z_2} k^*/k^{*3}$. There are two possibilities:

- $ab = a'b', \quad [a][b]^{-1} = [a'][b']^{-1}$;
- $ab = -a'b', \quad [a][b]^{-1} = [b'][a']^{-1}$.

In the first case, we have

$$[ab][a][b]^{-1} = [a'b'][a'][b']^{-1},$$

hence $[a^2] = [a'^2]$ and so $[a] = [a']$ as the group k^*/k^{*3} is of exponent 3. Thus, there exists $r \in k^*$ such that $a' = r^3 a$ and $b' = \frac{1}{r^3} b$. If we define $h \in GL(2, k)$ by

$$h \cdot x = rx, \quad h \cdot y = \frac{1}{r} y,$$

it is clear that $h \in Sl(2, k)$ and $h \cdot (g \cdot P) = g' \cdot P'$. Hence, P and P' are in the same $Sl(2, k)$-orbit.

In the second case, we have $[a^2] = [b'^2]$, $[a] = [b']$ and there exists $r \in k^*$ such that $b' = r^3 a$ and $a' = -\frac{1}{r^3} b$. If we define $h \in GL(2, k)$ by

$$h \cdot x = ry, \quad h \cdot y = -\frac{1}{r} x,$$

it is clear that $h \in Sl(2, k)$ and $h \cdot (g \cdot P) = g' \cdot P'$. Hence, P and P' are in the same $Sl(2, k)$-orbit and we have proved that $I_{\mathcal{O}_{[1]}} : \mathcal{O}_{[1]} \to k^* \times_{Z_2} k^*/k^{*3}$ separates $Sl(2, k)$-orbits.

To prove (ii), it remains to prove that $I_{\mathcal{O}_{[1]}} : \mathcal{O}_{[1]} \to k^* \times_{Z_2} k^*/k^{*3}$ is surjective. Let $[q, [\alpha]] \in k^* \times_{Z_2} k^*/k^{*3}$ and consider the binary cubic

$$P = \frac{1}{q\alpha} x^3 + q^2 \alpha y^3.$$

Then

$$I_{\mathcal{O}_{[1]}}(P) = \left[q, \left[\frac{1}{q\alpha} \right] [q^2\alpha] \right]^{-1} \right] = \left[q, \left[\frac{1}{q^3\alpha^2} \right] \right] = [q, [\alpha]].$$

and so $I_{\mathcal{O}_{[1]}} : \mathcal{O}_{[1]} \to k^* \times_{Z_2} k^*/k^{*3}$ is surjective. This completes the proof of (ii).

To prove (iii), recall that the representation $P = \lambda_1 \phi_1^3 + \lambda_2 \phi_2^3$ is unique up to permutation. Then $g \cdot P = P$ leads to two cases:

- $g \cdot (\lambda_1 \phi_1^3) = \lambda_1 \phi_1^3$ and $g \cdot (\lambda_2 \phi_2^3) = \lambda_2 \phi_2^3$;
- $g \cdot (\lambda_1 \phi_1^3) = \lambda_2 \phi_2^3$ and $g \cdot (\lambda_2 \phi_2^3) = \lambda_1 \phi_1^3$.

In the first case, $g \cdot \phi_i = j_i \phi_i$ where $j_i^3 = 1$ and since $g \in \mathrm{Sl}(2, k)$, we must have $j_1 j_2 = 1$. In the second case, there exist $r, s \in k^*$ such that $g \cdot \phi_1 = r\phi_2$, $g \cdot \phi_2 = s\phi_1$, $\lambda_1 r^3 = \lambda_2$, $\lambda_2 s^3 = \lambda_1$ and $rs = -1$. Hence, $(rs)^3 = 1$ and $rs = -1$ which is impossible and this case does not occur. $\qquad\square$

Properties of orbit space

We will use the parameterization

$$I_{\mathcal{O}_{[1]}} : \mathcal{O}_{[1]}/\mathrm{Sl}(2, k) \longleftrightarrow k^* \times_{Z_2} k^*/k^{*3}.$$

to study orbit space. The parameter space has two natural maps

$$sq : k^* \times_{Z_2} k^*/k^{*3} \to k^{*2}, \quad sq([q, \alpha]) = q^2, \tag{27}$$

and

$$t : k^* \times_{Z_2} k^*/k^{*3} \to (k^*/k^{*3})/Z_2, \quad t([q, \alpha]) = [\alpha] \tag{28}$$

corresponding to projection onto the orbit spaces of the two factors. We then have the following diagram:

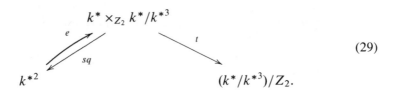

$$\tag{29}$$

The map

$$sq : k^* \times_{Z_2} k^*/k^{*3} \to k^{*2} \tag{30}$$

is the fibration associated with the principal Z_2-fibration

$$k^* \to k^{*2}$$

and the action of Z_2 on k^*/k^{*3} by inversion. Since Z_2 acts by automorphisms, the fiber $sq^{-1}(q^2)$ over any point $q^2 \in k^{*2}$ has a natural group structure

$$[q, \alpha] \times [q, \beta] = [q, \alpha\beta] \tag{31}$$

independent of the choice of square root q of q^2. Taking the identity at each point, we get a canonical section $e : k^{*2} \to k^* \times_{Z_2} k^*/k^{*3}$ of (30) given by

$$e(q^2) = [q, 1] \tag{32}$$

but, although each fiber is a group isomorphic to k^*/k^{*3}, the fibration (30) is not in general isomorphic to the product

$$k^{*2} \times k^*/k^{*3} \to k^{*2}.$$

To translate the above features of orbit space into more concrete statements about binary cubics over k, note that the map sq is essentially the quartic Q_n since for all $P \in \mathcal{O}_{[1]}$,

$$sq(I_{\mathcal{O}_{[1]}}(P)) = Q_n(P).$$

Theorem 3.35. *Let $M \in k^{*2}$, let*

$$\mathcal{O}_M = \{P \in S^3(k^{2^*}) : Q_n(P) = M\}$$

and let $\mathcal{O}_M/Sl(2,k)$ be the space of $Sl(2,k)$-orbits in $\mathcal{O}_M$.

(i) *The map $I_{\mathcal{O}_{[1]}} : \mathcal{O}_{[1]} \to k^* \times_{Z_2} k^*/k^{*3}$ induces a bijection*

$$\mathcal{O}_M/Sl(2,k) \longleftrightarrow sq^{-1}(M)$$

and, by pullback of (31), a group structure on $\mathcal{O}_M/Sl(2,k)$.
(ii) *As groups, $\mathcal{O}_M/Sl(2,k) \cong k^*/k^{*3}$.*
(iii) *Let $q \in k^*$ be a square root of M. The identity element of $\mathcal{O}_M/Sl(2,k)$ is characterized by:*

$$Sl(2,k) \cdot P = 1 \Leftrightarrow P \text{ is reducible over } k \Leftrightarrow I_{\mathcal{O}_{[1]}}(P) = [q,1].$$

Proof. Parts (i) and (ii) follow from the discussion above. Part (iii) follows from Theorem 3.37(i) and equation (32). $\qquad\square$

Remark 3.36. From the Corollary, it follows that if the classical discriminant is a nonzero square there is a unique $Sl(2,k)$ orbit consisting of reducible polynomials. We remove the 'square' restriction in Corollary 3.48. In particular, over an algebraically closed field, there is only one orbit of fixed nonzero discriminant.

To finish this section, we briefly discuss the map $t : k^* \times_{Z_2} k^*/k^{*3} \to (k^*/k^{*3})/Z_2$ in diagram (29) given by

$$t([q,\alpha]) = [\alpha].$$

This a fibration with fiber k^* outside the identity coset $[1]$ but

$$t^{-1}([1]) = e(k^{*2})$$

is a 'singular fiber'. There is a k^*-action:

$$\lambda \cdot [q, \alpha] = [\lambda q, \alpha], \tag{33}$$

which maps fibers of sq to fibers of sq:

$$sq([q', \alpha']) = sq([q, \alpha]) \Rightarrow sq(\lambda \cdot [q', \alpha']) = sq(\lambda \cdot [q, \alpha]),$$

and whose orbits are exactly the fibers of t:

$$t([q', \alpha']) = t([q, \alpha]) \Leftrightarrow \exists \lambda \in k^* \text{ s.t. } [q', \alpha'] = \lambda \cdot [q, \alpha].$$

Isotropy for this action is given by: $Isot_{k^*}([q, \alpha]) = \begin{cases} 1 & \text{if } \alpha \neq 1 \\ \{\pm 1\} & \text{if } \alpha = 1. \end{cases}$

It would be interesting to interpret these features of orbit space in terms of the original binary cubics. Conversely, one can also identify actions on the orbits in terms of their orbit parameters. For example, the commutant of $Sl(2, k)$ in $Gl(S^3(k^{2^*}))$ acts on orbit space. This gives the action

$$\lambda \cdot' [q, \alpha] = [\lambda^2 q, \alpha]$$

of k^* on $k^* \times_{Z_2} k^*/k^{*3}$, which is the square of the action (33). Another example is obtained from $\Psi : S^3(k^{2^*}) \to S^3(k^{2^*})$ which, since it commutes with the action of $Sl(2, k)$, induces a map from $k^* \times_{Z_2} k^*/k^{*3}$ to itself. This is easily seen to be given by

$$[q, \alpha] \mapsto [-q^3, [q]\alpha], \tag{34}$$

where $[q]$ denotes the class of q in k^*/k^{*3}.

Reducibility and factorization

Theorem 3.37. *Let $P \in S$ and let $\{\phi_1, \phi_2\}$ be a basis of k^{2^*} such that $P = \lambda_1 \phi_1^3 + \lambda_2 \phi_2^3$ with $\lambda_1, \lambda_2 \in k^*$. Let $q \in k^*$ be a square root of $Q_n(P)$. The following are equivalent:*

(a) P is reducible over k.
(b) $\frac{\lambda_1}{\lambda_2}$ is a cube in k^*.
(c) $q\lambda_1$ is a cube in k^*.
(d) $q\lambda_2$ is a cube in k^*.
(e) There is a basis $\{\phi_1', \phi_2'\}$ of k^{2^*} such that $P = \frac{1}{q}(\phi_1'^3 + \phi_2'^3)$.

Proof. $(a) \Rightarrow (b)$: Suppose P is reducible over k. Then for all $g \in Sl(2, k)$,

$$g \cdot P = \lambda_1 (g \cdot \phi_1)^3 + \lambda_2 (g \cdot \phi_2)^3$$

is also reducible over k. Since ϕ_1, ϕ_2 form a basis of k^{2*}, we can choose g such that $g \cdot \phi_1 = x$ and $g \cdot \phi_2 = \rho y$ for some $\rho \in k^*$ so that

$$\lambda_1 x^3 + \lambda_2 \rho^3 y^3$$

is reducible over k. Hence, there exist $a, b, c, d, e \in k$ such that

$$\lambda_1 x^3 + \lambda_2 \rho^3 y^3 = (ax + by)(cx^2 + dxy + ey^2),$$

which gives the system

$$\lambda_1 = ac, \ 0 = ad + bc,$$

$$\lambda_2 \rho^3 = be, \ 0 = ae + bd.$$

Since λ_1 and λ_2 are nonzero, it follows that a, b, c, d, e are nonzero and, since $c = -\frac{ad}{b}$ and $e = -\frac{bd}{a}$ we get $\frac{\lambda_1}{\lambda_2} = (\rho \frac{a}{b})^3$.

$(b) \Rightarrow (a)$: Suppose $\frac{\lambda_1}{\lambda_2} = r^3$ with $r \in k^*$. Then

$$P = \lambda_2(r^3 \phi_1^3 + \phi_2^3) = \lambda_2(r\phi_1 + \phi_2)(r^2 \phi_1^2 + r\phi_1 \phi_2 + \phi_2^2) \tag{35}$$

and P is reducible over k.

$(b) \Leftrightarrow (c) \Leftrightarrow (d)$: Set $v_1 = q\lambda_1$ and $v_2 = q\lambda_2$. By Proposition 3.33, there exists $s \in k^*$ such that $v_1 v_2 = s^3$. Hence if any one of the three numbers $v_1, v_2, \frac{v_1}{v_2} = \frac{\lambda_1}{\lambda_2}$ is a cube so are the other two since formally

$$v_1 = \left(\frac{v_1}{\sqrt[3]{\frac{v_1}{v_2}} \sqrt[3]{v_1 v_2}} \right)^3, \quad v_2 = \left(\sqrt[3]{\frac{v_1}{v_2}} \frac{v_2}{\sqrt[3]{v_1 v_2}} \right)^3, \quad v_2 = \left(\frac{\sqrt[3]{v_1 v_2}}{\sqrt[3]{v_1}} \right)^3.$$

$(a) \Rightarrow (e)$: If P is reducible we have just proved that there exists $r \in k^*$ and $s \in k^*$ such that $\lambda_1 = \frac{1}{q} r^3$ and $\lambda_2 = \frac{1}{q} s^3$. Set $\phi_1' = r\phi_1$ and $\phi_2' = s\phi_2$. Then

$$P = \lambda_1 \phi_1^3 + \lambda_2 \phi_2^3 = \frac{1}{q} \left(\phi_1'^3 + \phi_2'^3 \right),$$

which proves (e).

$(e) \Rightarrow (a)$: Evident since $\phi'_1 + \phi'_2$ divides $\phi_1'^3 + \phi_2'^3$. $\qquad\square$

Corollary 3.38. *Let $P \in S$ be reducible and let $\{\phi_1', \phi_2'\}$ be a basis of k^{2*} such that $P = \frac{1}{q}(\phi_1'^3 + \phi_2'^3)$.*

(a) If -3 is not a square in k, then

$$P = \frac{1}{q}(\phi_1' + \phi_2')(\phi_1'^2 - \phi_1' \phi_2' + \phi_2'^2)$$

and $\phi_1'^2 - \phi_1' \phi_2' + \phi_2'^2$ is irreducible over k.

(b) If −3 is a square in k, then

$$P = \frac{1}{q}(\phi_1' + \phi_2')(j\phi_1' + j^{-1}\phi_2')(j^2\phi_1' + j^{-2}\phi_2'), \qquad (36)$$

where $j = \frac{1}{2}(-1 + \sqrt{-3})$. The factors of P are pairwise independent.

To a certain extent, we can normalize bases of k^{2*} satisfying Theorem 3.27(e).

Corollary 3.39. *Let $P \in S$.*

(a) *P is reducible iff there is a basis $\{\phi_1', \phi_2'\}$ of k^{2*} such that $P = \frac{1}{q}(\phi_1'^3 + \phi_2'^3)$ and $\Omega(\phi_1', \phi_2') = q$.*

(b *If $\{\phi_1', \phi_2'\}$ and $\{\phi_1'', \phi_2''\}$ are two bases of k^{2*} satisfying (a), there exists a cube root of unity $j \in k^*$ such that $\phi_1'' = j\phi_1'$ and $\phi_2'' = j^{-1}\phi_2'$.*

Proof. Choose a basis $\{\phi_1, \phi_2\}$ of k^{2*} and $\lambda_1, \lambda_2 \in k^*$ such that $P = \lambda_1\phi_1^3 + \lambda_2\phi_2^3$ and let $q = \lambda_1\lambda_2\Omega(\phi_1, \phi_2)^3$. If P is reducible, by Theorem 3.27, there exists $r \in k^*$ such that $\lambda_1 = \frac{1}{q}r^3$. Set $s = \frac{q}{r\Omega(\phi_1, \phi_2)}$, $\phi_1' = r\phi_1$ and $\phi_2' = s\phi_2$. Then $\Omega(\phi_1', \phi_2') = q$ and

$$s^3 = \left(\frac{q}{r\Omega(\phi_1, \phi_2)}\right)^3 = \frac{1}{r^3}\left(\frac{q}{\Omega(\phi_1, \phi_2)}\right)^3 = \frac{1}{q\lambda_1}(q\lambda_1)(q\lambda_2) = q\lambda_2.$$

Hence,

$$P = \lambda_1\phi_1^3 + \lambda_2\phi_2^3 = \frac{1}{q}\left(\phi_1'^3 + \phi_2'^3\right).$$

In the classical literature on cubics, this is called the Viète Substitution.

Conversely, if there is a basis $\{\phi_1', \phi_2'\}$ of k^{2*} such that $P = \frac{1}{q}(\phi_1'^3 + \phi_2'^3)$, then $\phi_1' + \phi_2'$ divides P and P is reducible.

To prove (b), note first that by Theorem 3.27(a), we have either $\phi_1''^3 = \phi_1'^3$ and $\phi_2''^3 = \phi_2'^3$ or $\phi_1''^3 = \phi_2'^3$ and $\phi_2''^3 = \phi_1'^3$.

In the first case, by unique factorization, there exist cube roots of unity j_1, j_2 such that $\phi_1'' = j_1\phi_1'$, $\phi_2'' = j_2\phi_2'$ and $j_1j_2 = 1$. This is exactly what we want to prove.

In the second case, there exist cube roots of unity j_1, j_2 such that $\phi_1'' = j_1\phi_2'$, $\phi_2'' = j_2\phi_1'$ and $j_1j_2 = -1$. This is impossible since $(j_1j_2)^3 = 1$. $\qquad\square$

Explicit formulae for $I_{\mathcal{O}_{[1]}}$ and Cardano–Tartaglia formulae

Proposition 3.40. *Let $P = ax^3 + 3bx^2y + 3cxy^2 + dy^3$ be an element of $\mathcal{O}_{[1]}$, let $q \in k^*$ be a square root of $Q_n(P)$ and define α, β, γ and δ in k by*

$$\mu(P) = \begin{pmatrix} (ad - bc) & 2(bd - c^2) \\ 2(b^2 - ac) & -(ad - bc) \end{pmatrix} = \begin{pmatrix} \alpha & \beta \\ \gamma & \delta \end{pmatrix}.$$

Then $P = \lambda_1 \phi_1^3 + \lambda_2 \phi_2^3$ *and* $I_{\mathcal{O}_{[1]}}(P) = [q, [\lambda_1][\lambda_2]^{-1}]$ *where:*

(i) *If* $\beta = \gamma = 0$,

$$\lambda_1 = a, \ \phi_1 = x,$$
$$\lambda_2 = d, \ \phi_2 = y, \qquad ad = q.$$

(ii) *If* $\gamma \neq 0$,

$$\lambda_1 = \frac{1}{2q}(\alpha + q)a + \frac{\gamma}{2q}b, \ \phi_1 = x - \left(\frac{\alpha - q}{\gamma}\right)y,$$
$$\lambda_2 = -\frac{1}{2q}(\alpha - q)a - \frac{\gamma}{2q}b, \ \phi_2 = x - \left(\frac{\alpha + q}{\gamma}\right)y, \qquad \Omega(\phi_1, \phi_2) = -\frac{2q}{\gamma}.$$

(iii) *If* $\beta \neq 0$,

$$\lambda_1 = \frac{\beta}{2q}c - \frac{1}{2q}(\alpha - q)d, \ \phi_1 = \left(\frac{\alpha + q}{\beta}\right)x + y,$$
$$\lambda_2 = -\frac{\beta}{2q}c + \frac{1}{2q}(\alpha + q)d, \ \phi_2 = \left(\frac{\alpha - q}{\beta}\right)x + y, \qquad \Omega(\phi_1, \phi_2) = \frac{2q}{\beta}.$$

If $P \in \mathcal{O}_{[1]}$ is reducible, we can use these formulae together with Theorem 3.37 and Corollary 3.38 to get an explicit formula for a linear factor of P in terms of the coefficients of P, a square root q of $Q_n(P)$ and a cube root r of $q\lambda_1$. Recall that the existence of a cube root of $q\lambda_1$ in k is a necessary and sufficient condition for P to be reducible over k.

Proposition 3.41. *Let* $P = ax^3 + 3bx^2y + 3cxy^2 + dy^3 \in \mathcal{O}_{[1]}$ *be reducible, let* $q \in k^*$ *be a square root of* $Q_n(P)$ *and suppose* $ad \neq 0$.

(i) *If* $\beta = \gamma = 0$, *let* r *be a cube root of* qa *and let* $s = \frac{q}{r}$. *Then*

$$rx + sy$$

divides P.

(ii) *If* $\gamma \neq 0$, *let* r *be a cube root of* $(\alpha + q)a + \gamma b$ *and let* $s = -\frac{\gamma}{r}$. *Then*

$$x + \left(\frac{r - s + b}{a}\right)y$$

divides P.

(iii) *If* $\beta \neq 0$, *let* r *be a cube root of* $\beta c - (\alpha - q)d$ *and let* $s = \frac{\beta}{r}$. *Then*

$$\left(\frac{s - r + c}{d}\right)x + y$$

divides P.

Proof. Since P is reducible, there exists a basis ϕ'_1, ϕ'_2 of k^{2*} such that $P = \frac{1}{q}(\phi'^3_1 + \phi'^3_2)$ (cf Theorem 3.37) and then $\phi'_1 + \phi'_2$ divides P. As shown in the proof of Corollary 3.39(a), we can take $\phi'_1 = r\phi_1$ and $\phi'_2 = s\phi_2$ where r is a cube root of $q\lambda_1$, $s = \frac{q}{r\Omega(\phi_1,\phi_2)}$ and $\phi_1, \phi_2, \lambda_1$ are given by Proposition 3.40. The explicit formulae in the three cases are:

(a) $\beta = \gamma = 0$: r is a cube root of qa, $rs = q$ and $\phi'_1 = rx$, $\phi'_2 = sy$;

(b) $\gamma \neq 0$: r is a cube root of $\frac{(\alpha+q)a+\gamma b}{2}$, $s = -\frac{\gamma}{2r}$ and

$$\phi'_1 = rx + \frac{1}{2s}(\alpha - q)y, \quad \phi'_2 = sx + \frac{1}{2r}(\alpha + q)y;$$

(c) $\beta \neq 0$: r is a cube root of $\frac{\beta c - (\alpha-q)d}{2}$, $s = \frac{\beta}{2r}$ and

$$\phi'_1 = \frac{1}{2s}(\alpha + q)x + ry, \quad \phi'_2 = \frac{1}{2r}(\alpha - q)x + sy.$$

Calculating $\phi'_1 + \phi'_2$ in the first case obviously gives (i). In the second case we have

$$\phi'_1 + \phi'_2 = (r + s)x + \left(\frac{1}{2s}(\alpha - q) + \frac{1}{2r}(\alpha + q) \right) y$$

$$= (r + s)x + \left(\frac{1}{2sa}(-2s^3 - \gamma b) + \frac{1}{2ra}(2r^3 - \gamma b) \right) y \quad (37)$$

since $r^3 = q\lambda_1 = \frac{(\alpha+q)a+\gamma b}{2}$ and $s^3 = q\lambda_2 = \frac{-(\alpha-q)a-\gamma b}{2}$. Simplifying the coefficient of y we get

$$\frac{1}{2sa}(-2s^3 - \gamma b) + \frac{1}{2ra}(2r^3 - \gamma b) = \frac{1}{a}\left(r^2 - s^2 - \frac{b\gamma}{2}\left(\frac{1}{r} + \frac{1}{s} \right) \right)$$

$$= (r + s)\frac{r - s + b}{a}$$

since $2rs = -\gamma$, and this implies (ii). Similarly, (iii) follows from (c). □

As an application of the above results, consider the homogeneous Cardano–Tartaglia polynomial

$$P = x^3 + pxy^2 + qy^3$$

over a field k of characteristic not 2 or 3. Assume $p \neq 0$ and $q \neq 0$ so that factorizing P is a nontrivial problem. Then

$$\mu(P) = \begin{pmatrix} q & -2\frac{p^2}{9} \\ -2\frac{p}{3} & -q \end{pmatrix}, \quad Q_n(P) = \left(q^2 + 4\frac{p^3}{27} \right).$$

To be able to apply our approach we assume $Q_n(P)$ has a square root in k^* which we denote $\sqrt{q^2 + 4\frac{p^3}{27}}$. Then by Theorem 3.37 and Proposition 3.40(ii), P is reducible iff

$$\frac{q}{2} + \sqrt{\frac{q^2}{4} + \frac{p^3}{27}} \quad \text{or} \quad -\frac{q}{2} + \sqrt{\frac{q^2}{4} + \frac{p^3}{27}}$$

has a cube root in k.

If this is the case, then Proposition 3.41 (ii) implies that $x + (r - s)y$ divides P, where r is a cube root of $\frac{q}{2} + \sqrt{\frac{q^2}{4} + \frac{p^3}{27}}$ and s is the cube root $\frac{p}{3r}$ of $-\frac{q}{2} + \sqrt{\frac{q^2}{4} + \frac{p^3}{27}}$. Hence, with the obvious notation,

$$\frac{p}{3\left(\sqrt[3]{\frac{q}{2} + \sqrt{\frac{q^2}{4} + \frac{p^3}{27}}} \right)} - \sqrt[3]{\frac{q}{2} + \sqrt{\frac{q^2}{4} + \frac{p^3}{27}}}$$

is a root of the inhomogeneous cubic $x^3 + px^2 + q$ and this is the classical Cardano–Tartaglia formula. If $k = \mathbb{R}$, this can be written

$$s - r = \sqrt[3]{-\frac{q}{2} + \sqrt{\frac{q^2}{4} + \frac{p^3}{27}}} - \sqrt[3]{\frac{q}{2} + \sqrt{\frac{q^2}{4} + \frac{p^3}{27}}}$$

since cube roots are unique.

3.4 Symplectic Covariants and Sums of Coprime Cubes in Quadratic Extensions

In this article, we have until now considered only binary cubics P such that $Q_n(P)$ is a square in k. In this section, we will study binary cubics P such that $Q_n(P)$ is a square in a fixed quadratic extension of k.

Let $\hat{k}$ be a quadratic extension of k. Recall that since $\text{char}(k) \neq 2$, the extension $\hat{k}/k$ is Galois and the Galois group $\text{Gal}(\hat{k}/k)$ is isomorphic to $\mathbb{Z}_2$. The Galois group $\mathrm{Gal}(\hat{k}/k)$ acts naturally on any space over $\hat{k}$ obtained by base extension of a space over k and its fixed point set is the original space over k. We always denote the action of the generator of $\mathrm{Gal}(\hat{k}/k)$ by $x \mapsto \bar{x}$ and we denote by $\hat{\Omega}$ and $\hat{\omega}$, respectively, the symplectic forms on $\hat{k}^{2*}$ and $S^3(\hat{k}^{2*})$ obtained by base extension of Ω and ω. The quartic on $S^3(\hat{k}^{2*})$ obtained by base extension of Q_n, will be denoted $\widehat{Q_n}$ and we set

$$\widehat{\mathcal{O}}_{[1]} = \{P \in S^3(\hat{k}^{2*}) : \widehat{Q_n}(P) \in \hat{k}^{*2}\}.$$

Finally, let $\mathrm{Im}\,\hat{k} = \{\lambda \in \hat{k} : \bar{\lambda} = -\lambda\}$ and let $\widehat{T} \subseteq S^3(\hat{k}^{2*})$ be the set of nonzero binary cubics over $\hat{k}$, which have a triple root over $\hat{k}$.

Remark 3.42. Note that $(\mathrm{Im}\,\hat{k}^*)^2 \subseteq k^*$ is the inverse image under $k^* \to k^{*2}$ of a single nontrivial square class in k^*/k^{*2}. Conversely, a nontrivial square class in k^*/k^{*2} determines up to isomorphism a quadratic extension of k with this property.

This notation out of the way, we make a symplectic definition

$$\mathcal{O}(\hat{k}) = \{P \in S^3(k^{2*}) : \hat{k} \text{ is a splitting field of } x^2 - Q_n(P)\}$$

and an algebraic definition

$$S(\hat{k}) = \{P \in S^3(k^{2^*}) : \exists T \in \widehat{T} \text{ s.t. } P = T + \bar{T} \text{ with } T, \bar{T} \text{ coprime}\}.$$

Proposition 3.43. $\mathcal{O}(\hat{k}) = S(\hat{k})$.

Proof. Let $P \in \mathcal{O}(\hat{k})$. Then $Q_n(P)$ has two square roots in $\hat{k}$ but no square roots in k since $\hat{k}$ is a splitting field of $x^2 - Q_n(P)$. By Theorem 3.27, there exists $T_1, T_2 \in \widehat{T}$ such that $P = T_1 + T_2$ and the square roots of $Q_n(P)$ are $\pm \hat{\omega}(T_1, T_2)$. Since $\bar{P} = P$ and since T_1 and T_2 are unique up to permutation, we have either $\bar{T}_1 = T_1$ and $\bar{T}_2 = T_2$ or $\bar{T}_1 = T_2$ and $\bar{T}_2 = T_1$. In the first case,

$$\overline{\hat{\omega}(T_1, T_2)} = \hat{\omega}(\bar{T}_1, \bar{T}_2) = \hat{\omega}(T_1, T_2),$$

so $\hat{\omega}(T_1, T_2) \in k$ and $Q_n(P)$ has a square root in k which is a contradiction. Hence, $P = T_1 + \bar{T}_1$. To prove that T_1 and $\bar{T}_1$ are coprime, write $T_1 = \lambda \alpha^3$ where $\lambda \in \hat{k}$ and $\alpha \in \hat{k}^{2*}$. Then, by unique factorization, T_1 and $\bar{T}_1$ are not coprime iff α and $\bar{\alpha}$ are proportional. But then $\hat{\omega}(T_1, \bar{T}_1) = 0$ and $Q_n(P) = 0$ has a square root in k. Hence, T_1 and $\bar{T}_1$ are coprime and $P \in S(\hat{k})$.

To prove inclusion in the opposite direction, suppose $P \in S(\hat{k})$ and let $P = T + \bar{T}$ with $T, \bar{T}$ coprime and $T \in \widehat{T}$. Note that $P \neq 0$ since otherwise T and $\bar{T}$ would not be coprime. By Theorem 3.27, we have $Q_n(P) = (\hat{\omega}(T, \bar{T}))^2$ and $Q_n(P)$ has two square roots $\pm \hat{\omega}(T, \bar{T})$ in $\hat{k}$. Let $T = \lambda \alpha^3$ where $\lambda \in \hat{k}^*$ and $\alpha \in \hat{k}^{2*}$.

As we saw above, T and $\bar{T}$ are coprime implies α and $\bar{\alpha}$ are not proportional, and this is equivalent to $\hat{\Omega}(\alpha, \bar{\alpha}) \neq 0$ since $\dim \hat{k}^{2*} = 2$. From

$$\hat{\omega}(T, \bar{T}) = \lambda \bar{\lambda}(\hat{\Omega}(\alpha, \bar{\alpha}))^3$$

it follows that $\hat{\omega}(T, \bar{T}) \neq 0$. On the other hand,

$$\overline{\hat{\omega}(T, \bar{T})} = \hat{\omega}(\bar{T}, T) = -\hat{\omega}(T, \bar{T})$$

and $\hat{\omega}(T, \bar{T})$ is pure imaginary. Hence, the square roots $\pm\hat{\omega}(T, \bar{T})$ of $Q_n(P)$ are not in k and $\hat{k}$ is a splitting field of $x^2 - Q_n(P)$.   □

Proposition 3.44 (Fibers of μ on $\mathcal{O}(\hat{k})$). *Let $X \in \mathfrak{sl}(2, k)$ be such that $-\det X \in (\operatorname{Im}\hat{k}^*)^2$ and $v_{\det X}(X) = [2]$. Let $q, \bar{q} \in \operatorname{Im}\hat{k}^*$ be its eigenvalues and let ϕ and $\bar{\phi}$ be corresponding eigenvectors in $\hat{k}^{2*}$.*

(i) There exists $a \in \hat{k}^*$ such that $a\bar{a}\,\Omega(\bar{\phi}, \phi)^3 = q$.
(ii)

$$\mu^{-1}(X) = \{ua\phi^3 + \bar{u}\bar{a}\,\bar{\phi}^3 : u \in \hat{k}^* \text{ and } u\bar{u} = 1\}.$$

Proof. Recall that $v_{\det X}(X) = [2]$ is a necessary and sufficient condition for X to be in the image of μ (cf Theorem 2.11). Since $\phi + \bar{\phi}$ is not an eigenvector of X, we have

$$[2] = [\Omega(\phi + \bar{\phi}, X \cdot \phi + X \cdot \bar{\phi})] = [-2q\Omega(\phi, \bar{\phi})].$$

Hence, there exists $\alpha \in \hat{k}^*$ such that $\alpha\bar{\alpha} = q\Omega(\bar{\phi}, \phi)$ and then $a = \frac{q^2}{\alpha^3}$ is a solution of (i).

By Corollary 3.31, the fiber of the $\hat{k}$-moment map $\hat{\mu} : S^3(\hat{k}^{2*}) \to \mathfrak{sl}(2, \hat{k})$ is

$$\hat{\mu}^{-1}(X) = \left\{ c\bar{\phi}^3 + \frac{q}{c\Omega(\bar{\phi}, \phi)^3}\phi^3 : c \in \hat{k}^* \right\}$$

and hence

$$\mu^{-1}(X) = \left\{ c\bar{\phi}^3 + \frac{q}{c\Omega(\bar{\phi}, \phi)^3}\phi^3 : c \in \hat{k}^*, \bar{c} = \frac{q}{c\Omega(\bar{\phi}, \phi)^3} \right\}.$$

This together with (i) implies (ii).   □

Orbit parameters for $\mathcal{O}(\hat{k})$

It is clear that $\mathcal{O}(\hat{k})$ is stable under the action of $\mathrm{Sl}(2, k)$ and in this section we will give a parameterization of the space of orbits.

Let $P \in \mathcal{O}(\hat{k})$. Then, since $Q_n(P) \in \hat{k}^{*2}$, the $Sl(2, \hat{k})$ orbit of P regarded as a binary cubic over $\hat{k}$ is entirely determined by $I_{\widehat{\mathcal{O}}_{[1]}}(P)$, where

$$I_{\widehat{\mathcal{O}}_{[1]}} : \widehat{\mathcal{O}}_{[1]} \to \hat{k}^* \times_{Z_2} \hat{k}^* / \hat{k}^{*3}$$

is the $Sl(2, \hat{k})$-invariant function defined in Theorem 3.34. Recall that to calculate $I_{\widehat{\mathcal{O}}_{[1]}}(P)$, we choose $\lambda \in \hat{k}^*$ and $\alpha \in \hat{k}^{2*}$ such that

$$P = \lambda\alpha^3 + \bar{\lambda}\bar{\alpha}^3$$

and then by definition,

$$I_{\widehat{\mathcal{O}}_{[1]}}(P) = [\hat{\omega}(\lambda\alpha^3, \overline{\lambda\alpha}), [\lambda\bar{\lambda}^{-1}]]. \tag{38}$$

The square roots $\pm\hat{\omega}(\lambda\alpha^3, \overline{\lambda}\bar{\alpha}^3)$ of $Q_n(P)$ are pure imaginary since

$$\overline{\hat{\omega}(\lambda\alpha^3, \overline{\lambda_1}\bar{\alpha}^3)} = \hat{\omega}(\overline{\lambda}\bar{\alpha}^3, \lambda\alpha^3) = -\hat{\omega}(\lambda\alpha^3, \overline{\lambda_1}\bar{\alpha}^3),$$

and the class $[\lambda\bar{\lambda}^{-1}]$ of $\lambda\bar{\lambda}^{-1}$ in the group $\hat{k}^* / \hat{k}^{*3}$ satisfies

$$[\lambda\bar{\lambda}^{-1}]\overline{[\lambda\bar{\lambda}^{-1}]} = 1.$$

It follows that

$$I_{\widehat{\mathcal{O}}_{[1]}}(P) \in \operatorname{Im} \hat{k}^* \times_{Z_2} U(\hat{k}^* / \hat{k}^{*3}),$$

where

$$U(\hat{k}^* / \hat{k}^{*3}) = \{\alpha \in \hat{k}^* / \hat{k}^{*3} \text{ s.t. } \alpha\bar{\alpha} = 1\}.$$

is the 'unitary' group of $\hat{k}^* / \hat{k}^{*3}$. Note that the Z_2 action on $\operatorname{Im} \hat{k}^* \times U(\hat{k}^* / \hat{k}^{*3})$ is precisely the natural action of $\operatorname{Gal}(\hat{k}/k)$.

Theorem 3.45. *Let* $I_{\widehat{\mathcal{O}}_{[1]}} : \mathcal{O}(\hat{k}) \to \operatorname{Im} \hat{k}^* \times_{Z_2} U(\hat{k}^* / \hat{k}^{*3})$ *be defined by* (38) *above.*

(i) Let $P, P' \in \mathcal{O}(\hat{k})$. *Then*

$$Sl(2, k) \cdot P' = Sl(2, k) \cdot P \iff I_{\widehat{\mathcal{O}}_{[1]}}(P') = I_{\widehat{\mathcal{O}}_{[1]}}(P).$$

(ii) The map $I_{\widehat{\mathcal{O}}_{[1]}}$ induces a bijection

$$\mathcal{O}(\hat{k})/Sl(2,k) \longleftrightarrow \operatorname{Im}\hat{k}^* \times_{Z_2} U(\hat{k}^*/\hat{k}^{*3}).$$

(iii) The isotropy group of $P \in \mathcal{O}(\hat{k})$ is isomorphic to

$$\left\{ \begin{pmatrix} \lambda & 0 \\ 0 & \frac{1}{\lambda} \end{pmatrix} \in Sl(2,\hat{k}) : \lambda^3 = 1,\ \lambda\bar{\lambda} = 1 \right\}.$$

Proof. (i): The function $I_{\widehat{\mathcal{O}}_{[1]}} : \mathcal{O}(\hat{k}) \to \operatorname{Im}\hat{k}^* \times_{Z_2} U(\hat{k}^*/\hat{k}^{*3})$ is $Sl(2,k)$-invariant since it is by definition the restriction of an $Sl(2,\hat{k})$-invariant function on a larger space.

To prove $I_{\widehat{\mathcal{O}}_{[1]}}$ separates orbits, suppose $I_{\widehat{\mathcal{O}}_{[1]}}(P') = I_{\widehat{\mathcal{O}}_{[1]}}(P)$. Writing $P = \lambda\alpha^3 + \bar{\lambda}\bar{\alpha}^3$ and $P' = \lambda'\alpha'^3 + \bar{\lambda}'\bar{\alpha}'^3$, there exists $\sigma \in Z_2$ such that

$$\left(\hat{\omega}\left(\lambda'\alpha'^3,\ \bar{\lambda}'\bar{\alpha}'^3\right), \left[\lambda'\bar{\lambda}'^{-1}\right] \right) = \sigma \cdot \left(\hat{\omega}\left(\lambda\alpha^3,\ \bar{\lambda}\bar{\alpha}^3\right), \left[\lambda\bar{\lambda}^{-1}\right] \right) \qquad (39)$$

and, permuting cube terms if necessary, we can suppose without loss of generality that σ is the identity. Then, equation (39) implies

$$\hat{\omega}\left(\lambda'\alpha'^3,\ \bar{\lambda}'\bar{\alpha}'^3\right) = \hat{\omega}\left(\lambda\alpha^3,\ \bar{\lambda}\bar{\alpha}^3\right), \qquad \left[\lambda'\bar{\lambda}'^{-1}\right] = \left[\lambda\bar{\lambda}^{-1}\right] \qquad (40)$$

or equivalently,

$$\lambda'\bar{\lambda}'\hat{\omega}\left(\alpha'^3,\ \bar{\alpha}'^3\right) = \lambda\bar{\lambda}\hat{\omega}\left(\alpha^3,\ \bar{\alpha}^3\right), \qquad \left[\lambda'\bar{\lambda}'^{-1}\right] = \left[\lambda\bar{\lambda}^{-1}\right],$$

which by (2) is equivalent to

$$\lambda'\bar{\lambda}'\,\widehat{\Omega}\left(\alpha',\ \bar{\alpha}'\right)^3 = \lambda\bar{\lambda}\,\widehat{\Omega}\left(\alpha,\ \bar{\alpha}\right)^3, \qquad \left[\lambda'\bar{\lambda}'^{-1}\right] = \left[\lambda\bar{\lambda}^{-1}\right]. \qquad (41)$$

Taking classes in $\hat{k}^*/\hat{k}^{*3}$, we get

$$[\lambda'\bar{\lambda}'] = [\lambda\bar{\lambda}], \qquad \left[\lambda'\bar{\lambda}'^{-1}\right] = \left[\lambda\bar{\lambda}^{-1}\right]$$

and multiplying the two equations gives

$$[\lambda'^2] = [\lambda^2].$$

From this, it follows that $[\lambda'] = [\lambda]$ since the cube of any element in $\hat{k}^*/\hat{k}^{*3}$ is the identity.

Let now $\xi \in \hat{k}^*$ be such that

$$\lambda' = \xi^3 \lambda.$$

Substituting in the first equation of (41), we get

$$\widehat{\Omega}(\xi\alpha', \overline{\xi\alpha'})^3 = \widehat{\Omega}(\alpha, \overline{\alpha})^3,$$

which means

$$\widehat{\Omega}(\xi\alpha', \overline{\xi\alpha'}) = j\, \widehat{\Omega}(\alpha, \overline{\alpha})$$

for some $j \in \hat{k}$ such that $j^3 = 1$. The conjugate of this equation is

$$-\widehat{\Omega}(\xi\alpha', \overline{\xi\alpha'}) = -\bar{j}\, \widehat{\Omega}(\alpha, \overline{\alpha})$$

and hence $\bar{j} = j$.

Define $g \in \mathrm{Gl}(2, \hat{k})$ by

$$g \cdot \alpha = j\xi\alpha', \quad g \cdot \overline{\alpha} = j\bar{\xi}\,\overline{\alpha'}.$$

Then g commutes with conjugation by definition, and preserves $\widehat{\Omega}$ since

$$\widehat{\Omega}(g \cdot \alpha, g \cdot \overline{\alpha}) = j^2\widehat{\Omega}(\xi\alpha', \overline{\xi\alpha'}) = j^3\widehat{\Omega}(\alpha, \overline{\alpha}) = \widehat{\Omega}(\alpha, \overline{\alpha}).$$

Hence, $g \in \mathrm{Sl}(2, k)$. Furthermore,

$$g \cdot P = \lambda(g \cdot \alpha)^3 + \bar{\lambda}(\overline{g \cdot \alpha})^3 = \lambda(j\xi\alpha')^3 + \bar{\lambda}(j\bar{\xi}\,\overline{\alpha'})^3 = \lambda'\alpha'^3 + \bar{\lambda}'\overline{\alpha'}^3 = P',$$

which shows that P and P' are in the same $\mathrm{Sl}(2, k)$-orbit. This proves (i).

To prove (ii), we only have to show that $I_{\widehat{\mathcal{O}}_{[1]}}$ is surjective since by (i), the function $I_{\widehat{\mathcal{O}}_{[1]}}$ induces an injection $\mathcal{O}(\hat{k})/\mathrm{Sl}(2, k) \hookrightarrow \mathrm{Im}\,\hat{k}^* \times_{\mathbb{Z}_2} U(\hat{k}^*/\hat{k}^{*3})$.

Let $(q, s) \in \mathrm{Im}\,\hat{k}^* \times U(\hat{k}^*/\hat{k}^{*3})$. First, pick $\lambda \in \hat{k}^*$ such that

$$[\lambda] = \bar{s}. \tag{42}$$

Since $[\lambda\bar{\lambda}] = s\bar{s} = 1$, we know $\lambda\bar{\lambda}$ is a cube in $\hat{k}^*$ but in fact, since $\hat{k}^*/k$ is a quadratic extension and $\lambda\bar{\lambda} \in k$, this implies that there exists $r \in k^*$ such that

$$\lambda\bar{\lambda} = r^3. \tag{43}$$

Now let

$$\alpha = -\frac{q}{2r}\hat{x} + \hat{y}$$

(where $\hat{x}, \hat{y} \in \hat{k}^{2*}$ are the base extensions of $x, y \in k^{2*}$) and let

$$P = \frac{\lambda}{q}\alpha^3 - \frac{\bar{\lambda}}{q}\bar{\alpha}^3. \tag{44}$$

This is a binary cubic of the form $T + \bar{T}$ where $T \in \hat{T}$. We are now going to show that $P \in \mathcal{O}(\hat{k})$ and that $I_{\widehat{\mathcal{O}}_{[1]}}(P) = [q, s]$.

Note first that

$$\widehat{\Omega}(\alpha, \bar{\alpha}) = \widehat{\Omega}\left(-\frac{q}{2r}\hat{x} + \hat{y}, -\overline{\left(\frac{q}{2r}\right)}\hat{x} + \hat{y}\right) = -\frac{q}{2r} + \overline{\left(\frac{q}{2r}\right)} = -\frac{q}{r},$$

so $\widehat{\Omega}(\alpha, \bar{\alpha}) \neq 0$ which means α and $\bar{\alpha}$ are not proportional. Hence, α^3 and $\bar{\alpha}^3$ are coprime and $P \in \mathcal{O}(\hat{k})$.

Next, we have

$$\hat{\omega}(\alpha^3, \bar{\alpha}^3) = \widehat{\Omega}(\alpha, \bar{\alpha})^3 = -\frac{q^3}{r^3} \tag{45}$$

and

$$\hat{\omega}\left(\frac{\lambda}{q}\alpha^3, -\frac{\bar{\lambda}}{q}\bar{\alpha}^3\right) = -\left(\frac{1}{q}\right)^2 \lambda\bar{\lambda}\hat{\omega}(\alpha^3, \bar{\alpha}^3) = q \tag{46}$$

using equations (43) and (45). Finally, it follows from (42) that

$$\left[\frac{\lambda}{q}\left(\frac{\bar{\lambda}}{q}\right)^{-1}\right] = [\lambda\bar{\lambda}^{-1}] = \bar{s}s^{-1} = s^{-1}s^{-1} = s^{-2} = s. \tag{47}$$

Hence, putting together equations (38), (44), (46) and (47), we get

$$I_{\widehat{\mathcal{O}}_{[1]}}(P) = [q, s]$$

and this proves that $I_{\widehat{\mathcal{O}}_{[1]}} : \mathcal{O}(\hat{k}) \to \mathrm{Im}\,\hat{k}^* \times_{Z_2} U(\hat{k}^*/\hat{k}^{*3})$ is surjective.

Part (iii) follows from Theorem 3.34 (iii).   □

Corollary 3.46. *Let* $P, P' \in \mathcal{O}(\hat{k})$. *Then*

$$Sl(2, k) \cdot P' = Sl(2, k) \cdot P \iff Sl(2, \hat{k}) \cdot P' = Sl(2, \hat{k}) \cdot P.$$

Proof. Both properties are equivalent to $I_{\widehat{\mathcal{O}}_{[1]}}(P) = I_{\widehat{\mathcal{O}}_{[1]}}(P')$ by the above theorem and Theorem 3.34.   □

Properties of orbit space

The parameter space

$$\mathrm{Im}\,\hat{k}^* \times_{Z_2} U(\hat{k}^*/\hat{k}^{*3})$$

for $\mathrm{Sl}(2,k)$ orbits in $\mathcal{O}(\hat{k})$ is very analogous to the parameter space

$$\hat{k}^* \times_{Z_2} k^*/k^{*3}$$

for $\mathrm{Sl}(2,k)$ orbits in $\mathcal{O}_{[1]}$ that we gave in Theorem 3.34. Its main features can best be summarized in the diagram

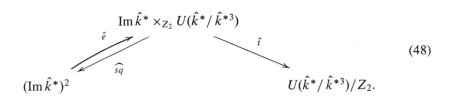

$$(48)$$

The map

$$\widehat{sq} : \operatorname{Im}\hat{k}^* \times_{Z_2} U(\hat{k}^*/\hat{k}^{*3}) \to (\operatorname{Im}\hat{k}^*)^2 \tag{49}$$

given by

$$\widehat{sq}([q,\alpha]) = q^2$$

is the fibration associated with the principal Z_2-fibration

$$\operatorname{Im}\hat{k}^* \to (\operatorname{Im}\hat{k}^*)^2$$

and the action of Z_2 on $U(\hat{k}^*/\hat{k}^{*3})$ by conjugation. Since Z_2 acts by automorphisms, the fiber $\widehat{sq}^{-1}(q^2)$ over any point $q^2 \in (\operatorname{Im}\hat{k}^*)^2$ has a natural group structure

$$[q,u_1] \times [q,u_2] = [q,u_1 u_2] \tag{50}$$

independent of the choice of square root q of q^2. Taking the identity at each point, we get a canonical section $\hat{e} : (\operatorname{Im}\hat{k}^*)^2 \to \operatorname{Im}\hat{k}^* \times_{Z_2} U(\hat{k}^*/\hat{k}^{*3})$ of (49) given by

$$\hat{e}(q^2) = [q,1] \tag{51}$$

but, although each fiber is a group isomorphic to $U(\hat{k}^*/\hat{k}^{*3})$, the fibration (49) is not in general isomorphic to the product

$$(\operatorname{Im}\hat{k}^*)^2 \times U(\hat{k}^*/\hat{k}^{*3}) \to (\operatorname{Im}\hat{k}^*)^2.$$

To translate the above features of orbit space into more concrete statements about binary cubics over k, note that the map $\widehat{sq}$ is essentially the quartic Q_n since for all $P \in \mathcal{O}(\hat{k})$,

$$\widehat{sq}\left(I_{\widehat{\mathcal{O}}_{[1]}}(P)\right) = Q_n(P).$$

Theorem 3.47. *Let $M \in (\operatorname{Im} \hat{k}^*)^2$, let*

$$\mathcal{O}_M = \{P \in S^3(k^{2^*}) : Q_n(P) = M\}$$

and let $\mathcal{O}_M/Sl(2,k)$ be the space of $Sl(2,k)$-orbits in $\mathcal{O}_M$.

(i) The map $I_{\widehat{\mathcal{O}}_{[1]}} : \mathcal{O}(\hat{k}) \to \operatorname{Im} \hat{k}^* \times_{Z_2} U(\hat{k}^*/\hat{k}^{*3})$ induces a bijection

$$\mathcal{O}_M/Sl(2,k) \longleftrightarrow \widehat{sq}^{-1}(M)$$

and, by pullback of (50), a group structure on $\mathcal{O}_M/Sl(2,k)$.
(ii) As groups, $\mathcal{O}_M/Sl(2,k) \cong U(\hat{k}^*/\hat{k}^{*3})$.
(iii) The identity element of $\mathcal{O}_M/Sl(2,k)$ is characterized by:

$$Sl(2,k) \cdot P = 1 \Leftrightarrow P \text{ is reducible over } k.$$

Proof. Parts (i) and (ii) follow from the discussion above. To prove (iii), first note that P is reducible over k iff P is reducible over $\hat{k}$ since P is cubic and $\hat{k}/k$ is a quadratic extension. By Theorem 3.35(iii), P is reducible over $\hat{k}$ iff $I_{\widehat{\mathcal{O}}_{[1]}}(P) = [q, 1]$, where $q \in \hat{k}$ is a square root of M, and by equation (51), this is the identity element of $\mathcal{O}_M/Sl(2,k)$. $\qquad\Box$

Corollary 3.48. *Let $P, P' \in S^3(k^{2^*})$ be reducible binary cubics such that $Q_n(P) = Q_n(P')$ is nonzero. Then there exists $g \in Sl(2,k)$ such that $P' = g \cdot P$.*

Proof. Suppose $Q_n(P) = Q_n(P') = M$. If $M \in k^{*2}$, the result follows from Theorem 3.35(iii). If $M \in k^*$ is not a square, one can find a quadratic extension $\hat{k}$ of k such that $M \in (\operatorname{Im} \hat{k}^*)^2$. The result then follows from Theorem 3.47 (iii). $\qquad\Box$

To finish this section, we briefly discuss the map $\hat{t} : \operatorname{Im} \hat{k}^* \times_{Z_2} U(\hat{k}^*/\hat{k}^{*3}) \to U(\hat{k}^*/\hat{k}^{*3})/Z_2$ in diagram (48) given by

$$\hat{t}([q, \alpha]) = [\alpha].$$

This is a fibration with fiber $\operatorname{Im} \hat{k}^*$ outside the identity coset [1] but

$$\hat{t}^{-1}([1]) = \hat{e}(k^{*2})$$

is a 'singular fiber'. There is a k^*-action:

$$\lambda \cdot [q, \alpha] = [\lambda q, \alpha], \tag{52}$$

which maps fibers of $\widehat{sq}$ to fibers of $\widehat{sq}$:

$$\widehat{sq}([q', \alpha']) = \widehat{sq}([q, \alpha]) \Rightarrow \widehat{sq}(\lambda \cdot [q', \alpha']) = \widehat{sq}(\lambda \cdot [q, \alpha]),$$

and whose orbits are exactly the fibers of $\hat{t}$:

$$\hat{t}([q', \alpha']) = t([q, \alpha]) \Leftrightarrow \exists \lambda \in k^* \text{ s.t. } [q', \alpha'] = \lambda \cdot [q, \alpha].$$

Isotropy for this action is given by: $Isot_{k^*}([q, \alpha]) = \begin{cases} 1 & \text{if } \alpha \neq 1 \\ \{\pm 1\} & \text{if } \alpha = 1. \end{cases}$

It would be interesting to interpret these features of the orbit space in terms of the original binary cubics.

4   Parameter Spaces for Gl$(2, k)$-Orbits

We have seen that the Sl$(2, k)$-orbits in

$$\mathcal{O}_{[1]} = \left\{ P \in S^3(k^{2*}) : Q_n(P) \in k^{*2} \right\}$$

are parameterized by

$$k^* \times_{\mathbb{Z}_2} k^* / k^{*3}$$

and that if $\hat{k}$ is a quadratic extension of k, the Sl$(2, k)$-orbits in

$$\mathcal{O}(\hat{k}) = \left\{ P \in S^3(k^{2*}) : Q_n(P) \in (\text{Im}\,\hat{k}^*)^2 \right\}$$

are parameterized by

$$\text{Im}\,\hat{k}^* \times_{\mathbb{Z}_2} U(\hat{k}^*/\hat{k}^{*3}).$$

The group $GL(2, k)$ also acts on binary cubics and since

$$Q_n(g \cdot P) = (\det g)^{-6} Q_n(P) \quad \forall g \in \text{Gl}(2, k), \forall P \in S^3(k^{2*}),$$

the spaces $\mathcal{O}_{[1]}$ and $\mathcal{O}(\hat{k})$ are stable under Gl$(2, k)$.

In general, if Gl$(2, k)$ acts on a space X, there is a map

$$X/\text{Sl}(2, k) \to X/\text{Gl}(2, k)$$

from the set of $Sl(2, k)$-orbits onto the set of $G1(2, k)$-orbits. The fibers of this map are the orbits of the k^*-action on $X/Sl(2, k)$ given by

$$\lambda * [x] = [\Lambda \cdot x], \tag{53}$$

where Λ is any element of $Gl(2, k)$ such that $\det \Lambda = \lambda$. Thus, to get parameter spaces for $\mathcal{O}_{[1]}/Gl(2, k)$ and $\mathcal{O}(\hat{k})/Gl(2, k)$ we need just to calculate the k^*-actions on $k^* \times_{Z_2} k^*/k^{*3}$ and $\text{Im}\,\hat{k}^* \times_{Z_2} U(\hat{k}^*/\hat{k}^{*3})$ corresponding to (53).

Lemma 4.1. *(i) Let k^* act on $k^* \times_{Z_2} k^*/k^{*3}$ by*

$$\lambda \cdot [\xi, \alpha] = [\lambda\xi, \alpha]$$

and let $I_{\mathcal{O}_{[1]}} : \mathcal{O}_{[1]} \to k^* \times_{Z_2} k^*/k^{*3}$ be defined by (23). Then

$$I_{\mathcal{O}_{[1]}}(g \cdot P) = (\det g)^{-3} \cdot I_{\mathcal{O}_{[1]}}(P) \qquad \forall P \in \mathcal{O}_{[1]}, \; \forall g \in Gl(2, k).$$

(ii) Let k^* act on $\text{Im}\,\hat{k}^* \times_{Z_2} U(\hat{k}^*/\hat{k}^{*3})$ by

$$\lambda \cdot [\xi, \alpha] = [\lambda\xi, \alpha]$$

and let $I_{\widehat{\mathcal{O}}_{[1]}} : \mathcal{O}(\hat{k}) \to \text{Im}\,\hat{k}^* \times_{Z_2} U(\hat{k}^*/\hat{k}^{*3})$ be defined by (38). Then

$$I_{\widehat{\mathcal{O}}_{[1]}}(g \cdot P) = (\det g)^{-3} \cdot I_{\mathcal{O}_{[1]}}(P) \qquad \forall P \in \mathcal{O}(\hat{k}), \; \forall g \in Gl(2, k).$$

Proof. To prove (i), since for any $P \in \mathcal{O}_{[1]}$ there exists $h \in Gl(2, k)$ and $a, b \in k^*$ such that

$$h \cdot P = ax^3 + by^3,$$

it is sufficient to prove that

$$I_{\mathcal{O}_{[1]}}(g \cdot (ax^3 + by^3)) = (\det g)^{-3} \cdot I_{\mathcal{O}_{[1]}}(ax^3 + by^3) \qquad \forall a, b \in k^*, \; \forall g \in Gl(2, k).$$

Consider

$$g' = \begin{pmatrix} \det g & 0 \\ 0 & 1 \end{pmatrix}.$$

Then $\det g' = \det g$, $g' \cdot x = \frac{1}{\det g} x$ and $g' \cdot y = y$. Hence $g'g^{-1} \in Sl(2, k)$,

$$I_{\mathcal{O}_{[1]}}(g \cdot (ax^3 + by^3)) = I_{\mathcal{O}_{[1]}}(g' \cdot (ax^3 + by^3))$$

and

$$I_{\mathcal{O}_{[1]}}(g' \cdot (ax^3 + by^3)) = I_{\mathcal{O}_{[1]}}\left(\frac{a}{(\det g)^3}x^3 + by^3\right) = \left[\frac{ab}{(\det g)^3}, [ab^{-1}]\right].$$

The result follows since $I_{\mathcal{O}_{[1]}}(ax^3 + by^3) = [ab, [ab^{-1}]]$.

Part (ii) follows from (i) applied to $\hat{k}$.   □

Corollary 4.2. *(i)* If $P \in \mathcal{O}_{[1]}$ and $\lambda \in k^*$ then

$$I_{\mathcal{O}_{[1]}}(\lambda * [P]) = \frac{1}{\lambda^3} \cdot I_{\mathcal{O}_{[1]}}([P]).$$

(ii) If $P \in \mathcal{O}(\hat{k})$ and $\lambda \in k^*$ then

$$I_{\widehat{\mathcal{O}}_{[1]}}(\lambda * [P]) = \frac{1}{\lambda^3} \cdot I_{\widehat{\mathcal{O}}_{[1]}}([P]).$$

Proof. Immediate from the lemma.   □

From this, we get the k^*-actions on the parameter spaces $k^* \times_{Z_2} k^*/k^{*3}$ and $\operatorname{Im}\hat{k}^* \times_{Z_2} U(\hat{k}^*/\hat{k}^{*3})$ corresponding to (53) : $\lambda \in k^*$ acts by multiplication by λ^{-3} on the first factor.

Hence, by the discussion above, the maps $I_{\mathcal{O}_{[1]}} : \mathcal{O}_{[1]} \to k^* \times_{Z_2} k^*/k^{*3}$ and $I_{\widehat{\mathcal{O}}_{[1]}} : \mathcal{O}(\hat{k}) \to \operatorname{Im}\hat{k}^* \times_{Z_2} U(\hat{k}^*/\hat{k}^{*3})$ induce bijections

$$\mathcal{O}_{[1]}/Gl(2,k) \longleftrightarrow (k^* \times_{Z_2} k^*/k^{*3})/k^{*3} = k^*/k^{*3} \times (k^*/k^{*3})/Z_2,$$

$$\mathcal{O}(\hat{k})/Gl(2,k) \longleftrightarrow (\operatorname{Im}\hat{k}^* \times_{Z_2} U(\hat{k}^*/\hat{k}^{*3}))/k^{*3} = (\operatorname{Im}\hat{k}^*)/k^{*3} \times U(\hat{k}^*/\hat{k}^{*3})/Z_2.$$

To summarize, we have proved the

Theorem 4.3. *(a)* Define $\pi : k^* \times_{Z_2} k^*/k^{*3} \to k^*/k^{*3} \times (k^*/k^{*3})/Z_2$ by $\pi([\xi, \alpha]) = ([\xi], [\alpha])$ and $J_{\mathcal{O}_{[1]}} : \mathcal{O}_{[1]} \to k^*/k^{*3} \times (k^*/k^{*3})/Z_2$ by $J_{\mathcal{O}_{[1]}} = \pi \circ I_{\mathcal{O}_{[1]}}$.

(i) Let $P, P' \in \mathcal{O}_{[1]}$ Then

$$Gl(2,k) \cdot P = Gl(2,k) \cdot P' \quad \Leftrightarrow \quad J_{\mathcal{O}_{[1]}}(P) = J_{\mathcal{O}_{[1]}}(P').$$

(ii) The map $J_{\mathcal{O}_{[1]}}$ induces a bijection

$$\mathcal{O}_{[1]}/Gl(2,k) \longleftrightarrow k^*/k^{*3} \times (k^*/k^{*3})/Z_2.$$

(b) Define $\hat{\pi} : \operatorname{Im}\hat{k}^* \times_{Z_2} U(\hat{k}^*/\hat{k}^{*3}) \to (\operatorname{Im}\hat{k}^*)/k^{*3} \times U(\hat{k}^*/\hat{k}^{*3})/Z_2$ by $\hat{\pi}([\xi, \alpha]) = ([\xi], [\alpha])$ and $J_{\widehat{\mathcal{O}}_{[1]}} : \mathcal{O}(\hat{k}) \to k^*/k^{*3} \times (k^*/k^{*3})/Z_2$ by $J_{\widehat{\mathcal{O}}_{[1]}} = \hat{\pi} \circ I_{\widehat{\mathcal{O}}_{[1]}}$.

(i) Let $P, P' \in \mathcal{O}(\hat{k})$. Then

$$Gl(2,k) \cdot P = Gl(2,k) \cdot P' \quad \Leftrightarrow \quad J_{\widehat{\mathcal{O}}_{[1]}}(P) = J_{\widehat{\mathcal{O}}_{[1]}}(P').$$

(ii) The map $J_{\widehat{\mathcal{O}}_{[1]}}$ induces a bijection

$$\mathcal{O}(\hat{k})/Gl(2,k) \longleftrightarrow (\text{Im}\,\hat{k}^*)/k^{*3} \times U(\hat{k}^*/\hat{k}^{*3})/Z_2.$$

Orbits spaces when k is a finite field of characteristic not 2 or 3

Let k be a finite field with q elements, not of characteristic 2 or 3. The following facts are well known:

- $k^*/k^{*2} \cong Z_2$ so up to isomorphism, there is only one quadratic extension of k and k^{*2} has $\frac{1}{2}(q-1)$ elements;
- if $q = 1 \bmod 3$, $k^*/k^{*3} \cong \mathbb{Z}/3\mathbb{Z}$;
- if $q = 2 \bmod 3$, $k^* = k^{*3}$;
- if $q = 1 \bmod 3$ and $\hat{k}/k$ is a quadratic extension, $U(\hat{k}^*/\hat{k}^{*3}) \cong 1$;
- if $q = 2 \bmod 3$ and $\hat{k}/k$ is a quadratic extension, $U(\hat{k}^*/\hat{k}^{*3}) \cong \mathbb{Z}/3\mathbb{Z}$.

These facts together with Theorem (3.34) and Theorem 4.3 immediately give the

Proposition 4.4. *Let k be a finite field with q elements, not of characteristic 2 or 3 and let $\hat{k}$ be a quadratic extension. Set*

$$\mathcal{O}_{[1]} = \left\{ P \in S^3(k^{2*}) : Q_n(P) \in k^{*2} \right\},$$

$$\mathcal{O}(\hat{k}) = \left\{ P \in S^3(k^{2*}) : Q_n(P) \in (\text{Im}\,\hat{k}^*)^2 \right\}.$$

(a) *If $q = 1 \bmod 3$, $\mathcal{O}_{[1]}$ is the union of $\frac{3}{2}(q-1)$ $Sl(2,k)$-orbits and $\mathcal{O}(\hat{k})$ is the union of $\frac{1}{2}(q-1)$ $Sl(2,k)$-orbits.*

(b) *If $q = 1 \bmod 3$, $\mathcal{O}_{[1]}$ is the union of 6 $Gl(2,k)$-orbits and $\mathcal{O}(\hat{k})$ is the union of 3 $Gl(2,k)$-orbits.*

(c) *If $q = 2 \bmod 3$, $\mathcal{O}_{[1]}$ is the union of $\frac{1}{2}(q-1)$ $Sl(2,k)$-orbits and $\mathcal{O}(\hat{k})$ is the union of $\frac{3}{2}(q-1)$ $Sl(2,k)$-orbits.*

(d) *If $q = 2 \bmod 3$, $\mathcal{O}_{[1]}$ is a $Gl(2,k)$-orbit and $\mathcal{O}(\hat{k})$ is the union of 2 $Gl(2,k)$-orbits.*

Proof. As examples, let us count the number of $Sl(2,k)$-orbits in $\mathcal{O}_{[1]}$ when $q = 1 \bmod 3$ and the number of $Gl(2,k)$-orbits in $\mathcal{O}(\hat{k})$ when $q = 2 \bmod 3$.

In the first case, by Theorem (3.34), the parameter space is $k^* \times_{Z_2} k^*/k^{*3}$ which, being a fiber bundle over k^{*2} with fiber k^*/k^{*3}, has $\frac{1}{2}(q-1) \times 3 = \frac{3}{2}(q-1)$ elements.

In the second case, by Theorem 4.3 , the parameter space is $(\mathrm{Im}\,\hat{k}^*)/k^{*3} \times U(\hat{k}^*/\hat{k}^{*3})/Z_2$ and this has $1 \times 2 = 2$ elements since Z_2 acts on $U(\hat{k}^*/\hat{k}^{*3})$ by inversion. $\qquad\square$

According to [10] (Proposition 5.6) at least part of the following corollary can be found in Dickson [7].

Corollary 4.5. *Let k be a finite field with q elements, not of characteristic 2 or 3. The number of $Sl(2, k)$-orbits of binary cubics with nonzero discriminant is $2(q-1)$. The number of $Gl(2, k)$-orbits of binary cubics with nonzero discriminant is 9 if $q = 1 \bmod 3$ and 3 if $q = 2 \bmod 3$.*

Proof. A binary cubic of nonzero discriminant is either in $\mathcal{O}_{[1]}$ or in $\mathcal{O}(\hat{k})$ since up to isomorphism, k has only one quadratic extension. Hence, the total number of $Sl(2, k)$-orbits with nonzero discriminant is the number of $Sl(2, k)$-orbits in $\mathcal{O}_{[1]}$ plus the number of $Sl(2, k)$-orbits in $\mathcal{O}(\hat{k})$. The same is true for $Gl(2, k)$-orbits and the result follows from Proposition 4.4. $\qquad\square$

5   A Symplectic Eisenstein Identity

The following identity is a symplectic generalization of the classical Eisenstein identity which, as we will see, is obtained from it in the special case when Q is the cube of a linear form. There is an analogous identity for the symplectic module associated with any Heisenberg graded Lie algebra ([15]).

Theorem 5.1. *Let $P, Q \in S^3(k^{2*})$. Then*

$$\omega(\Psi(P), Q)^2 - 9Q_n(P)\,\omega(P, Q)^2 = -\frac{9}{2}\omega(\mu(P)^{\otimes 3} \cdot Q, Q)$$

$$-\frac{9}{2}Q_n(P)\,\omega(\mu(P) \cdot Q, Q), \qquad (54)$$

where $\mu(P)^{\otimes 3}$ denotes the unique endomorphism of $S^3(k^{2*})$ satisfying $\mu(P)^{\otimes 3} \cdot (\alpha^3) = (\mu(P) \cdot \alpha)^3$ for all $\alpha \in k^{2*}$.

Proof. If $\mu(P) = 0$, then $\Psi(P) = 0$, $Q_n(P) = 0$ and all terms in the identity are zero.

If $\mu(P)$ is nilpotent nonzero, then $Q_n(P) = 0$ and there exists $g \in Sl(2, k)$ such that $g \cdot P = x^2 y$. Since the identity (54) is $Sl(2, k)$-invariant, we can suppose without loss of generality that $P = x^2 y$. Then, by calculation,

$$\Psi(P) = -\frac{2}{9}x^3, \quad \mu(P) = \frac{2}{9}\begin{pmatrix} 0 & 0 \\ 1 & 0 \end{pmatrix}$$

and so $\mu(P) \cdot x = 0$ and $\mu(P) \cdot y = -\frac{2}{9}x$. Let

$$Q = px^3 + 3rx^2y + 3sxy^2 + ty^3.$$

The LHS of (54) is

$$\omega\left(-\frac{2}{9}x^3, Q\right)^2 = \left(\frac{2}{9}\right)^2 t^2.$$

and the RHS of (54) is

$$-\frac{9}{2}\omega\left(\mu(P)^{\otimes 3} \cdot Q, Q\right) = -\frac{9}{2}\omega\left(-\left(\frac{2}{9}\right)^3 tx^3, Q\right) = \left(\frac{2}{9}\right)^2 t^2.$$

Thus, (54) holds if $\mu(P)$ is nilpotent nonzero.

To complete the proof of the proposition, it remains to prove (54) if $Q_n(P) \neq 0$. As the identity is independent of the field we may suppose that $Q_n(P)$ is a square in k^* and hence that $P \in \mathcal{O}_{[1]}$. Since the identity (54) is $\mathrm{Sl}(2, k)$-invariant, we can further suppose without loss of generality that

$$P = ax^3 + dy^3.$$

Then

$$Q_n(P) = a^2d^2, \quad \Psi(P) = 3ad(-ax^3 + dy^3), \quad \mu(P) = \begin{pmatrix} ad & 0 \\ 0 & -ad \end{pmatrix}$$

and so $\mu(P) \cdot x = -adx$ and $\mu(P) \cdot y = ady$. Let

$$Q = px^3 + 3rx^2y + 3sxy^2 + ty^3.$$

The LHS of (54) is

$$\omega(\Psi(P), Q)^2 - 9Q_n(P)\omega(P, Q)^2$$

$$= 9a^2d^2\left(\omega(-ax^3 + dy^3, Q)^2 - \omega(ax^3 + dy^3, Q)^2\right)$$

$$= -36a^3d^3\omega(x^3, Q)\omega(y^3, Q)$$

$$= 36a^3d^3pt. \tag{55}$$

On the other hand, the first term of the RHS of (54) is

$$-\frac{9}{2}\omega(\mu(P)^{\otimes 3} \cdot Q, Q) = -\frac{9}{2}a^3 d^3 \omega(-px^3 + 3rx^2 y - 3sxy^2 + ty^3, Q)$$

$$= -\frac{9}{2}a^3 d^3(-2pt - 6rs)$$

$$= 9a^3 d^3(pt + 3rs) \tag{56}$$

and the second term of the RHS of (54) is

$$-\frac{9}{2}Q_n(P)\omega(\mu(P) \cdot Q, Q) = -\frac{9}{2}a^3 d^3 \omega(-3px^3 - 3rx^2 y + 3sxy^2 + 3ty^3, Q)$$

$$= -\frac{9}{2}a^3 d^3(-6pt + 6rs)$$

$$= 27a^3 d^3(pt - rs). \tag{57}$$

The result follows from equations (55), (56) and (57).   □

To obtain the classical Eisenstein identity from this result, recall that one can use the symplectic form Ω on k^{2*} to define a $Sl(2, k)$-equivariant isomorphism $\tilde{\ }$: $k^2 \to k^{2*}$: if $v \in k^2$, we let $\tilde{v} \in k^{2*}$ be the unique linear form such that

$$\phi(v) = \Omega(\phi, \tilde{v}) \quad \forall \phi \in k^{2*}.$$

It then follows that

$$P(v) = \omega(P, \tilde{v}^3) \quad \forall P \in S^3(k^{2*}), \forall v \in k^2, \tag{58}$$

so that the operation of evaluating a binary cubic at a point of k^2 can be expressed in terms of the symplectic form ω on $S^3(k^{2*})$. One can also pullback Ω to get an $Sl(2, k)$-invariant symplectic form Ω_{k^2} on k^2:

$$\Omega_{k^2}(v, w) = \Omega(\tilde{v}, \tilde{w}) \quad \forall v, w \in k^2.$$

Corollary 5.2 (Classical Eisenstein identity). *Let $P \in S^3(k^{2*})$ and let $v \in k^2$.*

$$\Psi(P)(v)^2 - 9Q_n(P) P(v)^2 = -\frac{9}{2}\Omega_{k^2}(\mu(P) \cdot v, v)^3.$$

Proof. Setting $Q = \tilde{v}^3$ in (54) and using (58), we get

$$\Psi(P)(v)^2 - 9Q_n(P) P(v)^2 = -\frac{9}{2}\omega(\mu(P)^{\otimes 3} \cdot \tilde{v}^3, \tilde{v}^3)$$

$$-\frac{9}{2}Q_n(P)\omega(\mu(P) \cdot \tilde{v}^3, \tilde{v}^3). \tag{59}$$

The result follows from this since

$$\omega(\mu(P) \cdot \tilde{v}^3, \tilde{v}^3) = 3\omega((\mu(P) \cdot \tilde{v})\tilde{v}^2, \tilde{v}^3) = 0$$

$((\mu(P) \cdot \tilde{v})\tilde{v}^2$ has at least a double root at v) and

$$\omega(\mu(P)^{\otimes 3} \cdot \tilde{v}^3, \tilde{v}^3) = \Omega(\mu(P) \cdot \tilde{v}, \tilde{v})^3 = \Omega_{k^2}(\mu(P) \cdot v, v)^3. \qquad \square$$

Acknowledgements MJS was supported in part by the Math Research Institute, OSU. RJS was supported in part by NSF Grants DMS-0301133 and DMS-0701198.

References

1. M. BHARGAVA, Higher composition laws. I. A new view on Gauss composition, and quadratic generalizations, *Ann. of Math. (2)* **159** (2004), no. 1, 217–250.
2. N. BOURBAKI, Élements de mathématique, Fascicule XXIV, Livre II, Algèbre Chapitre 9, Hermann, Paris, 1959.
3. M. CAHEN AND L. SCHWACHHÖFER, Special symplectic connections and Poisson geometry. *Lett. Math. Phys.* **69** (2004), 115–137.
4. M. DEMAZURE, Automorphismes et déformations des variétés de Borel. *Invent. Math.* **39** (1977), no. 2, 179–186.
5. L. E. DICKSON, **History of the Theory of Numbers**, vol. III. Chelsea, 1952.
6. L. E. DICKSON, **Algebraic Invariants**, Mathematical Monographs, No. 14, John Wiley and Sons, New York, 1914.
7. L. E. DICKSON, **On Invariants and The Theory of Numbers**, The Madison Colloquim, Dover Publications, New York, 1966.
8. G. EISENSTEIN, Untersuchungen über die cubischen Formen mit zwei Variabeln, *J. Crelle* **27** (1844), 89–104 = Mathematische Werke, Band I, Chelsea Publ. Co., 1975, 10–25.
9. D. HAILE, On the Clifford algebra of a binary cubic form, *Amer. J. Math.* **106** (1984), 1269–1280.
10. J. W. HOFFMAN AND J.MORALES, Arithmetic of binary cubic forms, *Enseign. Math. (2)*, **46** (2000), no. 1-2, 61–94.
11. L. J. MORDELL, The diophantine equation $y^2 - k = x^3$, *Proc. Lond. Math. Soc. (2)*, **13**, (1913), 60-80.
12. L. J. MORDELL, **Diophantine equations**, Pure and Applied Mathematics, Vol. 30 Academic Press, London-New York, 1969.
13. J. NAKAGAWA, On the relations among the class numbers of binary cubic forms, *Invent. Math.* **134** (1998), no. 1, 101–138.
14. T. SHINTANI, On Dirichlet series whose coefficients are class numbers of integral binary cubic forms, *J. Math. Soc. Japan* **24**, (1972), 132-188.
15. M. J. SLUPINSKI AND R. J. STANTON, Symplectic geometry of Heisenberg graded Lie algebras, nearing completion.
16. D. J. WRIGHT, The adelic zeta function associated with the space of binary cubic forms, I: global theory, *Math. Ann.* **270** (1985), 503–534.

On the Restriction of Representations of SL(2, ℂ) to SL(2, ℝ)

B. Speh and T.N. Venkataramana

Abstract We prove that for a certain range of the continuous parameter, the complementary series representation of SL(2,ℝ) is a direct summand of the complementary series representations of SL(2,ℂ). For this, we construct a continuous "geometric restriction map" from the complementary series representations of SL(2,ℂ) to the complementary series representations of SL(2,ℝ). In the second part, we prove that the Steinberg representation σ of SL(2,ℝ) is a direct summand of the restriction of the Steinberg representation π of SL(2,ℂ). We show that σ does not contain any smooth vectors of π.

Keywords Complementary series representations • Restriction • Subgroup

Mathematics Subject Classification (2010): 22D10

1 Introduction

Let $G = SL(2, \mathbb{C})$ and denote by $B(\mathbb{C})$ the (Borel-)subgroup of upper triangular matrices in G, by $N(\mathbb{C})$ the subgroup of unipotent upper triangular matrices in G. Given an element $b = \begin{pmatrix} a & n \\ 0 & a^{-1} \end{pmatrix}$ of $B(\mathbb{C})$, write $\rho(b) = \mid a \mid^2$. The group $K = SU(2)$ is a maximal compact subgroup of G. Given a complex number u, we obtain a $(\mathfrak{g}, K)$–module π_u realized on the space of functions on G, which satisfy for all

B. Speh (✉)
Department of Mathematics, 310 Malott Hall Cornell University Ithaca, NY 14853-4201, USA
e-mail: speh@math.cornell.edu

T.N. Venkataramana
Tata Institute for Fundamental Research, Homi Bhabha Road, Mumbai 400005, India
e-mail: venky@math.tifr.res.in

B. Krötz et al. (eds.), *Representation Theory, Complex Analysis, and Integral Geometry,*   231
DOI 10.1007/978-0-8176-4817-6_9, © Springer Science+Business Media, LLC 2012

$b \in B(\mathbb{C})$ and all $g \in G(\mathbb{C})$ the formula

$$f(bg) = \rho(b)^{1+u} f(g)$$

and in addition are K-finite under the action of K by right translations. If $Re(u) > 0$, define the map $I_G(u) : \pi_u \to \pi_{-u}$ by the formula

$$(I_G(u)f)(x) = \int_{N(\mathbb{C})} dn f(w_0 nx).$$

The integral converges for $Re(u) > 0$. If u is real and $0 < u < 1$, then the pairing

$$\langle f, f \rangle_{\pi_u} = \int_K \overline{f}(k) I_G(u)(f)(k) dk$$

defines a positive definite G-invariant inner product on K-finite functions in π_u. The completion $\widehat{\pi}_u$ with respect to this inner product is the complementary series representation with continuous parameter u.

Given a complex number $u' \in \mathbb{C}$, denote by $\sigma_{u'}$ the representation of $(U(\mathfrak{h}), K_H)$, where $\mathfrak{h}$ is the Lie algebra of $H = SL(2, \mathbb{R})$, and $K_H = SO(2)$ is the maximal compact subgroup of H, defined as the space of complex valued right K_H-finite functions on H such that for all upper triangular matrices $b = \begin{pmatrix} a & n \\ 0 & a^{-1} \end{pmatrix}$ in H and all $h \in H$, we have $f(bh) = | a |^{1+u'} f(h)$. The character $| a |^2$ is the character $\rho(a)^2$, where ρ^2 is the sum of positive roots.

Denote by N_H the group of unipotent upper triangular matrices in H. If $Re(u') > 0$, we define the intertwining operator $I_H(u') : \sigma_{u'} \to \sigma_{-u'}$ as follows: for all $g \in H$, set

$$(I_H(u')f)(g) = \int_{N_H(\mathbb{R})} dn f(w_0 ng).$$

The integral is convergent if $Re(u') > 0$. Now for $f_H, g_H \in \sigma_{u'}$ and u' is real and $0 < u' < 1$, the pairing

$$\langle f_H, g_H \rangle_{\sigma_{u'}} = \int_{K_H} \overline{f}_H(k_H)(I_H(u')g_H)(k_H) dk_H$$

defines a positive definite H-invariant inner product on σ'_u. The completion is the complementary series representation $\widehat{\sigma}'_u$.

Theorem 1.1. *Let* $\frac{1}{2} < u < 1$ *and* $u' = 2u - 1$. *The complementary series representation* $\widehat{\sigma}_{u'}$ *of* $SL(2, \mathbb{R})$ *is a direct summand of the restriction of the complementary series representation* $\widehat{\pi}_u$ *of* $SL(2, \mathbb{C})$.

This theorem is proved by Mukunda [6] in 1968. In this paper, we give a different proof and we realize the projection map from $\widehat{\pi}_u$ to $\widehat{\sigma}_{2u-1}$ as a simple geometric map of sections of a line bundle on the flag varieties of $G = SL(2, \mathbb{C})$ and $H = SL(2, \mathbb{R})$. For a precise formulation and details, see Theorem 2.5 and its corollary. In a sequel to this article, we will use this idea to analyze the restriction of the complementary series representations of $SO(n, 1)$ [7].

Consider the **Steinberg representation** $\widehat{\pi} = Ind_B^G(\chi)$. Here, *Ind* refers to **unitary** induction from a unitary character χ of B. Given two functions $f, f' \in \pi$, the product $\phi = f \overline{f'}$ lies in $Ind_B^G(\rho)$. The G-invariant inner product on $\widehat{\pi}$ is defined by

$$\langle f, f' \rangle = \int_K (f\overline{f'})(k)dk.$$

Let π be the $(\mathfrak{g}, K)$-module of the Steinberg representation. We have the exact sequence of $(\mathfrak{g}, K)$–modules

$$0 \to \pi \to \pi_1 \to 1 \to 0.$$

The $(\mathfrak{h}, K_H)$– module σ of the Steinberg representation of SL(2, ℝ) is defined by the exact sequence

$$0 \to \sigma \to \sigma_1 \to 1 \to 0$$

and the completion $\widehat{\sigma}$ is a direct sum of 2 discrete series representations.

Theorem 1.2. *The restriction to H of $\widehat{\pi}$ contains the Steinberg Representation $\widehat{\sigma}$ of H as a direct summand.*

More precisely, the restriction is a sum of the holomorphic discrete series representation σ of H, its complex conjugate $\overline{\sigma}$, and a sum of two copies of $L^2(H/K \cap H)$, where $K \cap H$ is a maximal compact subgroup of H. By a theorem of T. Kobayashi (Theorem 4.2.6 in [4]), this implies that $\widehat{\sigma}$ does not contain any nonzero K–finite vectors in $\widehat{\pi}$. Using an explicit description of the functions in the subspace $\widehat{\sigma}$ we prove a stronger result.

Theorem 1.3. *The intersection*

$$\widehat{\sigma} \cap \widehat{\pi}^\infty = 0.$$

That is $\widehat{\sigma}$ does not contain any nonzero smooth vectors in $\widehat{\pi}$.

It is very important in the above theorems to consider a unitary representation of G (respectively H) and not only the unitary $(\mathfrak{g}, K)$-modules (respectively $(\mathfrak{h}, K_H)$-modules) as the following example shows.

Fix a semisimple noncompact real algebraic group G and let $C_c(G)$ denote the space of continuous complex valued functions on G with compact support. Let π denote an irreducible representation on a Hilbert space (which, we denote again by π) of G of the complementary series, which is unramified (i.e., fixed under a maximal compact subgroup K of G). Fix a nonzero K-invariant vector v in π.

Denote by $|| \ w \ ||_\pi$ the metric on the space π. Given $\phi \in C_c(G)$, we get a bounded operator $\pi(\phi)$ on π. Define a metric on $C_c(G)$ by setting

$$|| \ \phi \ ||^2 = || \ \pi(\phi)(v) \ ||^2_\pi + || \ \phi \ ||^2_{L^2},$$

where the latter is the L^2-norm of ϕ. The group G acts by left translations on $C_c(G)$ and preserves the above metric. Hence, it operates by unitary transformations on the completion (the latter is a Hilbert space) of this metric.

Proposition 1.4. *Under the foregoing metric, the completion of $C_c(G)$ is the direct sum of the Hilbert spaces*

$$\pi \oplus L^2(G).$$

The action of the group G on the direct sum, restricted to the subspace $C_c(G)$, is by left translations.

Note that the direct sum $\pi \oplus L^2(G)$ and $L^2(G)$ both share the same dense subspace $C_c(G)$ on which the G action is identical, namely by left translations, and yet the completions are different: $\pi \oplus L^2$ is the completion with respect to the new metric and $L^2(G)$ is the completion under the L^2-metric. We have therefore an example of two nonisomorphic unitary G-representations with an isomorphic dense subspace.

This is not possible in the case of **irreducible** unitary representations, as can be seen as follows.

The kernel to the map $\phi \mapsto \pi(\phi)v$ on $C_c(G)$ is just those functions, whose Fourier transform vanishes at a point on $\mathbb{C}$ (the latter is the space of not-necessarily unitary characters of $\mathbb{R}$). This is clearly dense in $C_c(G)$ and hence dense in $L^2(G)$. The restriction of the new metric to the kernel is simply the L^2-metric, and the kernel is dense in L^2. Therefore, the completion of the kernel gives all of L^2.

Since the map from $C_c(G)$ to the first factor π is nonzero, it follows that the completion of $C_c(G)$ cannot be only L^2. The irreducibility of π now implies that the completion must be $\pi \oplus L^2(G)$.

2 Complementary Series Representations for $SL(2,\mathbb{R})$ and $SL(2,\mathbb{C})$

2.1 Complementary Series Representations $\hat{\sigma}_{u'}$ for $SL(2,\mathbb{R})$

The space $\sigma_{u'}$ consists, by construction, of K_H-finite vectors and the restriction of $\sigma_{u'}$ to K_H is an injection; under this map, $\sigma_{u'}$ may be identified with trigonometric polynomials on K_H, which are *even*. The space of even trigonometric polynomials is spanned by the characters

$$\chi_l : \theta \mapsto e^{4\pi i l \theta} \qquad l \in \mathbb{Z}.$$

Each χ_l-eigenspace in $\pi_{u'}$ is one dimensional and has a unique vector $\chi_{l,u'}$ such that for all $k \in K_H$ we have $\chi_{l,u'}(k) = \chi_l(k)$. The intertwining operator $I_H(u')$ maps $\chi_{l,u'}$ into a multiple of $\chi_{l,-u'}$. After replacing $I_H(u')$ by a scalar multiple, we may assume that for the K_H- fixed vector $\chi_{0,u'}$

$$I_H(u')\chi_{0,u'} = \chi_{0,-u'}.$$

The normalized intertwining operator will, by abuse of notation, also be denoted $I_H(u')$. One computes that for all integers $l \neq 0$ and all $k \in K_H$,

$$I_H(u')\chi_{l,u'} = d_l(u')\chi_{l,-u'},$$

where

$$d_l(u') = \frac{(1 - u')(3 - u') \cdots (2 \mid l \mid -1 - u')}{(1 + u')(3 + u') \cdots (2 \mid l \mid -1 + u')} \tag{1}$$

and $d_0(u') = 1$. Note that if $c(u') = \frac{\Gamma((1-u')/2)}{\Gamma((1+u')/2)}$ then we have

$$d_l(u') = c(u')^{-1} \frac{\Gamma(\mid l \mid +(1 - u')/2)}{\Gamma(\mid l \mid +(1 + u')/2)} \tag{2}$$

We note that for $\chi_{l,u'}$ we have

$$\| \chi_{l,u'} \|^2_{\sigma_{u'}} = \langle \chi_{l,u'}, I_H(u')\chi_{l,u'} \rangle \tag{3}$$

$$= d_l(u') \| \chi_l \|^2_{L^2(K_H)}. \tag{4}$$

Therefore, the norm of $\chi_{l,-u'}$ in $\sigma_{-u'}$ is given by

$$\| \chi_{l,-u'} \|^2_{\sigma_{-u'}} = \frac{1}{d_l(u')} \| \chi_l \|^2_{L^2(K_H)}. \tag{5}$$

We have already noted that if $0 < u' < 1$, the pairing $\langle , \rangle$ on $\sigma_{u'}$ is positive definite. This easily follows from the formula (1) for $d_l(u')$, which shows that $d_l(u') > 0$, and the (3).

The space $Rep(\{\pm 1\} \backslash K_H) = \oplus_{l \geq 0} \sigma_l$, where $\sigma_l = \mathbb{C}\chi_l \oplus \mathbb{C}\chi_{-l}$ for $l \geq 1$ and $\sigma_0 = \mathbb{C}\chi_0$. The elements of σ_l may be thought of as the space of **Harmonic Polynomials** in the circle of degree $2l$.

Note that $d_l(u') < 1$. It can be shown, using Stirling's asymptotic formula for the Gamma function and (2), that for $|\, l \,| \to \infty$ there exists a constant C such that

$$d_l(u') \simeq C \frac{1}{|\, l \,|^{u'}}.$$

If $u' = 2u - 1$ and $l \geq 1$, then

$$d_l(u') = \frac{(1-u)(2-u)\cdots(l-u)}{(1+u)(2+u)\cdots(l+u)} \frac{(l+u)}{u}.$$

Define $\lambda_l(u)$ by the formula

$$d_l(u') = \lambda_l(u) \frac{l+u}{u}.$$

2.2 Complementary Series $\widehat{\pi}_u$ of $SL(2,\mathbb{C})$

For any $u \in \mathbb{C}$, the restriction of the $(\mathfrak{g}, K)$–module π_u to the maximal compact subgroup K is isomorphic to $Rep(T\backslash K)$, where T is the group of diagonal matrices in K, and $Rep(T\backslash K)$ denotes the space of representation functions on $T\backslash K$ on which K acts by right translations. It is known that as a representation of K, we have

$$Rep(T\backslash K) = \oplus_{m \geq 0} \rho_m,$$

where $\rho_m = Sym^{2m}(\mathbb{C}^2)$ is the $2m$-th symmetric power of $\mathbb{C}^2$, the standard two-dimensional representation of K; ρ_m is irreducible and occurs exactly once in $Rep(T\backslash K)$. The same decomposition holds if π_u is replaced by π_{-u}. The operator $I_G(u)$ may be normalized so that under the identification of the K-representations

$$\pi_u \simeq R(T\backslash K) \simeq \pi_{-u},$$

it acts on each ρ_m by the scalar

$$\lambda_m(u) = \frac{(1-u)(2-u)\cdots(m-u)}{(1+u)(2+u)\cdots(m+u)}.$$

Write $c_{\mathbb{C}}(u) = \frac{\Gamma(1-u)}{\Gamma(1+u)}$. Then we have

$$\lambda_m(u) = c_{\mathbb{C}}(u)^{-1} \frac{\Gamma(m+1-u)}{\Gamma(m+1+u)}. \tag{6}$$

Lemma 2.1. *Let* $\rho_m = Sym^{2m}(\mathbb{C}^2)$ *be the $2m$-th symmetric power of the standard representation of* $K = SU(2)$. *Let* $(\,,\,)$ *be a K-invariant inner product on* ρ_m *and* *v, w vectors in ρ_m of norm one with respect to (,) such that v is invariant under the diagonals T on K and the group $K_H = SO(2)$ acts by the character χ_l on the vector w. Then the formula*

$$| (v, w) | = \frac{2^m \Gamma(\frac{m-l+1}{2})\Gamma(\frac{m+l+1}{2})}{\sqrt{(m-l)!(m+l)!}}$$

holds.

Proof. The formula clearly does not depend on the K-invariant metric chosen, since any two invariant inner products are scalar multiples of each other. We will view elements of ρ_m as homogeneous polynomials of degree $2m$ with complex coefficients in two variables X and Y such that if $k = \begin{pmatrix} \alpha & \beta \\ -\bar{\beta} & \bar{\alpha} \end{pmatrix} \in SU(2)$, then k acts on X and Y by $k(X) = \alpha X - \bar{\beta}Y$ and $k(Y) = \beta X + \bar{\alpha}Y$. The vector $v' = X^m Y^m \in \rho_m$ is invariant under the diagonal subgroup T of $SU(2)$.

The subgroup $K_H = SO(2)$ is conjugate to T by the element $k_0 = \begin{pmatrix} \frac{1}{\sqrt{2}} & \frac{i}{\sqrt{2}} \\ \frac{i}{\sqrt{2}} & \frac{1}{\sqrt{2}} \end{pmatrix}$. That is, $SO(2) = k_0 T k_0^{-1}$. If $-m \le l \le m$, then the element $w'' = X^{m+l}Y^{m-l}$ is an eigenvector for T with eigencharacter $\chi_l : \theta \mapsto e^{4\pi i l \theta}$. Consequently, the vector $w' = k_0(w'')$ is an eigenvector of $SO(2)$ with eigencharacter χ_l.

If

$$f = \sum_{\mu=-m}^{m} a_\mu X^{m+\mu} Y^{m-\mu}, \quad g = \sum_{\mu=-m}^{m} b_\mu X^{m+\mu} Y^{m-\mu} \in \rho_m,$$

then the inner product

$$(f, g) = \sum_{\mu=-m}^{m} a_\mu \bar{b}_\mu (m+\mu)!(m-\mu)!$$

is easily shown to be K-invariant (see p. 44 of [8]). Therefore, the vectors

$$w = \frac{X^m Y^m}{m!}, \quad v = k_0 \left(\frac{X^{m+l} Y^{m-l}}{\sqrt{(m+l)!(m-l)!}} \right) \tag{7}$$

satisfy the conditions of Lemma 2.1. We compute

$$k_0(X^{m+l}Y^{m-l}) = \left(\frac{X+iY}{\sqrt{2}} \right)^{m+l} \left(\frac{iX+Y}{\sqrt{2}} \right)^{m-l}$$

$$= \left(\sum_{a=0}^{m+l} \binom{m+l}{a} X^a (iY)^{m+l-a} \right) \left(\sum_{b=0}^{m-l} \binom{m-l}{b} (iX)^b Y^{m-l-b} \right).$$

Using the fact that the vectors $X^{m+l}Y^{m-l}$ are orthogonal for varying l, we find that the inner product of $X^m Y^m$ with $k_0(X^{m+l}Y^{m-l})$ is the sum (over $a \le m + l$ and $b \le m - l$)

$$\sum_{a+b=m} \frac{(m!)^2}{2^m} i^{m+l-a} i^b \binom{m+l}{a} \binom{m-l}{b}.$$

Lemma 2.2 implies that the absolute value of this sum is equal to

$$\frac{1}{\pi} \frac{m!}{2^m} 4^m \Gamma\left(\frac{m+l+1}{2}\right) \Gamma\left(\frac{m-l+1}{2}\right). \tag{8}$$

if $m + l$ is even and 0 if $m + l$ is odd.

The Lemma follows from (7) and (8). $\square$

Lemma 2.2. *The equality*

$$\frac{m!}{2^m} \sum_{a+b=m} \binom{m+l}{a} \binom{m-l}{b} (-1)^b = \frac{1}{\pi} 2^m \Gamma\left(\frac{m+l+1}{2}\right) \Gamma\left(\frac{m-l+1}{2}\right)$$

holds if $m + l$ is even; the sum on the left-hand side is 0 if $m + l$ is odd.

Proof. If $f(z) = \sum a_k z^k$ is a polynomial with complex coefficients, then the coefficient a_m is given by the formula

$$a_m = \frac{1}{2\pi} \int_0^{2\pi} d\theta f(e^{i\theta}) e^{-im\theta}.$$

The sum Σ on the left-hand side of the statement of the Lemma is clearly ($\frac{m!}{2^m}$ times) the mth-coefficient of the polynomial

$$f(z) = (1+z)^{m+l}(1-z)^{m-l}.$$

We use the foregoing formula for the mth coefficient to deduce that

$$\Sigma = \frac{1}{2\pi} \int_0^{2\pi} d\theta e^{-im\theta} (1 + e^{i\theta})^{m+l} (1 - e^{i\theta})^{m-l}.$$

After a few elementary manipulations, the integral becomes

$$\frac{i^{m-l} 4^m}{\pi} \int_0^2 \pi d\theta \left(\cos\left(\frac{\theta}{2}\right) \sin\left(\frac{\theta}{2}\right)\right)^m \left(\frac{\sin\left(\frac{\theta}{2}\right)}{\cos\left(\frac{\theta}{2}\right)}\right)^l.$$

Substituting $t = \tan(\theta/2)$ the integral becomes

$$\frac{2i^{m-l}4^m}{\pi} \int_0^\infty dt \, \frac{t^{m-l}}{(1+t^2)^{m+1}}$$

and the latter, when multiplied by $\frac{m!}{2^m} = \frac{\Gamma(m+1)}{2^m}$, is the right side of the Lemma 2.2. □

We now collect some estimates for the Gamma function, which will be needed later.

Lemma 2.3. *If $Re(z) > 0$, then we have, as m tends to infinity, the asymptotic relation*

$$\Gamma(m+z) \simeq \text{Constant } m^{m+z-\frac{1}{2}} \cdot \frac{1}{e^m}.$$

In particular, as m tends to infinity through integers,

$$m! = \Gamma(m+1) \simeq \text{Constant } m^{m+\frac{1}{2}} \frac{1}{e^m}.$$

The formula for the inner product in Lemma 2.1 is unchanged if we replace l by $-l$. We may therefore assume that $l \geq 0$. Let $m \geq 0$ and $0 \leq l \leq m$. Put $m = k+l$. From Lemmas 2.3 and 2.1 we obtain (notation as in Lemma 2.1), as m tends to infinity and l is arbitrary, the asymptotic

$$|(v,w)| \simeq \frac{\text{Constant } 2^{k+l}}{(k+2l+1)^{\frac{k+2l+(1/2)}{2}}(k+1)^{\frac{k+(1/2)}{2}}} \left(\frac{k+2l+1}{2}\right)^{\frac{k+2l+1}{2}} \left(\frac{k+1}{2}\right)^{\frac{k}{2}}$$

$$\simeq \frac{\text{Constant}}{(k+2l+1)^{1/4}(k+1)^{1/4}}.$$

Moreover, the constant is independent of l.
This proves:

Lemma 2.4. *Let $m \geq 0$ be an integer and $(\ ,\ )$ a $SU(2)$-invariant inner product on the representation $\rho_m = Sym^{2m}(\mathbb{C}^2)$. Let $0 \leq l \leq m$ and put $m = k+l$. Let v_m a vector of norm 1 in ρ_m invariant under the diagonals T in $SU(2)$ and $w_{m,l} \in \rho_m$ a vector of norm 1 on which $SO(2)$ acts by the character χ_l. We have the following asymptotic as $m = k+l$ tends to infinity:*

$$|(v_m, w_{m,l})| \simeq \frac{\text{Constant}}{(k+2l+1)^{1/4}(k+1)^{1/4}}.$$

Given $m \geq 0$ and $-m \leq l \leq m$, define the function for $k \in K = SU(2)$ by the formula

$$\psi_{m,l}(k) = (v_m, \rho_m(k)w_{m,l}).$$

The functions $\psi_{m,l}$ form a complete orthogonal set for $Rep(T\backslash K)$. The norm of $\psi_{m,l}$ with respect to the L^2 norm on functions K, is, by the Orthogonality Relations for matrix coefficients of ρ_m, equal to $\sqrt{2m+1}$.

Notation: If ψ is a function on K in $Rep(T\backslash K)$, denote by $\|\psi\|_K^2$ the integral (dk is the Haar measure on K)

$$\int_K |\psi(k)|^2 \, dk.$$

Define similarly the number $\|\phi\|_{K_H}^2$ for $\phi \in Rep(K_H)$, where $K_H = SO(2)$.

2.3 The Restriction of $\hat{\pi}_u$ to $SL(2,\mathbb{R})$

The restriction of the function $\psi_{m,l}$ to $K_H = SO(2)$ is, by the choice of the vector $w_{m,l}$, a multiple of the character χ_l: for $k_H \in K_H$, we have

$$\psi_{m,l}(k_H) = \psi_{m,l}(1)\chi_l(k_H).$$

By Lemma 2.4, we have, for $k + l = m$ tending to infinity, the asymptotic

$$|\psi_{m,l}(1)|^2 \simeq \frac{\text{Constant}}{\sqrt{(k+2l+1)(k+1)}}. \tag{9}$$

Let $\frac{1}{2} < u < 1$ and let π_{-u} be the $(\mathfrak{g}, K)$- module of the complementary series representation of $G = SL(2,\mathbb{C})$ as before. Set $u' = 2u-1$. Then $0 < u' < 1$. If $\sigma_{-u'}$ is the $(\mathfrak{h}, K_H)$-module of the complementary series representation of $SL(2,\mathbb{R})$ as before, the restriction of the functions (sections) in π_{-u} on $G/B(\mathbb{C})$ to the subspace $H/B(\mathbb{R})$ lies in $\sigma_{-u'}$, as is easily seen. Denote by

$$\text{res} : \pi_{-u} \to \sigma_{-u'}$$

this restriction of sections.

Note that if $\psi \in \rho_m \subset Rep(T\backslash K) \simeq \pi_{-u}$ (the latter isomorphism is of K modules), then

$$\|\psi\|_{\pi_{-u}}^2 = \frac{1}{\lambda_l(u)} \|\psi\|_K^2. \tag{10}$$

Similarly, if $\phi \in \mathbb{C}\chi_l \subset Rep(\{\pm 1\}\backslash K_H) \simeq \sigma_{-u'}$ (the last isomorphism is of K_H-modules), then

$$\|\phi\|_{\sigma_{-u'}}^2 = \frac{1}{d_l(u')} \|\phi\|_{K_H}^2.$$

Moreover, from (6), (2) and the Stirling approximation for the Gamma function (Lemma 2.3), we have the asymptotic

$$\lambda_m(u) \simeq \frac{\text{Constant}}{m^{2u}}, \quad d_l(u') \simeq \frac{\text{Constant}}{|l|^{2u-1}}, \tag{11}$$

as m and $|l|$ tends to infinity.

Theorem 2.5. *Let* $\frac{1}{2} < u < 1$. *The map*

$$\text{res} : \pi_{-u} \to \sigma_{-(2u-1)}$$

is a continuous map of the unitary for $(\mathfrak{g}, K)$–, *respectively* $(\mathfrak{h}, K_H)$-*modules of the complementary series representations.*

Proof. We must prove the existence of a constant C such that for all $\psi \in \pi_{-u}$, the estimate

$$\| \psi \|^2_{\pi_{-u}} \leq C \| \text{res}(\psi) \|^2_{\sigma_{-(2u-1)}} .$$

The map *res* is equivariant for the action of H and in particular, for the action of K_H. The orthogonality of distinct eigenspaces for K_H implies that we need to only prove this estimate when ψ is an eigenvector for the action of K_H; however, the constant C must be proved to be independent of the eigencharacter.

Assume then that ψ is an eigenvector for K_H with eigencharacter χ_l. The function ψ is a linear combination of the functions $\psi_{m,l}$ ($m \geq |l|$). Write

$$\psi = \sum_{m \geq |l|} x_m \psi_{m,l},$$

where the sum is over a finite set of the m's; the finite set could be arbitrarily large.

The orthogonality of $\psi_{m,l}$ and the equalities in (10) imply

$$\| \psi \|^2_{\pi_{-u}} = \sum_{m \geq |l|} |x_m|^2 \| \psi_{m,l} \|^2_{\pi_{-u}} = \sum |x_m|^2 \frac{1}{\lambda_m(u)} \| \psi_{m,l} \|^2_K .$$

We therefore get, for $\psi \in \pi_{-u}$,

$$\| \psi \|^2_{\pi_{-u}} = \sum_{m \geq |l|} |x_m|^2 \frac{1}{(2m+1)\lambda_m(u)}. \tag{12}$$

We now compute $\mathrm{res}(\psi)$ and its norm. Since ψ is an eigenvector for K_H with eigencharacter χ_l, we have

$$\mathrm{res}(\psi) = \psi(1)\chi_l = \left(\sum_{m \geq |l|} x_m \psi_{m,l}(1) \right) \chi_l.$$

Therefore,

$$\| \mathrm{res}(\psi) \|_{\sigma_{-(2u-1)}}^2 = \left| \left(\sum x_m \psi_{m,l}(1) \right) \right|^2 \frac{1}{d_l(2u-1)}.$$

The Cauchy–Schwartz inequality implies

$$\| \mathrm{res}(\psi) \|_{\sigma_{-(2u-1)}}^2$$

$$\leq \left(\sum | x_m |^2 \frac{1}{\lambda_m(u)(2m+1)} \right) \left(\sum (2m+1)\lambda_m(u) | \psi_{m,l}(1) |^2 \right) \frac{1}{d_l(u')}.$$

Assume for convenience that $l \geq 0$. Put $k = m + l$. Then $k \geq 0$. The estimate (9) and the equality (12) imply that (write σ for $\sigma_{-(2u-1)}$ and π for π_{-u}),

$$\| \mathrm{res}(\psi) \|_{\sigma}^2 \leq \| \psi \|_{\pi}^2 \left(\sum_{k \geq 0} \frac{2k + 2l + 1}{\sqrt{(k+2l+1)(k+1)}} \lambda_{k+l}(u) \frac{1}{d_l(u')} \right).$$

Let Σ denote the sum in brackets in the above equation. To prove Theorem 2.5, we must show that Σ is bounded above by a constant independent of l. We now use the asymptotic (11) to get a constant C such that

$$\Sigma \leq C \sum_{k \geq 0} \frac{2k + 2l + 1}{\sqrt{(k+2l+1)(k+1)}} \frac{l^{2u-1}}{(k+l)^{2u}}.$$

This is a *decreasing* series in k and therefore bounded above by the sum of the $k = 0$ term and the integral

$$\int_0^{\infty} dk \, \frac{2k + 2l + 1}{\sqrt{(k+2l+1)(k+1)}} \frac{l^{2u-1}}{(k+l)^{2u}}.$$

We first compute the $k = 0$ term: this is

$$\frac{2l+1}{\sqrt{(2l+1)}} \frac{l^{2u-1}}{l^{2u}} \leq \frac{2}{\sqrt{2l+1}},$$

which therefore tends to 0 for large l and is bounded for all l.

To estimate the integral, we first change the variable from k to kl. The integral becomes

$$\int_0^\infty l\,dk\,\frac{2kl+2l+1}{\sqrt{(kl+2l+1)(kl+1)}}\,\frac{l^{2u-1}}{(kl+l)^{2u}}$$

$$\leq \int_0^\infty dk\,\frac{2k+3}{\sqrt{(k+2)(k)}}\,\frac{1}{(k+1)^{2u}},$$

and since $2u > 1$, the latter integral is finite (and is independent of l).

We have therefore checked that both the $k = 0$ term and the integral are bounded by constants independent of l and this proves Theorem 2.5.   □

Corollary 1. *Let $\frac{1}{2} < u < 1$. If $u' = 2u - 1$, then $\widehat{\sigma}_{u'}$ is a direct summand of $\widehat{\pi}_u$ restricted to $SL(2,\mathbb{R})$.*

Proof. We may replace π_u and $\sigma_{u'}$ by the isomorphic (and isometric) modules π_{-u} and $\sigma_{-u'}$. By Theorem 2.5, the restriction map $\pi_{-u} \rightarrow \sigma_{-u'}$ is continuous and extends to the completions. Hence, $\widehat{\pi}_{-u}$ is, as a representation of $SL(2,\mathbb{R})$, the direct sum of the kernel of this restriction map and of $\widehat{\sigma}_{-u'}$. This completes the proof.   □

Remark 1. This corollary is proved in [6]; the proof in this paper is that the "abstract" projection map is realized as a simple geometric map of sections of a line bundle on the flag varieties of $G = SL(2, \mathbb{C})$ and $H = SL(2, \mathbb{R})$.

3 Branching Laws for the Steinberg Representation

Let $G = SL_2(\mathbb{C})$ and $H = SL_2(\mathbb{R})$. and let $\widehat{\pi}$ be the Steinberg Representation of G (For a definition see Sect. 1.).

3.1 The Representation $\tilde{\pi}_0$ and a G-Invariant Linear Form

Consider the representation $\tilde{\pi}_0 = ind_B^G(\rho^2)$. In this equality, *ind* refers to **nonunitary** induction and $\tilde{\pi}_o$ is the space of all continuous complex valued functions on G such that for all $g \in G$ and $man \in MAN = B$, we have

$$\phi(mang) = \rho^2(a)\phi(g).$$

Here, ρ^2 is the product of all the positive roots of the split torus A occurring in the Lie algebra of the unipotent radical N of B and M is a maximal compact subgroup of the centralizer of A in G.

Now, $\tilde{\pi}_0$ a nonunitary representation and has a G-invariant linear form L defined on it as follows. The map $C_c(G) \rightarrow \tilde{\pi}_0$ given by integration with respect to a **left** invariant Haar measure on B is surjective. Given an element $\phi \in \tilde{\pi}_0$ select any function $\phi^* \in C_c(G)$ in the preimage of ϕ and define $L(\phi)$ as the integral of ϕ^* with respect to the Haar measure on G. This is well defined (i.e., independent of the function ϕ^* chosen) and yields a linear form L. Moreover, if a function $\phi \in \tilde{\pi}_0$ is a positive function on G, then $L(\phi)$ is positive.

Under the action of the subgroup H on the G-space G/B, the space G/B has three disjoint orbits: the upper half plane $\mathbb{H}^+$, the lower half plane $\mathbb{H}^-$ and the space $H/B \cap H$. The upper and lower half planes form open orbits. Given a function $\phi \in C_c(\mathbb{H}^+)$, we may view it as a function in π_0 as follows. The restriction of the character ρ^2 to the maximal compact subgroup of H is trivial; therefore, the restriction of any element of $\tilde{\pi}_0$ to H yields a function on $\mathbb{H}^+$ and also on $\mathbb{H}^-$. Conversely, given $\phi \in C_c(\mathbb{H}^+)$, extend ϕ by zero outside $\mathbb{H}^+$; we get a function, which we will again denote by ϕ, in $\tilde{\pi}_0$. The linear form L applied to $C_c(\mathbb{H}^+)$ yields a positive linear functional, which is H-invariant. Hence, the positive linear functional L is a Haar measure on $\mathbb{H}^+$, respectively on $\mathbb{H}^-$.

3.2   The Metric on the Steinberg Representation of G

Consider the Steinberg representation $\tilde{\pi} = Ind_B^G(\chi)$. Here, Ind refers to **unitary** induction from a unitary character χ of B and again we consider only continuous functions. Given two functions $f, f' \in \tilde{\pi}$, the product $\phi = f\overline{f'}$ ($\overline{f'}$ is the complex conjugate of f') lies in $\tilde{\pi}_0$. The linear form L applied to ϕ gives a pairing

$$\langle f, f' \rangle = L(f\overline{f'})$$

on $\tilde{\pi}$ which is clearly G-invariant. This is the G-invariant inner product on $\tilde{\pi}$.

Given a compactly supported function f on H, which under the left action of $K \cap H$ acts via the restriction of a character χ to $K \cap H$, we extend it by zero to an element of π. Then the inner product $< f, f >$ is, by the conclusion of the last paragraph in (2.1), just the Haar integral on H applied to the function $\mid f \mid^2 \in C_c(\mathbb{H}^+)$. Consequently, the metric on $\tilde{\pi}$ restricted to $C_c(H) \cap \tilde{\pi}$ is just the restriction of the L^2-metric on $C_c(H)$.

Remark 2. We know that the Steinberg representation $\hat{\pi}$ of G is tempered and is induced by a unitary character from the Borel subgroup of upper triangular matrices. The tempered dual of G does not contain isolated points, since G does not have discrete series representations. Moreover, the entire tempered dual is automorphic [3]. Consequently, the Steinberg representation, which has nontrivial $(\mathfrak{g}, K)$– cohomology, is not isolated in the automorphic dual of G.

3.3   Decomposition of the Steinberg Representation $\widehat{\pi}$

Proposition 3.1. *The restriction to H of $\widehat{\pi}$ contains the Steinberg representation of H. More precisely, the restriction is a sum of the Steinberg representation $\widehat{\sigma}$ of H, and a sum of two copies of $L^2(H/K \cap H)$, where $K \cap H$ is a maximal compact subgroup of H.*

Proof. The Steinberg Representation $\widehat{\pi}$ is unitarily induced from a **unitary** character χ of the Borel Subgroup $B = B(\mathbb{C})$ of upper triangular matrices in $G = SL_2(\mathbb{C})$. Recall that the group H has three orbits the space G/B; the upper half plane $\mathbb{H}^+$, the lower half plane $\mathbb{H}^-$ and the projective line $\mathbf{P}^1(\mathbb{R})$ over $\mathbb{R}$. The first two are open orbits and $\mathbf{P}^1(\mathbb{R})$ has zero measure in G/B. From this, it is clear from Sect. 3.2, that $\widehat{\pi}$ is the direct sum of $L^2(\mathbb{H}^+, \chi_{K\cap H})$ and $L^2(\mathbb{H}^-, \chi^*_{K\cap H})$, where the subscript denotes the restriction of the character χ to the subgroup $K \cap H$ and χ^* denotes the complex conjugate of χ.

The representations χ and χ^* are such that their restrictions to $K \cap H$ are minimal K-types of holomorphic, respectively, antiholomorphic discrete series representations of $H = SL(2, \mathbb{R})$. The space $L^2(\mathbb{H}^+, \chi_{K\cap H}) \oplus L^2(\mathbb{H}^-, \chi^*_{K\cap H})$ is therefore a direct sum of the Steinberg representation $\widehat{\sigma}$ and 2 copies of the full unramified tempered spectrum, since any unramified representation contains $\chi_{K\cap H}$ and $\chi^*_{K\cap H}$ as a $K \cap H$-types.

The Proposition now follows immediately.   □

Remark 3. The Steinberg representation $\widehat{\pi}$ is unitarily induced from the unitary character χ. Thus, it is nonunitarily induced from the character $\delta_\mathbb{C}\chi$ whose restriction to $B(\mathbb{R})$ is $\delta_\mathbb{R}^2$. Here, $\delta_\mathbb{R}^2$ denotes the character by which the split torus $S(\mathbb{R})$ acts on the Lie algebra of the unipotent radical of $B(\mathbb{R})$. Similarly, $\delta_\mathbb{C}^2$ denotes the **square** of the character by which the split real torus in $S(\mathbb{C})$ acts on the complex Lie algebra of the unipotent radical of $B(\mathbb{C})$.

The proposition was proved by restricting $\widetilde{\pi}$ to the open orbits; we may instead restrict $\widetilde{\pi}$ to the **closed** orbit $G(\mathbb{R})/B(\mathbb{R})$. We thus get a surjection of $\widetilde{\pi}$ onto the space of $K \cap H$-finite sections of the line bundle on $G(\mathbb{R})/B(\mathbb{R})$, which is induced from the character $\delta_\mathbb{R}^2$ on $B(\mathbb{R})$.

The latter representation contains the trivial representation as a quotient. We have therefore obtained that the trivial representation is a quotient of the restriction of $\widetilde{\pi}$ to the subgroup $SL_2(\mathbb{R})$. This shows that there is a mapping of the $(\mathfrak{h}, K \cap H)$-modules of the restriction of π to $\mathfrak{h}$ onto the trivial module of H; however, this map does not extend to a map of the corresponding Hilbert spaces, since the Howe–Moore Theorem implies that the matrix coefficients of $\widehat{\pi}$ restricted to the noncompact subgroup H must tend to zero at infinity.

3.4 A Generalization

Suppose that $G_1 = SO(2m + 1, 1)$ and $H_1 = SO(2m, 1)$ and let $\widehat{\Pi}_m$ be the unitary irreducible representation of G_1, which has nonzero cohomology in degree m, and vanishing cohomology in lower degrees. Then $\widehat{\Pi}_m$ is a tempered representation [1]. Let $\widehat{\Sigma}_m$ be the unitary representation of H_1 which has nontrivial cohomology in degree m and vanishing cohomology below that. Then $\widehat{\Sigma}_m$ is a discrete series representation [1]. Following the proof of Proposition 3.1, we obtain the following proposition.

Proposition 3.2. *The representation $\widehat{\Sigma}_m$ is a direct summand of the restriction of the G_1–representation $\widehat{\Pi}_m$ to H_1.*

Remark 4. If $G_1 = SO(2m + 1, 1)$, then G_1 hs no compact Cartan subgroup, and hence $L^2(G_1)$ does not have discrete spectrum. Let Γ be an arithmetic (congruence) subgroup of G. The notion of "automorphic spectrum" of G_1 with respect to the $\mathbb{Q}$-structure associated with Γ was defined by Burger and Sarnak. [3] Since all the tempered dual of G is automorphic [3], it follows that the representation $\widehat{\Pi}_m$ is not isolated in the automorphic spectrum of G_1. Thus, representations with nontrivial cohomology may not be isolated in the automorphic dual.

3.5 Functions in $\sigma \subset \widehat{\pi}$

Denote by σ the space of $K \cap H$-finite functions in the Steinberg representation $\widehat{\sigma}$ of $SL_2(\mathbb{R})$. By Proposition 3.1, this space of functions restricts trivially to the lower half plane. Moreover, in the space of L^2-functions on the upper half plane, the representation $\widehat{\sigma}$ occurs with multiplicity one. In this subsection, we describe explicitly, elements in σ viewed as functions on the upper half plane.

We will now replace $H = SL_2(\mathbb{R})$ with the subgroup $SU(1, 1)$ of $G = SL_2(\mathbb{C})$. Since $SU(1, 1)$ is conjugate to H, this does not affect the statement and proof of Proposition 3.1. The upper and lower half planes are then replaced, respectively, by the open unit ball in $\mathbb{C}$ and the complement of the closed unit ball in $\mathbb{P}^1(\mathbb{C})$. With this notation, elements of σ are now thought of as functions on $SU(1, 1)$ with the equivariance property

$$f(ht) = \chi(t)f(h) \forall t \in K \cap H, \forall h \in SU(1, 1).$$

Some functions in of σ are explicitly described in [5] (Chap. IX, Sect. 2, Theorem 1 in p. 181 of Lang with $m = 2$). The eigenvectors of $K \cap H$ in one summand of σ are

$$\phi_{2+2r} = \alpha^{-2} \left(\frac{\beta}{\alpha} \right)^r,$$

with $r = 0, 1, 2, \cdots$. In this formula, an element of $SU(1, 1)$ is of the form

$$\begin{pmatrix} \alpha & \beta \\ \bar{\beta} & \bar{\alpha} \end{pmatrix},$$

with $\alpha, \beta \in \mathbb{C}$ such that

$$| \alpha |^2 - | \beta |^2 = 1.$$

These functions span the $(\mathfrak{h}, K \cap H)$-modules D.

Furthermore, the function ϕ_2 vanishes on the complement of the closed disc. That is, if $g = \begin{pmatrix} a & b \\ c & d \end{pmatrix}$ with $| \frac{c}{d} | > 1$, then $\phi_2(g) = 0$.

It follows from the last two paragraphs that if $g = \begin{pmatrix} a & b \\ c & d \end{pmatrix} \in SL_2(\mathbb{C})$, then one of the following two conditions hold:

Proposition 3.3. *If* $| \frac{d}{c} | < 1$, *then for any matrix* $h = \begin{pmatrix} \alpha & \beta \\ \bar{\beta} & \bar{\alpha} \end{pmatrix} \in SU(1, 1)$ *with* $(\infty)h = (\infty)g$ *(the inequality satisfied by g ensures that there exists an h with this property), we have*

$$\phi_2(g) = \alpha^{-2} \frac{1}{| d |^2 - | c |^2}.$$

If $| \frac{d}{c} | > 1$, *then* $\phi_2(g) = 0$.

Proof. The points on the open unit disc are obtained as translates of the point at infinity by an element of $SU(1, 1)$. Therefore, if $\frac{d}{c}$ has modulus less than one, there exists an element $h \in SU(1, 1)$ such that $(\infty)g = \frac{d}{c} = \infty(h)$. This means that

$$g = \begin{pmatrix} u & n \\ 0 & u^{-1} \end{pmatrix} h$$

for some element $b = \begin{pmatrix} u & n \\ 0 & u^{-1} \end{pmatrix} \in SL_2(\mathbb{C})$ (elements of type b form the isotropy subgroup of G at infinity).

The intersection of the isotropy at infinity with $SU(1, 1)$ is the space of diagonal matrices whose entries have absolute value one. Therefore, we may assume that the entry u above of the matrix b is real and positive. Then it follows that

$$\chi\delta(b) = u^2 = \frac{1}{| d |^2 - | c |^2},$$

and this proves the first part of the proposition.

The second part was already proved, as we noted that the restriction of the functions in the discrete series representations to the complement of the closed unit disc vanishes.   □

Consider the decomposition

$$\widehat{\pi} = \widehat{\sigma} \oplus L^2(K \cap H \backslash H) \oplus L^2(K \cap H \backslash H)$$

of $\widehat{\pi}$ as a representation of the group H and recall that $\widehat{\sigma}$ is a direct sum of discrete series representations $\widehat{D} \oplus \widehat{\overline{D}}$. It can be proved that the space π^∞ of smooth vectors for the action of $G = SL_2(\mathbb{C})$ is simply the space of smooth functions on G which lie in $\widehat{\pi}$, by proving the corresponding statement for the maximal compact subgroup $K = SU(2)$ of G. A natural question that arises is whether the $(\mathfrak{h}, K \cap H)$-module σ contains any smooth vectors in $\widehat{\pi}$. We answer this in the negative.

Proposition 3.4. *The intersection*

$$\widehat{\sigma} \cap \pi^\infty = 0.$$

That is, $\widehat{\sigma}$ does not contain any nonzero smooth vectors in $\widehat{\pi}$.

Proof. We will show the proposition for $\widehat{D}$. The proof for $\widehat{\overline{D}}$ is similar.

The intersection in the proposition is stable under H and hence under the maximal compact subgroup $K \cap H$. If the intersection is nonzero, then it contains nonzero $K \cap H$-finite vectors. The space D of $K \cap H$-finite vectors is irreducible as a $(\mathfrak{h}, K \cap H)$-module. Therefore, the space of smooth vectors in $\widehat{D}$ contains all of D and in particular, contains the function $f = \phi_2$ introduced above. That is, the function ϕ_2 is smooth on G (and hence on K).

We will now view ϕ_2 as a function on the group

$$SO(2) = \left\{ k_\theta = \begin{pmatrix} \cos \theta & \sin \theta \\ -\sin \theta & \cos \theta \end{pmatrix} : 0 \le \theta \le 2\pi \right\}.$$

If $\left| \frac{\cos \theta}{-\sin \theta} \right| < 1$, then there exists a real number t such that

$$\frac{\cos \theta}{-\sin \theta} = \frac{\cosh(t)}{\sinh(t)}.$$

By Proposition 3.3,

$$\phi_2(k_\theta) = \alpha^{-2} u^{-2} = \cosh(t)^{-2} \frac{1}{\cos^2 \theta - \sin^2 \theta}.$$

Moreover, it follows from the fact that $h = bg$ (in the notation of Proposition 3.3) that $u^{-1} \cosh(t) = \cos\theta$ and hence that $\cosh^2(t)u^{-2} = \cos\theta^{-2}$. We have then:

$$\phi_2(k_\theta) = \frac{1}{\cos^2\theta}$$

if $0 < \theta < \pi/4$ and 0 if $\pi/4 < \theta < \pi/2$. This contradicts the smoothness of ϕ_2 as a function of θ and proves the Proposition. 3.4.   □

Remark 5. The Proposition shows in particular that although the **completion** of the Steinberg module of $SL(2, \mathbb{C})$ contains discretely the completion of the Steinberg module of $SL(2, \mathbb{R})$, this decomposition does not hold at the level of K-finite vectors. This also follows from the results of Kobayashi (see [4]),

Acknowledgments B. Speh was partially supported by NSF grant DMS 0070561. T.N Venkataramana was supported by the J.C Bose Fellowship for the period 2008-2013.

References

1. [BW] A. Borel and N.Wallach, *Continuous cohomology, discrete groups and representations of reductive groups* Edition 2, AMS Bookstore, 2000.
2. [B] C. Boyer, On the Supplementary series of $SO_0(p, 1)$, J. Mathematical Phys., **14**, No. 5, (1973), 609-617.
3. [BS] M. Burger and P. Sarnak, Ramanujan Duals (II), Invent. Math. **106** (1991), no. 1, 1-11.
4. [K] T. Kobayashi, Restrictions of Unitary Representations of Real Reductive Groups, *Lie Theory, Unitary Representations and Compactifications of Symmetric Spaces*, Progress in Mathematics 229, Birkhauser 2005.
5. [L] S. Lang, $SL(2, \mathbb{R})$, Addison Wesley Publishing Company, Reading, Mass.-London-Amsterdam, 1975.
6. [M] N. Mukunda, Unitary Representations of the Lorentz Groups: reduction of the supplementary series under a non-compact subgroup, J. Mathematical Phys., **9** (1968), 417-431.
7. [SV] B. Speh and T.N. Venkataramana, Discrete Components of some complementary series, to appear in Forum Mathematicum
8. [S] M. Suguira, Unitary Representations and Harmonic Analysis, Kodansha Ltd., Tokyo, 1975.

Asympotics of Spherical Functions For Large Rank: An Introduction

Jacques Faraut

Abstract *We present the scheme developed by Okounkov and Olshanski for studying asymptotics of spherical functions on a compact symmetric space as the rank goes to infinity. The method is explained in the special case of the unitary group, and results are stated in the general case.*

Keywords Spherical function • Symmetric space • Schur function • Jack polynomial • Jacobi polynomial

Mathematics Subject Classification (2010): 43A90, 43A75, 53C35, 33C52

This paper has been written following a talk given as an introduction to the work of Okounkov and Olshanski about asymptotics of spherical functions for compact symmetric spaces as the rank goes to infinity. This topic belongs to the asymptotic harmonic analysis, *i.e.*, the study of the asymptotics of functions related to the harmonic analysis on groups or homogeneous spaces as the dimension goes to infinity. Such questions have been considered before, for instance, by Krein and Schoenberg for Euclidean spaces, spheres and real hyperbolic spaces, which are Riemannian symmetric spaces of rank one. The behavior is very different when the rank is unbounded, and new phenomenons arise in that case.

In this introductory paper, we present the scheme developed by Okounkov and Olshanski for studying limits of spherical functions on a compact symmetric space

J. Faraut (✉)
Institut de Mathématiques de Jussieu, Université Pierre et Marie Curie, 4 place Jussieu, case 247, 75252 Paris cedex, France
e-mail: faraut@math.jussieu.fr

B. Krötz et al. (eds.), *Representation Theory, Complex Analysis, and Integral Geometry*,   251
DOI 10.1007/978-0-8176-4817-6_10, © Springer Science+Business Media, LLC 2012

$G(n)/K(n)$ as the rank n goes to infinity. These limits are identified as spherical functions for the Olshanski spherical pair (G, K), with

$$G = \bigcup_{n=1}^{\infty} G(n), \quad K = \bigcup_{n=1}^{\infty} K(n).$$

We will explain results and methods in the special case of the unitary groups $U(n)$. This amounts to studying asymptotics of Schur functions. The proof uses a binomial formula for Schur functions involving shifted Schur functions. This presentation is based on two papers: [Okounkov-Olshanski, 1998c], for the type A, and [Okounkov-Olshanski, 2006], for the type BC. The case of the unitary groups have been considered by Vershik and Kerov, following a slightly different method ([1982]).

In Sect. 5, we present without proof general results by Okounkov and Olshanski for series of classical compact symmetric spaces, and finally in Sect. 6 we consider the cases for which there is a determinantal formula for the spherical functions.

1 Olshanski Spherical Pairs

Let us recall first what is a spherical function for a Gelfand pair. A pair (G, K), where G is a locally compact group, and K a compact subgroup, is said to be a *Gelfand pair* if the convolution algebra $L^1(K \backslash G / K)$ of K-biinvariant integrable functions on G is commutative. Fix now a Gelfand pair (G, K). A *spherical function* is a continuous function φ on G which is K-biinvariant, $\varphi(e) = 1$, and satisfies the functional equation

$$\int_K \varphi(xky)\alpha(dk) = \varphi(x)\varphi(y) \quad (x, y \in G),$$

where α is the normalized Haar measure on the compact group K. The characters χ of the commutative Banach algebra $L^1(K \backslash G / K)$ are of the form

$$\chi(f) = \int_G f(x)\varphi(x)m(dx),$$

where φ is a bounded spherical function (m is a Haar measure on the group G, which is unimodular since (G, K) is a Gelfand pair).

If the spherical function φ is of positive type (i.e., positive definite), there is an irreducible unitary representation $(\pi, \mathcal{H})$ with $\dim \mathcal{H}^K = 1$, where $\mathcal{H}^K$ denotes the subspace of K-invariant vectors in $\mathcal{H}$, such that

$$\varphi(x) = (u|\pi(x)u),$$

with $u \in \mathcal{H}^K$, $\|u\| = 1$. The representation $(\pi, \mathcal{H})$ is unique up to equivalence. An irreducible unitary representation $(\pi, \mathcal{H})$ with $\dim \mathcal{H}^K = 1$ is said to be *spherical*, and the set Ω of equivalence classes of spherical representations will be called the *spherical dual* for the pair (G, K). Equivalently Ω is the set of spherical functions of positive type. We will denote the spherical functions of positive type for the Gelfand pair (G, K) $\varphi(\lambda; x)$ $(\lambda \in \Omega, x \in G)$.

Consider now an increasing sequence of Gelfand pairs $\big(G(n), K(n)\big)$:

$$G(n) \subset G(n+1), \quad K(n) \subset K(n+1), \quad K(n) = G(n) \cap K(n+1),$$

and define

$$G = \bigcup_{n=1}^{\infty} G(n), \quad K = \bigcup_{n=1}^{\infty} K(n).$$

We say that (G, K) is an *Olshanski spherical pair*. A *spherical function* for the Olshanski spherical pair (G, K) is a continuous function φ on G, $\varphi(e) = 1$, which is K-biinvariant and satisfies

$$\lim_{n \to \infty} \int_{K(n)} \varphi(xky)\alpha_n(dk) = \varphi(x)\varphi(y) \quad (x, y \in G),$$

where α_n is the normalized Haar measure on $K(n)$. As in the case of a Gelfand pair, if φ is a spherical function of positive type, there exists a spherical representation $(\pi, \mathcal{H})$ of G (*i.e.*, irreducible, unitary, with $\dim \mathcal{H}^K = 1$) such that

$$\varphi(x) = \big(u | \pi(x)u\big),$$

with $u \in \mathcal{H}^K$, $\|u\| = 1$. In the same way, the spherical dual Ω is identified with the set of spherical functions of positive type. Such a function will be written $\varphi(\omega; x)$ $(\omega \in \Omega, x \in G)$.

On Ω, seen as the set of spherical functions of positive type, we will consider the topology of uniform convergence on compact sets.

We will consider the following question. Let Ω_n be the spherical dual for the Gelfand pair $\big(G(n), K(n)\big)$, and let us write a spherical function of positive type for $\big(G(n), K(n)\big)$ as $\varphi_n(\lambda, x)$ $(\lambda \in \Omega_n, x \in G(n))$. For which sequences $(\lambda^{(n)})$, with $\lambda^{(n)} \in \Omega_n$, does there exist $\omega \in \Omega$ such that

$$\lim_{n \to \infty} \varphi_n(\lambda^{(n)}; x) = \varphi(\omega; x) \quad (x \in G) \, ?$$

In the cases we will consider, there is, for each n, a map

$$T_n : \Omega_n \to \Omega,$$

such that, if

$$\lim_{n\to\infty} T_n(\lambda^{(n)}) = \omega,$$

for the topology of Ω, then

$$\lim_{n\to\infty} \varphi_n(\lambda^{(n)}; x) = \varphi(\omega; x).$$

It is said that $(\lambda^{(n)})$ is a *Vershik–Kerov sequence*.

2   The Unitary Group

For a compact group U, we consider the pair

$$G = U \times U, \quad K = \{(u, u) \mid u \in U\} \simeq U.$$

Then, $G/K \simeq U$. A K-biinvariant function f on G is identified to a central function φ on U by

$$f(x, y) = \varphi(xy^{-1}).$$

The convolution algebra $L^1(K\backslash G/K)$ is isomorphic to the convolution algebra $L^1(U)_{\text{central}}$ of central integrable functions on U, which is commutative. Hence, (G, K) is a Gelfand pair. We will say that a continuous central function φ is spherical if $\varphi(e) = 1$, and

$$\int_U \varphi(xuyu^{-1})\alpha(du) = \varphi(x)\varphi(y) \quad (x, y \in U),$$

where α is the normalized Haar measure on U. In fact, it amounts to saying that the corresponding function f on G is spherical for the Gelfand pair (G, K).

If $(\pi, \mathcal{H})$ is an irreducible representation of U, then the normalized character

$$\varphi(u) = \frac{\chi_\pi(u)}{\chi_\pi(e)}, \quad \chi_\pi(u) = \text{tr}\big(\pi(u)\big),$$

is a spherical function, and all spherical functions are of that form. Hence the spherical dual Ω for the pair (G, K) is the dual $\hat{U}$ of the compact group U.

For $U = U(n)$, the unitary group, the spherical dual $\Omega_n = \widehat{U(n)}$ is identified to the set of signatures

$$\lambda = (\lambda_1, \ldots, \lambda_n), \ \lambda_i \in \mathbb{Z}, \ \lambda_1 \geq \cdots \geq \lambda_n.$$

The character χ_λ of an irreducible representation in the class λ is given by a Schur function. Define, for $t = (t_1, \ldots, t_n) \in (\mathbb{C}^*)^n$, $\alpha = (\alpha_1, \ldots, \alpha_n) \in \mathbb{Z}^n$,

$$A_\alpha(t) = \det(t_j^{\alpha_i}).$$

For a signature λ, the Schur function s_λ is given by

$$s_\lambda(t) = \frac{A_{\lambda+\delta}(t)}{V(t)},$$

where $\delta = (n-1, n-2, \ldots, 1, 0)$, $V(t) = A_\delta(t)$ is the Vandermonde determinant:

$$V(t) = \prod_{i<j}(t_i - t_j).$$

For a diagonal matrix, $u = \operatorname{diag}(t_1, \ldots, t_n)$,

$$\chi_\lambda(u) = s_\lambda(t).$$

3 The Infinite Dimensional Unitary Group

The infinite dimensional unitary group $U(\infty)$ is defined as

$$U(\infty) = \bigcup_{n=1}^{\infty} U(n).$$

One associates to $U(\infty)$ the following inductive limit of Gelfand pairs:

$$G(n) = U(n) \times U(n), \quad K(n) = \{(u, u) \mid u \in U(n)\},$$

$$G = \bigcup_{n=1}^{\infty} G(n) = U(\infty) \times U(\infty),$$

$$K = \bigcup_{n=1}^{\infty} K(n) = \{(u, u) \mid u \in U(\infty)\}.$$

Let us first state the following result by Voiculescu [1976]. Consider a power series

$$\Phi(t) = \sum_{m=0}^{\infty} c_m t^m,$$

with

$$c_m \geq 0, \quad \Phi(1) = \sum_{m=0}^{\infty} c_m = 1, \quad |t| \leq 1.$$

Define the function φ on $U(\infty)$ by

$$\varphi(g) = \det \Phi(g).$$

This means that the function φ is central, and, if $g = \text{diag}(t_1, \ldots, t_n, 1, \ldots)$, then

$$\varphi(g) = \Phi(t_1) \ldots \Phi(t_n).$$

Theorem 3.1 (Voiculescu, 1976). *The function φ is of positive type if and only if Φ has the following form:*

$$\Phi(t) = e^{\gamma(t-1)} \prod_{k=1}^{\infty} \frac{1 + \beta_k(t-1)}{1 - \alpha_k(t-1)},$$

with

$$\alpha_k \geq 0, \ 0 \leq \beta_k \leq 1, \ \gamma \geq 0, \ \sum_{k=1}^{\infty} (\alpha_k + \beta_k) < \infty.$$

We propose to call such a function a *Voiculescu function*. Let Ω_0 be the set of triples $\omega = (\alpha, \beta, \gamma)$ as above. We will write

$$\Phi(t) = \Phi(\omega; t),$$

and consider on Ω_0 the topology corresponding to the uniform convergence of the functions $\Phi(\omega; \cdot)$ on the unit circle. This topology can be expressed in terms of the parameters α, β, γ as follows: for a continuous function u on $\mathbb{R}$, put

$$L_u(\omega) = \sum_{k=1}^{\infty} \alpha_k u(\alpha_k) + \sum_{k=1}^{\infty} \beta_k u(-\beta_k) + \gamma u(0).$$

Then the topology of Ω_0 coincides with the initial topology defined by the functions L_u (i.e., the coarser topology for which all the functions L_u are continuous).

The Voiculescu function $\Phi(\omega; t)$ is meromorphic in t, with poles $1 + \frac{1}{\alpha_k}$. It is holomorphic in the disc $|t| < r$, with $r = 1 + \inf \frac{1}{\alpha_k}$. Its logarithmic derivative is holomorphic near 1:

$$\frac{d}{dz} \log \Phi(\omega; 1 + z) = \sum_{m=0}^{\infty} a_m z^m,$$

with

$$a_0 = \gamma + \sum_{k=1}^{\infty} \alpha_k + \sum_{k=1}^{\infty} \beta_k,$$

$$a_m = \sum_{k=1}^{\infty} \alpha_k^{m+1} + (-1)^m \sum_{k=1}^{\infty} \beta_k^{m+1}, \quad m \geq 1.$$

Observe that

$$a_m = L_{u_m}(\omega) \quad \text{with} \quad u_m(s) = s^m.$$

Theorem 3.2. *The spherical functions of positive type on $U(\infty)$ are the following ones:*

$$\varphi(\omega^+, \omega^-; g) = \det \Phi(\omega^+; g) \det \Phi(\omega^-; g^{-1}),$$

with ω^+, $\omega^- \in \Omega_0$.

[Vershik-Kerov, 1982], [Boyer, 1983].

Hence, the spherical dual of the Olshanski spherical pair (G, K) associated to $U(\infty)$ is the set $\Omega = \Omega_0 \times \Omega_0$ of pairs (ω^+, ω^-).

We will now describe the sequences of signatures $(\lambda^{(n)})$ with

$$\lambda^{(n)} = \left(\lambda_1^{(n)}, \ldots, \lambda_n^{(n)} \right) \in \Omega_n,$$

for which there exists $\omega = (\omega^+, \omega^-)$ such that

$$\lim_{n \to \infty} \varphi_n(\lambda^{(n)}; g) = \varphi(\omega^+, \omega^-; g).$$

We will first consider the case of positive signatures. We say that a signature λ is positive if the numbers λ_i are ≥ 0, and we will denote by Ω_n^+ the set of positive signatures in Ω_n. One defines the Frobenius parameters $a = (a_i)$ and $b = (b_i)$ of a positive signature λ as follows:

$$a_i = \lambda_i - i \quad \text{if } \lambda_i > i, \ a_i = 0 \text{ otherwise,}$$

$$b_j = \lambda'_j - j + 1 \quad \text{if } \lambda'_j > j - 1, \ b_j = 0 \text{ otherwise,}$$

where λ' is the transpose signature. For instance, if $\lambda = (6, 4, 4, 2, 1)$, then $a = (5, 2, 1, 0, 0)$, $b = (5, 3, 1, 0, 0)$.

We define the map

$$T_n : \Omega_n^+ \to \Omega_0, \quad \lambda \mapsto \omega = (\alpha, \beta, \gamma),$$

by

$$\alpha_k = \frac{a_k}{n}, \quad \beta_k = \frac{b_k}{n}, \quad \gamma = 0.$$

Theorem 3.3. *Let* $\lambda^{(n)} = (\lambda_1^{(n)}, \dots, \lambda_n^{(n)})$ *be a sequence of positive signatures. Assume that*

$$\lim_{n \to \infty} T_n(\lambda^{(n)}) = \omega,$$

for the topology of Ω_0. *Then, for* $g \in U(\infty)$,

$$\lim_{n \to \infty} \varphi_n(\lambda^{(n)}; g) = \det \Phi(\omega; g),$$

uniformly on each $U(k)$.

[Vershik-Kerov, 1982], [Okounkov-Olshanski, 1998c].

Example. For two numbers $p, k \in \mathbb{N}$ with $p \geq k$, consider the positive signature

$$\lambda = (p, \dots, p, 0, \dots),$$

where p is repeated k times. The Young diagram of λ is a rectangle with sides p and k. The Frobenius parameters are $a = (a_i)$ with

$$a_i = p - i \quad \text{if } i \leq k, \quad a_i = 0 \quad \text{if } i > k,$$

and $b = (b_j)$ with

$$b_j = k - j + 1 \quad \text{if } j \leq k, \quad b_j = 0 \quad \text{if } j > k.$$

Observe that

$$\sum a_i + \sum b_j = kp.$$

For a continuous function u on $\mathbb{R}$,

$$L_u(T_n(\lambda)) = \sum_{i=1}^{k} \frac{p-i}{n} u\left(\frac{p-i}{n}\right) + \sum_{j=1}^{k} \frac{k-j+1}{n} u\left(-\frac{k-j+1}{n}\right).$$

Consider now two sequences $(p^{(n)})$ and $(k^{(n)})$, and let $(\lambda^{(n)})$ be the corresponding sequence of signatures. Assume that

$$p^{(n)} \sim \sqrt{n}, \quad k^{(n)} \sim \sqrt{n}.$$

Then

$$\lim_{n \to \infty} L_u(T_n(\lambda^{(n)})) = u(0).$$

This means that

$$\lim_{n\to\infty} T_n\big(\lambda^{(n)}\big) = \omega,$$

with $\omega = (0,0,1)$, i.e. $\alpha_k = 0$, $\beta_k = 0$, $\gamma = 1$. Therefore

$$\lim_{n\to\infty} \varphi_n(\lambda^{(n)}; g) = \det\big(\exp(g - I)\big) = e^{\operatorname{tr}(g-I)}.$$

We consider now the general case. To a signature λ, one associates two positive signatures λ^+ and λ^-: if

$$\lambda_1 \geq \cdots \lambda_p \geq 0 \geq \lambda_{p+1} \geq \cdots \geq \lambda_n,$$

then

$$\lambda^+ = (\lambda_1, \ldots, \lambda_p, 0, \ldots), \quad \lambda^- = (-\lambda_n, \ldots, -\lambda_{p+1}, 0, \ldots).$$

One adds as many zeros as necessary to get positive signatures λ^+, λ^- in Ω_n^+. Then we define the map

$$T_n : \Omega_n \to \Omega = \Omega_0 \times \Omega_0$$

by extending the map T_n previously defined:

$$T_n(\lambda) = \big(T_n(\lambda^+), T_n(\lambda^-)\big).$$

Theorem 3.4. *Let $(\lambda^{(n)})$ be a sequence of signatures, with $\lambda^{(n)} \in \Omega_n$. Assume that*

$$\lim_{n\to\infty} T_n(\lambda^{(n)}) = \omega = (\omega^+, \omega^-).$$

Then, for $g \in U(\infty)$,

$$\lim_{n\to\infty} \varphi_n(\lambda^{(n)}; g) = \det \Phi(\omega^+; g) \det \Phi(\omega^-; g^{-1})$$

uniformly on each $U(k)$.

We will prove Theorem 3.3 in Sect. 5. For the proof of Theorem 3.4, see [Okounkov-Olshanski, 1998c], and also [Faraut, 2008]. The proof of Theorem 3.3 will involve a binomial formula for Schur functions.

4   Binomial Formula for Schur Functions

We will use a formula for Schur expansions due to Hua ([Hua, 1963], Theorem 1.2.1).

Proposition 4.1 (Hua's formula). *Consider n power series:*

$$f_i(w) = \sum_{m=0}^{\infty} c_m^{(i)} w^m,$$

which are convergent for $|w| < r$ for some $r > 0$. Define the function F on $\mathbb{C}^n$ by

$$F(z) = F(z_1, \ldots, z_n) = \frac{\det(f_i(z_j))}{V(z)} \quad |z_j| < r.$$

Then F admits the following Schur expansion:

$$F(z) = \sum_{m_1 \geq \cdots \geq m_n \geq 0} a_\mathbf{m} s_\mathbf{m}(z),$$

with

$$a_\mathbf{m} = \det(c_{m_j+n-j}^{(i)}).$$

In particular

$$\lim_{z_1,\ldots,z_n \to 0} \frac{\det f_i(z_j)}{V(z)} = F(0) = a_0 = \det(c_{n-j}^{(i)}).$$

For a positive signature $\mathbf{m} = (m_1, \ldots, m_n)$, the *shifted Schur function* $s_\mathbf{m}^*$ is defined, for a signature $\lambda = (\lambda_1, \ldots, \lambda_n)$ by

$$s_\mathbf{m}^* = \frac{\det([\lambda_i + \delta_i]_{m_j + \delta_j})}{\det([\lambda_i + \delta_i]_{\delta_j})},$$

where $\delta_i = n - i$, and

$$[a]_k = a(a-1) \ldots (a-k+1).$$

The functions $s_\mathbf{m}^*(\lambda)$ are shifted symmetric functions. The ordinary Schur function $s_\mathbf{m}(x)$ is symmetric, i.e.,

$$s_\mathbf{m}(\ldots, x_i, x_{i+1}, \ldots) = s_\mathbf{m}(\ldots, x_{i+1}, x_i, \ldots),$$

while the shifted Schur function $s_\mathbf{m}^*(\lambda)$ satisfies

$$s_\mathbf{m}^*(\ldots, \lambda_i, \lambda_{i+1}, \ldots) = s_\mathbf{m}^*(\ldots, \lambda_{i+1} - 1, \lambda_i + 1, \ldots).$$

The algebra of symmetric functions is denoted by Λ, and the algebra of shifted symmetric functions will be denoted by Λ^*. (See [Okounkov-Olshanski, 1998a] and [1998b].)

Theorem 4.2 (Binomial formula).

$$\frac{s_\lambda(1 + z_1, \ldots, 1 + z_n)}{s_\lambda(1, \ldots, 1)} = \sum_{m_1 \geq \cdots \geq m_n \geq 0} \frac{\delta!}{(\mathbf{m} + \delta)!} s_{\mathbf{m}}^*(\lambda) s_{\mathbf{m}}(z).$$

For $n = 1$ this is nothing but the classical binomial formula:

$$(1 + z)^\lambda = \sum_{m=0}^{\infty} \frac{1}{m!} [\lambda]_m w^m.$$

Proof. The theorem is a straightforward application of Hua's formula (Proposition 4.1) in the case

$$f_i(w) = (1 + w)^{\lambda_i + \delta_i} = \sum_{m=0}^{\infty} \frac{1}{m!} [\lambda_i + \delta_i]_m w^m.$$

One observes that

$$s_\lambda(1, \ldots, 1) = \frac{V(\lambda + \delta)}{V(\delta)} = \frac{\det([\lambda_i + \delta_i]_{\delta_j})}{\delta!}. \qquad \square$$

If λ is a positive signature, then $s_{\mathbf{m}}^*(\lambda) = 0$ if $\mathbf{m} \not\subseteq \lambda$, and

$$\frac{s_\lambda(1 + z_1, \ldots, 1 + z_n)}{s_\lambda(1, \ldots, 1)} = \sum_{\mathbf{m} \subseteq \lambda} \frac{\delta!}{(\mathbf{m} + \delta)!} s_{\mathbf{m}}^*(\lambda) s_{\mathbf{m}}(z).$$

If, in Theorem 4.2, one takes $z_1 = z, z_2 = 0, \ldots, z_n = 0$, then one obtains Lemma 3 in [Vershik-Kerov, 1982]:

$$\frac{s_\lambda(1 + z, 1, \ldots, 1)}{s_\lambda(1, \ldots, 1)} = 1 + \sum_{m=1}^{\infty} \frac{1}{n(n + 1) \ldots (n + m - 1)} h_m^*(\lambda) z^m.$$

The shifted complete symmetric function $h_m^*(\lambda)$ is denoted by $\Phi_m(\lambda)$ in [Vershik-Kerov, 1982]. By using the fact that the value of a determinant does not change when adding to a column a linear combination of the other ones, one obtains, with $\ell_i = \lambda_i + n - i$,

$$h_m^*(\lambda) = \frac{1}{V(\ell)} \begin{vmatrix} [\ell_1]_{m+n-1} & [\ell_1]_{n-2} & \cdots & 1 \\ [\ell_2]_{m+n-1} & [\ell_2]_{n-2} & \cdots & 1 \\ \vdots & \vdots & & \vdots \\ [\ell_n]_{m+n-1} & [\ell_n]_{n-2} & \cdots & 1 \end{vmatrix}$$

$$= \frac{1}{V(\ell)} \begin{vmatrix} [\ell_1]_{m+n-1} & \ell_1^{n-2} & \cdots & 1 \\ [\ell_2]_{m+n-1} & \ell_2^{n-2} & \cdots & 1 \\ \vdots & \vdots & & \vdots \\ [\ell_n]_{m+n-1} & \ell_n^{n-2} & \cdots & 1 \end{vmatrix}.$$

By expanding now $[x]_{m+n-1}$ in powers of x:

$$[x]_{m+n-1} = x(x-1)\ldots(x-m-n+2)$$
$$= \sum_{k=0}^{m} e_{m-k}\big(0, -1, \ldots, -(m+n-2)\big)x^{k+n-1}$$
$$+ \text{ terms of degree } < n-1,$$

where e_k is the k-th elementary symmetric function, one obtains the formula from Lemma 3 in [Vershik-Kerov, 1982]:

$$h_m^*(\lambda) = \sum_{k=0}^{m} e_{m-k}\big(0, -1, \ldots, -(m+n-2)\big)h_k(\ell).$$

5   Proof of Theorem 3.3

We follow the method of proof of [Okounkov-Olshanski, 1998c].

(a) *The morphism $\Lambda \to C(\Omega_0)$*

One defines an algebra morphism $\Lambda \to C(\Omega_0)$ which maps a symmetric function f to a continuous function $\tilde{f}$ on Ω_0. Since the power sums

$$p_m(x_1, \ldots, x_n, \ldots) = \sum_i x_i^m$$

generate Λ as an algebra, this morphism is uniquely determined by their images $\widetilde{p_m}$. One puts, for $\omega = (\alpha, \beta, \gamma) \in \Omega_0$, with $\alpha = (\alpha_k)$, $\beta = (\beta_k)$,

$$\widetilde{p_1}(\omega) = \sum_{k=1}^{\infty} \alpha_k + \sum_{k=1}^{\infty} \beta_k + \gamma,$$

$$\widetilde{p_m}(\omega) = \sum_{k=1}^{\infty} \alpha_k^m + (-1)^{m-1} \sum_{k=1}^{\infty} \beta_k^m \quad (m \geq 2).$$

The functions $\widetilde{p_m}$ are continuous on Ω_0. In fact, as we saw above, $\widetilde{p_m}(\omega) = L_u(\omega)$, with $u(s) = s^{m-1}$ $(m \geq 1)$.

Proposition 5.1. *The functions $\widetilde{h_m}(\omega)$ are the Taylor coefficients of the Voiculescu function $\Phi(\omega; t)$ at $t = 1$: for $z \in \mathbb{C}$, $|z| < r = \inf \frac{1}{\alpha_k}$,*

$$\Phi(\omega; 1 + z) = \sum_{m=0}^{\infty} \widetilde{h_m}(\omega) z^m.$$

Proof. One starts from the generating function of the complete symmetric functions h_m:

$$H(x; z) = \sum_{m=0}^{\infty} h_m(x) z^m = \prod_{j=1}^{n} \frac{1}{1 - x_j z}.$$

Its logarithmic derivative is given by

$$\frac{d}{dz} \log H(x; z) = \sum_{m=0}^{\infty} p_{m+1}(x) z^m.$$

On the other hand, as we saw in Sect. 3,

$$\frac{d}{dz} \log \Phi(\omega; 1 + z) = \sum_{m=0}^{\infty} \widetilde{p_{m+1}}(\omega) z^m.$$

Therefore, the coefficients $c_m(\omega)$ defined by

$$\Phi(\omega; 1 + z) = \sum_{m=0}^{\infty} c_m(\omega) z^m,$$

are images, by the morphism $f \mapsto \tilde{f}$, of the complete symmetric functions h_m: $c_m(\omega) = \widetilde{h_m}(\omega)$. □

Corollary 5.2. *For $z = (z_1, \ldots, z_n) \in \mathbb{C}^n$, $|z_j| < r$,*

$$\prod_{j=1}^{n} \Phi(\omega; 1 + z_j) = \sum_{m_1 \geq \cdots \geq m_1 \geq 0} \widetilde{s_{\mathbf{m}}}(\omega) s_{\mathbf{m}}(z).$$

Proof. Observe that the statement of Proposition 5.1 can be written as

$$\tilde{H}(\omega; z) = \Phi(\omega; 1 + z),$$

and apply the morphism $f \mapsto \tilde{f}$ to both sides of the Cauchy identity

$$\prod_{j=1}^{n} H(x; z_j) = \prod_{i,j=1}^{n} \frac{1}{1 - x_i z_j} = \sum_{m_1 \geq \cdots \geq m_n \geq 0} s_{\mathbf{m}}(x) s_{\mathbf{m}}(z).$$

(b) *Asymptotics of shifted symmetric functions*

Proposition 5.3. *Consider a sequence $(\lambda^{(n)})$ of positive signatures with $\lambda^{(n)} \in \Omega_n^+$, and let $\omega \in \Omega_0$. Assume that, for the topology of Ω_0,*

$$\lim_{n \to \infty} T_n(\lambda^{(n)}) = \omega.$$

Then, for every shifted symmetric function $f^* \in \Lambda^*$

$$\lim_{n \to \infty} \frac{1}{n^m} f^*(\lambda^{(n)}) = \tilde{f}(\omega),$$

where m is the degree of f^*, and f is the homogeneous part of degree m in f^*.

We will prove the statement in the special case $f^* = q_m^*$:

$$q_m^*(\lambda) = \sum_{i \geq 1} ([\lambda_i - i + 1]_m - [-i + 1]_m).$$

The function $q_m^*(\lambda)$ is shifted symmetric of degree m and the homogeneous part of degree m is equal to the Newton power sum $p_m(\lambda)$. Since the functions $q_m^*(\lambda)$ generate Λ^* as an algebra, the statement of the proposition will be proven.

Lemma 5.4. *Let $a = (a_i)$, $b = (b_j)$ be the Frobenius parameters of the positive signature λ. Then*

$$q_m^*(\lambda) = \sum_{i \geq 1} [1 + a_i]_m - \sum_{j \geq 1} [1 - b_j]_m.$$

Proof of Proposition 5.3. Let $a^{(n)} = (a_i^{(n)})$ and $b^{(n)} = (b_j^{(n)})$ be the Frobenius parameters of the positive signature $\lambda^{(n)}$, and $\omega = (\alpha, \beta, \gamma) \in \Omega_0$, with $\alpha = (\alpha_k)$, $\beta = (\beta_k)$. By assumption, for every continuous function u on $\mathbb{R}$,

$$\lim_{n \to \infty} L_u(T_n(\lambda^{(n)})) = L_u(\omega),$$

or

$$\lim_{n\to\infty}\left(\sum_{i\geq1}\frac{a_i^{(n)}}{n}u\left(\frac{a_i^{(n)}}{n}\right)+\sum_{j\geq1}\frac{b_j^{(n)}}{n}u\left(-\frac{b_j^{(n)}}{n}\right)\right)$$

$$=\sum_{k=1}^{\infty}\alpha_k u(\alpha_k)+\sum_{k=1}^{\infty}\beta_k u(-\beta_k)+\gamma u(0).$$

Consider the sequence of the functions

$$u_n(s)=\frac{1}{n^m s}[ns+1]_m.$$

Then

$$L_{u_n}\left(T_n(\lambda^{(n)})\right)=\frac{1}{n^m}q_m^*(\lambda^{(n)}).$$

On the other hand, the sequence $u_n(s)$ converges to the function $u(s)=s^{m-1}$ uniformly on compacts sets in $\mathbb{R}$, and

$$L_u(\omega)=\widetilde{p_m}(\omega).$$

It follows that

$$\lim_{n\to\infty}\frac{1}{n^m}q_m^*(\lambda^{(n)})=\widetilde{p_m}(\omega).\qquad\qquad\square$$

(c) *End of the proof of Theorem 3.3*

To finish the proof, one applies the following:

Proposition 5.5. *Let ψ_n be a sequence of C^∞-functions on the torus $\mathbb{T}^k$ of positive type, with $\psi_n(0)=1$, and ψ an analytic function in a neighborhood of 0. Assume that, for every $\alpha=(\alpha_1,\ldots,\alpha_k)\in\mathbb{N}^k$,*

$$\lim_{n\to\infty}\partial^\alpha\psi_n(0)=\partial^\alpha\psi(0).$$

Then ψ has an analytic extension to $\mathbb{T}^k$, and ψ_n converges to ψ uniformly on $\mathbb{T}^k$.

For the proof, see for instance [Faraut, 2008], Proposition 3.11.
We consider a sequence of positive signatures $(\lambda^{(n)})$ such that

$$\lim_{n\to\infty}T_n(\lambda^{(n)})=\omega.$$

Put, with $t_j = e^{i\theta_j}$,

$$\psi_n(t_1, \ldots, t_k) = \varphi_n\big(\lambda^{(n)}; \mathrm{diag}(t_1, \ldots, t_k, 1, \ldots)\big),$$

$$\psi(t_1, \ldots, t_k) = \prod_{j=1}^{k} \Phi(\omega; t_j).$$

By Theorem 4.2,

$$\psi_n(1 + z_1, \ldots, 1 + z_k) = \sum_{m_k \geq \cdots \geq m_1 \geq 0} \frac{\delta!}{(\mathbf{m} + \delta)!} s_{\mathbf{m}}^*(\lambda^{(n)}) s_{\mathbf{m}}(z_1, \ldots, z_k).$$

Then, by Proposition 5.3,

$$\lim_{n \to \infty} \frac{1}{n^{|\mathbf{m}|}} s_{\mathbf{m}}^*(\lambda^{(n)}) = \widetilde{s}_{\mathbf{m}}(\omega),$$

and, by Corollary 5.2,

$$\sum_{m_1 \geq \cdots \geq m_k} \widetilde{s}_{\mathbf{m}}(\omega) s_{\mathbf{m}}(z_1, \ldots, z_k) = \prod_{j=1}^{k} \Phi(\omega; 1 + z_k) = \psi(1 + z_1, \ldots, 1 + z_k).$$

Finally, observing that

$$\frac{(\mathbf{m} + \delta)!}{\delta!} \sim n^{|\mathbf{m}|} \quad (n \to \infty),$$

we obtain, by Proposition 5.5,

$$\lim_{n \to \infty} \psi_n(t_1, \ldots, t_k) = \psi(t_1, \ldots, t_k),$$

uniformly on $\mathbb{T}^k$. In fact, the Taylor coefficients of ψ_n, as a function on $\mathbb{T}^k$, are finite linear combinations of the coefficients in the Schur expansion of $\psi_n(1 + z_1, \ldots, 1 + z_n)$. □

6 Inductive Limits of Compact Symmetric Spaces

One knows that if G/K is a Riemannian symmetric space, then (G, K) is a Gelfand pair. Let $G(n)/K(n)$ be a compact symmetric space of rank n, and

$$\mathfrak{g}(n) = \mathfrak{k}(n) + \mathfrak{p}(n)$$

be a Cartan decomposition of the Lie algebra $\mathfrak{g}(n)$ of $G(n)$. Fix a Cartan subspace $\mathfrak{a}(n) \subset \mathfrak{p}(n)$, $\mathfrak{a}(n) \simeq \mathbb{R}^n$, and put $A(n) = \exp \mathfrak{a}(n) \simeq \mathbb{T}^n$. Let $\mathcal{R}_n$ denote the system of restricted roots for the pair $(\mathfrak{a}(n)_{\mathbb{C}}, \mathfrak{g}(n)_{\mathbb{C}})$.

(a) *Classical series of type A*

We consider one of the following series of compact symmetric spaces.

$G(n)$	$K(n)$	d
$U(n)$	$O(n)$	1
$U(n) \times U(n)$	$U(n)$	2
$U(2n)$	$Sp(n)$	4

The system $\mathcal{R}_n$ of restricted roots is of type A_{n-1}. For a suitable basis $(e_1, \ldots, e_n)$ of $\mathfrak{a}(n)$, the restricted roots are

$$\alpha_{ij} = \varepsilon_i - \varepsilon_j \quad (i \neq j)$$

$((\varepsilon_1, \ldots, \varepsilon_n)$ is the dual basis), with multiplicities $d = 1, 2, 4$.

These symmetric spaces appear as Shilov boundaries of bounded symmetric domains of tube type. In particular, the symmetric space $U(n)/O(n)$ can be seen as the space of symmetric unitary $n \times n$ matrices. The subgroup $A(n)$ can be taken as the subgroup of unitary diagonal matrices. The space $U(n)/O(n)$ can also be seen as the Lagrangian manifold $\Lambda(n)$, the manifold of n-Lagrangian subspaces in $\mathbb{R}^{2n}$.

The spherical dual Ω_n of the Gelfand pair $(G(n), K(n))$ is parametrized by signatures

$$\lambda = (\lambda_1, \ldots, \lambda_n), \ \lambda_i \in \mathbb{Z}, \ \lambda_1 \geq \cdots \geq \lambda_n.$$

The restricted highest weight of the spherical representation corresponding to λ is $\sum_{i=1}^n \lambda_i \varepsilon_i$.

The restriction to $A(n) \simeq \mathbb{T}^n$ of the spherical function $\varphi_n(\lambda; x)$ is a normalized Jack function: for $a = (t_1, \ldots, t_n)$,

$$\varphi_n(\lambda; a) = \frac{J_\lambda(t_1, \ldots, t_n; \alpha)}{J_\lambda(1, \ldots, 1; \alpha)},$$

with $\alpha = \frac{2}{d}$. For $d = 2$, it is a Schur function. (See [Stanley, 1989] for definition and properties of Jack functions, and also [Macdonald, 1995], Sect. VI.10.)

The Jack functions are orthogonal with respect to the following inner product:

$$(P|Q) = \int_{\mathbb{T}^n} P(t)\overline{Q(t)}|V(t)|^d \beta(dt),$$

where β is the normalized Haar measure on $\mathbb{T}^n$. With $t_j = e^{i\theta_j}$,

$$|V(t)|^d = \prod_{j<k} 4 \left| \sin \frac{\theta_j - \theta_k}{2} \right|^d,$$

$$\beta(\mathrm{d}t) = \frac{1}{(2\pi)^n} \mathrm{d}\theta_1 \ldots \mathrm{d}\theta_n.$$

We consider now the Olshanski spherical pair (G, K) with

$$G = \bigcup_{n=1}^{\infty} G(n), \quad K = \bigcup_{n=1}^{\infty} K(n).$$

We state without proof the main results by Okounkov-Olshanski ([1998c]). The spherical dual for the pair (G, K) is, as in the case of the infinite dimensional unitary group, parametrized by a pair $\omega = (\omega^+, \omega^-)$, i.e., $\Omega = \Omega_0 \times \Omega_0$. For $\omega \in \Omega_0$, $\omega = (\alpha, \beta, \gamma)$, with $\alpha = (\alpha_k)$, $\beta = (\beta_k)$, define

$$\Phi^{(d)}(\omega; t) = e^{\gamma(t-1)} \prod_{k=1}^{\infty} \frac{1 + \beta_k(t-1)}{\left(1 - \frac{2}{d}\alpha_k(t-1)\right)^{\frac{d}{2}}} \quad (t \in \mathbb{T}).$$

For $d = 2$, it is the Voiculescu function we considered in Sect. 3.

Theorem 6.1. *The spherical functions of positive type, for the Olshanski spherical pair (G, K), are given, for $a = (t_1, \ldots, t_n, 1, \ldots) \in A \simeq \mathbb{T}^{(\infty)}$, by*

$$\varphi(\omega; a) = \prod_{j=1}^{n} \Phi^{(d)}(\omega^+; t_j) \Phi^{(d)}\left(\omega^-; \frac{1}{t_j}\right),$$

with $\omega = (\omega^+, \omega^-) \in \Omega$.

One defines the map $T_n : \Omega_n \to \Omega = \Omega_0 \times \Omega_0$ as in the case of the unitary groups (see Sect. 3).

Theorem 6.2. *Let $(\lambda^{(n)})$ be a sequence of signatures with $\lambda^{(n)} \in \Omega_n$. If*

$$\lim_{n\to\infty} T_n(\lambda^{(n)}) = \omega = (\omega^+, \omega^-),$$

then, with $a = (t_1, \ldots, t_k, 1, \ldots) \in A$,

$$\lim_{n\to\infty} \varphi_n(\lambda^{(n)}; a) = \prod_{j=1}^{k} \Phi^{(d)}(\omega^+; t_j) \Phi^{(d)}\left(\omega^-; \frac{1}{t_j}\right).$$

Since there is no simple formula for the Jack functions for $\alpha \neq 1$, the proof for $d \neq 2$ is more difficult than in the case of the unitary groups. However, it follows the same lines. The first step is a binomial formula for the normalized Jack functions.

(b) *Classical series of type BC*

We consider the following series of compact symmetric spaces.

	$G(n)$	$K(n)$	$\mathcal{R}_n$	d	p	q
1	$O(2n) \times O(2n)$	$O(2n)$	D_n	2	0	0
2	$O(2n+1) \times O(2n+1)$	$O(2n+1)$	B_n	2	2	0
3	$Sp(n) \times Sp(n)$	$Sp(n)$	C_n	2	0	2
4	$Sp(n)$	$U(n)$	C_n	1	0	1
5	$O(4n)$	$U(2n)$	C_n	4	0	1
6	$O(4n+2)$	$U(2n+1)$	BC_n	4	4	1
7	$O(2n+k)$	$O(n) \times O(n+k)$	BC_n	1	k	0
8	$U(2n+k)$	$U(n) \times U(n+k)$	BC_n	2	2k	1
9	$Sp(2n+k)$	$Sp(n) \times Sp(n+k)$	BC_n	4	4k	3

The possible roots and multiplicities are

α	$\pm\varepsilon_i \pm \varepsilon_j$	ε_i	$2\varepsilon_i$
m_α	d	p	q

Series 1, 2, and 3 are compact groups seen as symmetric spaces.

Series 4, 5, and 6 are compact Hermitian symmetric spaces.

Series 7, 8, and 9 are Grassmann manifolds: spaces of n-subspaces in $\mathbb{F}^{2n+k}$, with $\mathbb{F} = \mathbb{R}, \mathbb{C},$ or $\mathbb{H}$, $d = \dim_\mathbb{R}\mathbb{F}$, $p = dk$, $q = d - 1$. If $k = 0$, the root system $\mathcal{R}_n$ is of type C_n. The symmetric space $U(2n+k)/U(n) \times U(n+k)$ is Hermitian as well. For series 7, 8, and 9 the Cartan subgroup $A(n)$ can be taken as the group of the following matrices:

$$a(\theta) = \begin{pmatrix} \cos\frac{\theta}{2} & 0 & -\sin\frac{\theta}{2} \\ 0 & I_k & 0 \\ \sin\frac{\theta}{2} & 0 & \cos\frac{\theta}{2} \end{pmatrix},$$

with $\theta = (\theta_1, \dots, \theta_n)$, and

$$\cos\frac{\theta}{2} = \text{diag}\left(\cos\frac{\theta_1}{2}, \dots, \cos\frac{\theta_n}{2}\right), \quad \sin\frac{\theta}{2} = \text{diag}\left(\sin\frac{\theta_1}{2}, \dots, \sin\frac{\theta_n}{2}\right).$$

We assume that the multiplicities d, p, q don't depend on n. The spherical dual Ω_n is parametrized by positive signatures:

$$\lambda = (\lambda_1, \ldots, \lambda_n), \ \lambda_i \in \mathbb{N}, \ \lambda_1 \geq \cdots \geq \lambda_n \geq 0.$$

The restriction to $A(n) \simeq \mathbb{T}^n$ of the corresponding spherical function is a normalized Jacobi polynomial. (See Hypergeometric and Special Functions, by Heckman, in [Heckman-Schichtkrull,1994], for definition and properties of Jacobi polynomials associated to a root system.) For $a = (t_1, \ldots, t_n) \in A(n)$,

$$\varphi_n(\lambda; a) = \frac{\mathfrak{P}_\lambda(t_1, \ldots, t_n)}{\mathfrak{P}_\lambda(1, \ldots, 1)}.$$

The polynomials $\mathfrak{P}_\lambda$ are orthogonal with respect to the inner product

$$(P|Q) = \int_{\mathbb{T}^n} P(t)\overline{Q(t)}|D(t)|\beta(dt),$$

with, if $t_j = e^{i\theta_j}$,

$$D(t) = \prod_{i<j}\left(\sin\frac{\theta_i+\theta_j}{2}\right)^d\left(\sin\frac{\theta_i-\theta_j}{2}\right)^d \prod_{i=1}^n\left(\sin\frac{\theta_i}{2}\right)^p (\sin\theta_i)^q.$$

By putting $x_i = \cos\theta_i = \frac{1}{2}(t_i + t_i^{-1})$, the inner product is carried over an integral on $[-1,1]^n$ with the weight

$$\prod_{i<j}|x_i - x_j|^d \prod_{i=1}^n(1-x_i)^\alpha(1+x_i)^\beta,$$

with $\alpha = \frac{1}{2}(p+q-1), \beta = \frac{1}{2}(q-1)$. We will write P_λ for the Jacobi polynomial in the variables x_i:

$$P_\lambda(x_1, \ldots, x_n) = \mathfrak{P}_\lambda(t_1, \ldots, t_n), \quad x_i = \frac{1}{2}(t_i + t_i^{-1}).$$

As in Sect. 6(a), we define, for $\omega \in \Omega_0$,

$$\Phi^{(d)}(\omega; t) = e^{\gamma(t-1)} \prod_{k=1}^\infty \frac{1 + \beta_k(t-1)}{\left(1 - \frac{2}{d}\alpha_k(t-1)\right)^{\frac{d}{2}}} \quad (t \in \mathbb{T}).$$

Theorem 6.3. *The spherical dual for the pair* (G, K) *is parametrized by* Ω_0. *The spherical functions are given, for* $a = (t_1, \ldots, t_n, 1, \ldots) \in A \simeq \mathbb{T}^{(\infty)}$, *by*

$$\varphi(\omega; a) = \prod_{j=1}^{n} \Phi^{(d)}(\omega; t_j) \Phi^{(d)}\left(\omega; \frac{1}{t_j}\right),$$

with $\omega \in \Omega_0$.

One defines the map $T_n : \Omega_n \to \Omega_0$ as in the case of the unitary groups for positive signatures.

Theorem 6.4. *Let* $(\lambda^{(n)})$ *be a sequence of signatures, with* $\lambda^{(n)} \in \Omega_n$. *If*

$$\lim_{n \to \infty} T_n(\lambda^{(n)}) = \omega,$$

then, for $a = (t_1, \ldots, t_k, 1, \ldots)$,

$$\lim_{n \to \infty} \varphi_n(\lambda^{(n)}; a) = \prod_{j=1}^{k} \Phi^{(d)}(\omega; t_j) \Phi^{(d)}\left(\omega; \frac{1}{t_j}\right).$$

7 The Case $d = 2$. Determinantal Formula, Binomial Formula for Multivariate Jacobi Polynomials

In this last section, we will present, in case $d = 2$, a determinantal formula for the multivariate Jacobi polynomials, and then a binomial formula.

In their paper, Berezin and Karpelevič gave a determinantal formula for the spherical functions on the Grassmann manifolds $U(p + q)/U(p) \times U(q)$ ([1958], see also [Takahashi, 1977], [Hoogenboom, 1982]). In fact, such a determinantal formula exists in all cases with $d = 2$.

Let μ be a positive measure on $\mathbb{R}$ with infinite support and finite moments: for all $m \in \mathbb{N}$,

$$\int_{\mathbb{R}} |t|^m \mu(\mathrm{d}t) < \infty.$$

By orthogonalizing the monomials t^m, one obtains a sequence of orthogonal polynomials $p_m(t)$:

$$\int_{\mathbb{R}} p_\ell(t) p_m(t) \mu(\mathrm{d}t) = 0 \quad \text{if } \ell \neq m.$$

For a positive signature λ, define the multivariate polynomials P_λ

$$P_\lambda(x_1,\ldots,x_n) = \frac{\det\big(p_{\lambda_i+\delta_i}(x_j)\big)}{V(x)},$$

where λ is a positive signature, and, as above, $\delta = (n-1,\ldots,1,0)$. The symmetric polynomials P_λ are orthogonal with respect to the inner product

$$(P|Q) = \int_{\mathbb{R}^n} P(x_1,\ldots,x_n)\overline{Q(x_1,\ldots,x_n)}V(x_1,\ldots,x_n)^2\mu(dx_1)\ldots\mu(dx_n).$$

If the polynomials p_m are normalized such that

$$p_m(t) = t^m + \text{ lower order terms},$$

then

$$P_\lambda(x_1,\ldots,x_n) = s_\lambda(x_1,\ldots,x_n) + \text{ lower order terms}.$$

Consider now the measure μ on $\mathbb{R}$ given by

$$\int_{\mathbb{R}} f(t)\mu(dt) = \int_{-1}^{1} f(t)(1-t)^\alpha(1+t)^\beta dt,$$

with $\alpha,\beta > -1$. Then, the orthogonal polynomials with respect to this measure are the Jacobi polynomials $p_m(t) = p_m^{(\alpha,\beta)}(t)$. The multivariable polynomials $P_\lambda^{(\alpha,\beta)}$ given by, for $x = (x_1,\ldots,x_n)$,

$$P_\lambda^{(\alpha,\beta)}(x) = \frac{\det\big(p_{\lambda_i+\delta_i}^{(\alpha,\beta)}(x_j)\big)}{V(x)},$$

are orthogonal for the inner product

$$(P|Q) = \int_{[-1,1]^n} P(x)\overline{Q(x)} \prod_{i<j}(x_i-x_j)^2 \prod_{i=1}^{n}(1-x_i)^\alpha(1+x_i)^\beta dx_1\ldots dx_n,$$

and are, up to a constant factor, the Jacobi polynomials associated with the root system of type BC_n and the multiplicity (d,p,q), with $d=2$.

Normalized by the condition $p_m^{(\alpha,\beta)}(1) = 1$, the Jacobi polynomials $p_m^{(\alpha,\beta)}$ admit the following hypergeometric representation:

$$p_m^{(\alpha,\beta)}(t) = {}_2F_1\left(-m, m+\alpha+\beta+1; \alpha+1; \frac{1-t}{2}\right)$$

$$= \sum_{k=0}^{m} \frac{(-m)_k(m+\alpha+\beta+1)_k}{(\alpha+1)_k} \frac{1}{k!}\left(\frac{1-t}{2}\right)^k.$$

Let us introduce the notation

$$\sigma = \frac{\alpha+\beta+1}{2}, \quad \ell = m+\sigma,$$
$$[\ell,\sigma]_k = \left(\ell^2-\sigma^2\right)\dots\left(\ell^2-(\sigma+k-1)^2\right).$$

The binomial formula for the Jacobi polynomial $p_m^{(\alpha,\beta)}$ can be written as

$$p_m^{(\alpha,\beta)}(1+w) = \sum_{k=0}^{m} a_k^{(m)} w^k = \sum_{k=0}^{m} \frac{1}{k!}\frac{[\ell,\sigma]_k}{(\alpha+1)_k}\left(\frac{w}{2}\right)^k.$$

By Hua's formula,

$$P_\lambda^{(\alpha,\beta)}(1,\dots,1) = \det\left(a_{\delta_j}^{(\lambda_i+\delta_i)}\right) = 2^{-\frac{n(n-1)}{2}}\frac{1}{\delta!}\prod_{i=1}^{n}\frac{1}{(\alpha+1)_{\delta_i}}V(\ell_1^2,\dots,\ell_n^2),$$

with $\ell_i = \lambda_i + \delta_i + \sigma$. Since

$$\det\left([\ell_i,\sigma]_{\delta_j}\right) = V(\ell_1^2,\dots,\ell_n^2).$$

Theorem 7.1.

$$\frac{P_\lambda^{(\alpha,\beta)}(1+z_1,\dots,1+z_n)}{P_\lambda^{(\alpha,\beta)}(1,\dots,1)}$$

$$= \sum_{\mu\subseteq\lambda} 2^{-|\mu|}\frac{\delta!}{(\mu+\delta)!}\frac{\prod_{i=1}^{n}(\alpha+1)_{\delta_i}}{\prod_{i=1}^{n}(\alpha+1)_{\mu_i+\delta_i}}S_\mu^*(\lambda)s_\mu(z_1,\dots,z_n),$$

with

$$S_\mu^*(\lambda) = \frac{\det\left([\ell_i,\sigma]_{\mu_j+\delta_j}\right)}{V(\ell_1^2,\dots,\ell_n^2)}, \quad \ell_i = \lambda_i + \delta_i + \sigma.$$

Proof. This is once more an application of Hua's formula (Proposition 4.1). In the present case

$$f_i(w) = p^{(\alpha,\beta)}_{\lambda_i+\delta_i}(1+w) = \sum_{k=0}^{\lambda_i+\delta_i} a_k^{(\lambda_i+\delta_i)} w^k = \sum_{k=0}^{\lambda_i+\delta_i} \frac{1}{k!} \frac{[\ell_i,\sigma]_k}{(\alpha+1)_k} 2^{-k} w^k,$$

with $\ell_i = \lambda_i + \delta_i + \sigma$. Then we get

$$P_\lambda^{(\alpha,\beta)}(1+z_1,\ldots,1+z_n) = \sum_{\mu_1 \ge \cdots \ge \mu_n \ge 0} a_\mu s_\mu(z_1,\ldots,z_n),$$

with

$$a_\mu = \det\left(c_{\mu_j+\delta_j}^{(\lambda_i+\delta_i)}\right) = \frac{1}{(\mu+\delta)!} \frac{1}{\prod_{i=1}^n (\alpha+1)_{\mu_i+\delta_i}} \det\left([\ell_i,\sigma]_{\mu_j+\delta_j}\right).$$

Observe that, if $\mu \not\subseteq \lambda$, then $a_\mu = 0$.   □

References

F. BEREZIN AND F.I KARPELEVIČ (1958). Zonal spherical functions and Laplace operators on some symmetric spaces, *Dokl. Akad. Nauk USSR*, **118**, 9–12.

R.P. BOYER (1983). Infinite traces of AF-algebras and characters of $U(\infty)$, *J. Operator Theory*, **9**, 205–236.

J. FARAUT (2006). Infinite dimensional harmonic analysis and probability. *in Probability measures on groups: recent directions and trends*, (eds. S.G Dani and P. Graczyk), Tata Inst. Fund. Res., 179–254.

J. FARAUT (2008). Infinite Dimensional Spherical Analysis. *COE Lecture Note Vol. 10, Kyushu University.*

G. HECKMAN, H. SCHLICHTKRULL (1994). Harmonic analysis and special functions on symmetric spaces. *Academic Press.*

B. HOOGENBOOM (1982). Spherical functions and differential operators on complex Grassmann manifolds. *Ark. Mat.* **20**, 69–85.

L.K. Hua (1963). Harmonic analysis of functions of several variables in the classical domains. *American Mathematical Society.*

I.G. MACDONALD (1995). Symmetric functions and Hall polynomials. *Oxford Science Publications.*

A. OKOUNKOV AND G. OLSHANSKI (1998a). Shifted Schur functions, *St. Petersburg Math. J.* **9**, 239–300.

A. OKOUNKOV AND G. OLSHANSKI (1998b). Shifted Schur functions II. *in Kirillov's Seminar on Representation Theory (ed. G. Olshanski), Amer. Math. Soc. Translations 181 (2),* 245–271.

A. OKOUNKOV AND G. OLSHANSKI (1998c). Asymptotics of Jack polynomials as the number of variables goes to infinity, *Internat. Math. Res. Notices*, **13**, 641–682.

A. OKOUNKOV AND G. OLSHANSKI (2006). Limits of BC-type orthogonal polynomials as the number of variables goes to infinity, *Contemporary Mathematics*, **417**, 281–318.

G. OLSHANSKI, (1990). Unitary representations of infinite dimensional pairs (G, K) and the formalism of R. Howe. *in Representations of Lie groups and related topics (eds. A.M. Vershik, D.P. Zhelobenko), Adv. Stud. Contemp. Math. 7, Gordon and Breach.*

R.P. STANLEY (1989) Some combinatorial properties of Jack symmetric functions, *Adv. in Math.* **77** 76–115.

R. TAKAHASHI (1977). Fonctions sphériques zonales sur $U(n, n + k, \mathbb{F})$. *Séminaire d'Analyse Harmonique, Faculté des Sciences de Tunis*.

A. VERSHIK AND S. KEROV (1982) Characters and factor representations of the infinite unitary group, *Soviet Math. Dokl.*, **26 No 3**, 570–574.

D. VOICULESCU (1976). Représentations factorielles de type II_1 de $U(\infty)$, *J. Math. Pures Appl.*, **55**, 1–20.

CPSIA information can be obtained at www.ICGtesting.com
Printed in the USA
LVOW080418010212

266488LV00005B/95/P